Integrated Crop Management in Sustainable Agriculture

Integrated Crop Management in Sustainable Agriculture

Editors

Mubshar Hussain
Sami Ul-Allah
Shahid Farooq

MDPI • Basel • Beijing • Wuhan • Barcelona • Belgrade • Manchester • Tokyo • Cluj • Tianjin

Editors
Mubshar Hussain
Bahauddin Zakariya
University (BZU)
Multan
Pakistan

Sami Ul-Allah
University of Layyah
Layyah
Pakistan

Shahid Farooq
Harran University
Şanlıurfa
Turkey

Editorial Office
MDPI
St. Alban-Anlage 66
4052 Basel, Switzerland

This is a reprint of articles from the Special Issue published online in the open access journal *Agriculture* (ISSN 2077-0472) (available at: https://www.mdpi.com/journal/agriculture/special_issues/integrated_crop_management).

For citation purposes, cite each article independently as indicated on the article page online and as indicated below:

LastName, A.A.; LastName, B.B.; LastName, C.C. Article Title. *Journal Name* **Year**, *Volume Number*, Page Range.

ISBN 978-3-0365-8062-3 (Hbk)
ISBN 978-3-0365-8063-0 (PDF)

Contents

About the Editors

Mubshar Hussain

Prof. Dr. Mubshar Hussain has worked for the Bahauddin Zakariya University in Multan, Pakistan, since 2008. His major research areas include agronomy, conservation agronomy, and allelopathy, and he has published more than 200 aritcles in these fields. These publications include many highly cited papers on the adaptation of crops to water-limited environments, resource conservation techniques, biofortification of staple crops, and non-chemical weed control. Prof. Hussain's research on crop water relations and adaptation to dryland environments has covered a fundamental understanding of the response of crops to water deficits, temperatures, and nutrient deficits. Prof. Hussain has optimized several seed enhancement techniques to improve crop performance under sub-optimum field conditions and deliver micronutrients. Prof. Hussain has also developed non-chemical strategies, such as tillage and the inclusion of allelopathic crops in cropping systems, for weed management in field crops.

Sami Ul-Allah

Dr. Sami Ul-Allah is working as an Associate Professor of plant breeding and genetics at the University of Layyah, Layyah, Pakistan, and he has accumulated nine years of experience in teaching and research. Dr. Sami Ul-Allah completed his PhD from the University of Kassel, Germany, in 2013 and started his university career by joining the Department of Plant Breeding and Genetics as an Assistant Professor. After two years of service, he moved to Bahauddin Zakariya University, Layyah campus, where he was promoted to the role of Associate Professor of plant breeding and genetics in 2021. Dr. Sami Ul-Allah has published more than 100 research articles, review papers, and book chapters with world-leading journals and publishers. He has an accumulative impact factor of $\sim$200. His areas of research include crop improvement for biotic and abiotic stress and biofortification of food crops.

Shahid Farooq

Dr. Shahid Farooq completed his doctorate from Tokat Gaziosmanpaşa University, Turkey, in Weed Science in 2018 and has worked at Harran University, Şanlıurfa, Turkey, since 2018. Dr. Farooq's research interests include invasive plant species, species distribution models, cropping systems, and weed management. Dr. Farooq's research is focused on improving crop production through seed enhancements, the inoculation of bacteria, and the spatial and potential distribution of weeds under current and future climatic conditions. Dr. Farooq has published more than 70 papers in highly reputed journals, mainly focusing on crop production in stressful environments, tolerance of invasive species to benign environmental conditions, factors driving weed distribution, and the potential spread of weed species.

Preface to "Integrated Crop Management in Sustainable Agriculture"

Welcome to the world of Integrated Crop Management! In an era of rapidly changing climate, evolving agricultural practices, and an increasing global population, the need for sustainable and efficient crop production has never been more crucial. Integrated Crop Management (ICM) is an innovative approach that brings together the best of traditional wisdom and modern advancements to optimize agricultural systems and ensure food security for generations to come.

This book, "Integrated Crop Management" aims to provide a comprehensive guide to understanding and implementing the principles of ICM. It delves into the intricate web of interactions that exist within agroecosystems, exploring the relationships between soil health, crop nutrition, pest and disease management, water management, and the socioeconomic aspects of farming.

The concept of Integrated Crop Management goes beyond the traditional approach of relying solely on synthetic inputs or specific management practices. It embraces a holistic perspective, considering the entire agricultural system as an interconnected entity. By adopting this integrated approach, farmers can optimize resource utilization, reduce environmental impacts, enhance biodiversity, and improve the resilience of their crops to withstand various challenges.

This book is a result of a collaboration of experts in the fields of crop science, soil science, pest management, agronomy, and sustainable agriculture. Each chapter has been meticulously crafted to provide a balance of theoretical foundations, practical applications, and case studies from around the world. The aim is to empower farmers, researchers, extension workers, and policymakers with the knowledge and tools they need to successfully implement Integrated Crop Management.

Throughout the pages of this book, you will embark on a journey that explores the importance of soil health, the role of integrated pest management, the significance of precision agriculture, and the benefits of crop diversification. You will discover innovative techniques such as conservation agriculture, organic farming, agroforestry, and the integration of digital technologies into farming practices. Furthermore, you will gain insights into the socioeconomic aspects of ICM, including market trends, policy frameworks, and the potential for enhancing livelihoods in rural communities.

It is our hope that this book will serve as a valuable resource for anyone interested in sustainable agriculture and the future of crop production. By embracing the principles of Integrated Crop Management, we can work together to achieve a balance between productivity, profitability, and environmental stewardship. Our collective efforts can pave the way for a resilient and sustainable agricultural system that nourishes both people and the planet.

We would like to express our sincere gratitude to the contributors who have generously shared their expertise and experiences in the field of Integrated Crop Management. Their dedication and passion for sustainable agriculture have made this book possible. We also extend our gratitude to the readers for their interest in this important subject matter.

Mubshar Hussain, Sami Ul-Allah, and Shahid Farooq

Editors

Editorial

Integrated Crop Management in Sustainable Agriculture

Mubshar Hussain [1,2,*]**, Sami Ul-Allah** [3] **and Shahid Farooq** [4]

1 Department of Agronomy, Bahauddin Zakariya University, Multan 60800, Pakistan
2 School of Veterinary and Life Sciences, Murdoch University, 90 South Street, Murdoch, WA 6150, Australia
3 College of Agriculture, University of Layyah, Layyah 31200, Pakistan; samipbg@bzu.edu.pk
4 Department of Plant Protection, Faculty of Agriculture, Harran University, Sanlıurfa 63050, Turkey; shahid@harran.edu.tr
* Correspondence: mubashir.hussain@bzu.edu.pk or mubashiragr@gmail.com

Citation: Hussain, M.; Ul-Allah, S.; Farooq, S. Integrated Crop Management in Sustainable Agriculture. *Agriculture* **2023**, *13*, 954. https://doi.org/10.3390/agriculture13050954

Received: 20 April 2023
Accepted: 23 April 2023
Published: 26 April 2023

Integrated crop management (ICM) aims to balance economic, environmental, and social factors in crop production. The ICM involves different crop management practices and technologies to increase crop yields, reduce environmental damage, and sustain crop production. The ICM is a whole-systems approach based on knowledge and stresses the importance of understanding local ecosystems and changing management practices to be better suited to these ecosystems.

The rapid increase in global population has resulted in the construction of new residential colonies on fertile agricultural lands, resulting in a major decrease in the area devoted to crop production. In addition, abrupt changes in regional and global climates are resulting in various biotic and abiotic stresses, which make agricultural production more challenging. This situation requires sustainable agricultural production on the available cultivated lands. Several management options are employed to sustain crop production, including the creation of climate-resilient genotypes, integrated soil and crop management, etc. The ICM combines several eco-friendly practices to sustain agricultural production. These practices include seed priming [1], the application of organic and inorganic amendments [2–4], the use of plant growth promoting bacteria [1], the application of macro- and micronutrients [2,4], biofortification, the use of biopesticides [5], the use of high-yielding genotypes [6], alternative cropping systems [7–9], the conservation of natural enemies [10], etc. The use of these management measures improves soil health and crop yield. Nevertheless, the environment, soil type and fertility, and crop type have a significant impact on the advantages of ICM. Reduced soil quality, which favors insect pest infestation and lowers farm earnings, is a result of monocropping or the adoption of the same crop rotation strategy. Combining exhausting and restorative crops in crop rotation, intercropping, and relay cropping improved soil health, crop nutrition, crop yield, and net returns. Hence, the advantages of integrated crop management should not be evaluated only on the short-term yield response, but rather on a long-term basis. This Special Issue, entitled "Integrated Crop Management in Sustainable Agriculture," focuses on the impacts of ICM practices on soil health, crop productivity, and a reduction in the impacts of expected climate changes on crop production in a sustainable manner.

Implementing diverse crop rotations helps maintain soil fertility, reduce pest and disease pressures, and minimize the risk of monoculture-related issues [7,9]. Monitoring soil nutrient levels and applying fertilizers in a targeted, efficient manner ensures optimal crop nutrition while minimizing nutrient leaching and environmental contamination [2–4]. Selecting and breeding crop varieties with desirable traits, such as disease resistance [11] and drought tolerance, can improve overall crop performance and resilience [6].

The inclusion of high-yielding crops in existing cropping systems can increase profitability; however, it cannot affect the overall productivity of other crops grown in the system. For example, the inclusion of Bt cotton (*Gossypium hirsutum* L.) genotypes in Pakistan's cotton–wheat (*Triticum aestivum* L.) cropping system modified nutrient availability, while the productivity of winter crops following cotton remained unchanged. Nevertheless, the economic analysis revealed that the Bt cotton–wheat cropping system was more profitable than the conventional non-Bt cotton–wheat cropping system practiced in the country [7]. Hence, the use of high-yielding crops or crop varieties in the existing cropping systems can improve farm profits. Similarly, the inclusion of a leguminous crop, i.e., mungbean (*Vigna radiata* L.) in barley (*Hordeum vulgare* L.)-based cropping systems, improved soil health to a significant extent. Furthermore, including an allelopathic crop, i.e., sorghum (*Sorghum bicolor* L.), in the cropping systems lowered weed infestation in the barley crop [9]. Hence, improved crop rotations exert notable positive impacts on crop productivity and weed infestation. Furthermore, legume species significantly altered soil microorganisms and nitrogen cycling. Common beans (*Phaseolus vulgaris* L.) grown with chicken dung biochar resulted in a diversified prokaryotic community compared to other legume species [8].

Nutrient management is another strategy employed to improve crop production in diverse environments. The combined application of phosphorus and potassium improved sorghum's fodder yield and quality [2]. The application of nutrients recommended by the nutrient expert model and leaf color chart significantly increased the yield and net economic returns of various rice (*Oryza sativa* L.) genotypes. Therefore, these strategies could be used to improve rice productivity and nutrient use efficiency [4]. Biochar can increase crop yield and soil health when applied at an appropriate dose. The combined application of manure and biochar considerably improved the chemical properties of manure, plant biomass, and soil chemical parameters. Hence, manure characteristics, soil health, and plant biomass can be improved through the application of a high biochar dose along with manure [3].

Plant growth-promoting rhizobacteria are widely used to sustain crop productivity in stressful and benign environments. Furthermore, priming seeds before sowing with different compounds also improves crop productivity and quality. Seed priming with boron and seed inoculation with boron-tolerant bacteria improved dry matter accumulation, yield, and economic returns of chickpea (*Cicer arietinum* L.) sown in arid and semi-arid environments. Furthermore, foliar application of boron improved grain B concentration. Hence, different application methods can be used to achieve the desired results [1].

Genotypes play a significant role in sustainable agriculture. The selection of genotypes with higher yields and economic returns helps to attain higher profits. The genotypes can be selected through large-scale adaptability trials and the best-performing can be recommended for cultivation. The exotic genotypes can be recommended for cultivation after thorough testing [6]. Nevertheless, recommended genotypes require an optimum growing environment, and sowing density is one of the most important factors affecting growth and productivity. Optimizing sowing density in stressful and benign environments is a helpful approach to improving crop yield and productivity [12].

Growing environmental concerns necessitate the development of environmentally friendly pest management options. Plant-based pesticides are being developed to manage different pests in field crops. Encapsulation using synthetic zeolite, natural zeolite, and gelatin created an eco-friendly biopesticide from clove bud essential oil that could be used as a biopesticide to manage disease and pest infestation [5]. Similarly, preserving natural enemies through the safe use of pesticides could also lower their adverse impacts on the environment. Timely use of insecticides with low impacts on non-target organisms can preserve natural enemies [10].

Disease prediction models can predict the onset of disease and recommend suitable management strategies. The onset of the potato leaf roll virus was predicted by stepwise regression with a high degree of success. Furthermore, foliar application of salicylic acid in combination with acetamiprid proved the most effective treatment against the disease's incidence and its vector [11].

The use of any of these ICM strategies could be helpful in sustaining crop production. Several other strategies not reported here, i.e., nanoparticles, plant extracts, precision models, and climate-resilient genotypes, could also be used to improve crop productivity. We hope that studies related to these aspects will be published in the future.

Funding: This research received no external funding.

Conflicts of Interest: The authors declare no conflict of interest.

References

1. Mehboob, N.; Yasir, T.A.; Hussain, S.; Farooq, S.; Naveed, M.; Hussain, M. Osmopriming combined with boron-tolerant bacteria (*Bacillus sp.* MN54) improved the productivity of desi chickpea under rainfed and irrigated conditions. *Agriculture* **2022**, *12*, 1269. [CrossRef]
2. Atique-ur-Rehman; Qamar, R.; Altaf, M.M.; Alwahibi, M.S.; Al-Yahyai, R.; Hussain, M. Phosphorus and potassium application improves fodder yield and quality of sorghum in aridisol under diverse climatic conditions. *Agriculture* **2022**, *12*, 593. [CrossRef]
3. Hammerschmiedt, T.; Holatko, J.; Kucerik, J.; Mustafa, A.; Radziemska, M.; Kintl, A.; Malicek, O.; Baltazar, T.; Latal, O.; Brtnicky, M. Manure maturation with biochar: Effects on plant biomass, manure quality and soil microbiological characteristics. *Agriculture* **2022**, *12*, 314. [CrossRef]
4. Bhatta, R.D.; Paudel, M.; Ghimire, K.; Dahal, K.R.; Amgain, L.P.; Pokhrel, S.; Acharya, S.; Kandel, B.P.; Aryal, K.; Bishwas, K.C.; et al. Production and profitability of hybrid rice is influenced by different nutrient management practices. *Agriculture* **2021**, *12*, 4. [CrossRef]
5. Milićević, Z.; Krnjajić, S.; Stević, M.; Ćirković, J.; Jelušić, A.; Pucarević, M.; Popović, T. Encapsulated clove bud essential oil: A new perspective as an eco-friendly biopesticide. *Agriculture* **2022**, *12*, 338. [CrossRef]
6. Ashfaq, M.; Zhu, R.; Ali, M.; Xu, Z.; Rasheed, A.; Jamil, M.; Shakir, A.; Wu, X. Adaptation and high yield performance of honglian type hybrid rice in pakistan with desirable agricultural traits. *Agriculture* **2023**, *13*, 242. [CrossRef]
7. Marral, M.W.R.; Ahmad, F.; Ul-Allah, S.; Atique-ur-Rehman; Farooq, S.; Hussain, M. Influence of transgenic (Bt) cotton on the productivity of various cotton-based cropping systems in Pakistan. *Agriculture* **2023**, *13*, 276. [CrossRef]
8. Kimura, A.; Uchida, Y.; Madegwa, Y.M. Legume species alter the effect of biochar application on microbial diversity and functions in the mixed cropping system—Based on a pot experiment. *Agriculture* **2022**, *12*, 1548. [CrossRef]
9. Naeem, M.; Minhas, W.A.; Hussain, S.; Ul-Allah, S.; Farooq, M.; Farooq, S.; Hussain, M. Barley-based cropping systems and weed control strategies influence weed infestation, soil properties and barley productivity. *Agriculture* **2022**, *12*, 487. [CrossRef]
10. Nadeem, A.; Tahir, H.M.; Khan, A.A.; Idrees, A.; Shahzad, M.F.; Qadir, Z.A.; Akhtar, N.; Khan, A.M.; Afzal, A.; Li, J. Response of natural enemies toward selective chemical insecticides; used for the integrated management of insect pests in cotton field plots. *Agriculture* **2022**, *12*, 1341. [CrossRef]
11. Ali, Y.; Raza, A.; Aatif, H.M.; Ijaz, M.; Ul-Allah, S.; Rehman, S.u.; Mahmoud, S.Y.M.; Farrag, E.S.H.; Amer, M.A.; Moustafa, M. Regression modeling strategies to predict and manage potato leaf roll virus disease incidence and its vector. *Agriculture* **2022**, *12*, 550. [CrossRef]
12. Szczepanek, M.; Lemańczyk, G.; Nowak, R.; Graczyk, R. Response of Indian dwarf wheat and Persian wheat to sowing density and hydrothermal conditions of the growing seasons. *Agriculture* **2022**, *12*, 205. [CrossRef]

Article

Production and Profitability of Hybrid Rice Is Influenced by Different Nutrient Management Practices

Ram Datta Bhatta [1,*], Mahendra Paudel [2], Kishor Ghimire [1], Khem Raj Dahal [1], Lal Prasad Amgain [3], Samiksha Pokhrel [4], Samikshya Acharya [5], Bishnu Prasad Kandel [6], Krishna Aryal [7], Bishwas K. C. [1], Meena Pandey [8], Yadav KC [9], Samrat Paudel [10], Milan Subedi [7], Bhoj Raj Pant [11], Tirtha Raj Bajgai [12], Niranjan Koirala [13], Salama Mostafa Aboelenin [14], Mohamed Mohamed Soliman [15], Gaber EI-Saber Batiha [16] and Jitendra Upadhyaya [7,*]

[1]   Post Graduate Program, Institute of Agriculture and Animal Science, Tribhuvan University, Kathmandu 44600, Nepal; iaaskishor@gmail.com (K.G.); d.khemraj@ymail.com (K.R.D.); vswask754@gmail.com (B.K.C.)

[2]   University Institute of Agricultural Sciences, Chandigarh University, Gharua 140413, India; mahendrapaudelp@gmail.com

[3]   Faculty of Agriculture, Far Western University, Kailali 10900, Nepal; amgainlp@gmail.com

[4]   College of Natural Resource Management, Agriculture and Forestry University, Chitwan 44200, Nepal; pokhrelsamiksha980@gmail.com

[5]   Institute of Agriculture and Animal Science, Prithu Technical College, Tribhuvan University, Dang 22400, Nepal; acharyasamikshya727@gmail.com

[6]   Purbanchal Agriculture Campus, Agriculture and Forestry University, Jhapa 57200, Nepal; bkandel33@gmail.com

[7]   Institute of Agriculture and Animal Science, Rampur Campus, Chitwan 44200, Nepal; krish.aryal2014@gmail.com (K.A.); myln_subedi@hotmail.com (M.S.)

[8]   Paklihawa Campus, Institute of Agriculture and Animal Science, Tribhuvan University, Bhairahawa 32900, Nepal; pandeymeena999@gmail.com

[9]   Central Campus of Technology, Department of Food Technology, Tribhuvan University, Dharan 56700, Nepal; ykcdng504@gmail.com

[10]   Department of Biotechnology, Kathmandu University, Dhulikhel 45200, Nepal; samratpaudel92@gmail.com

[11]   Nepal Academy of Science and Technology, Khumaltar 00977, Nepal; bhojraj.pant@nast.gov.np

[12]   Minhas Microbrewery, Distillery and Winery, 1314 44 Ave NE, Calgary, AB T2E 6L6, Canada; tirraj@yahoo.com

[13]   Department of Natural Products Research, Dr. Koirala Research Institute for Biotechnology and Biodiversity, Kathmandu 44600, Nepal; Koirala.biochem@gmail.com

[14]   Biology Department, Turabah University College, Taif University, Al Hawiyah 21995, Saudi Arabia; s.aboelenin@tu.edu.sa

[15]   Clinical Laboratory Sciences Department, Turabah University College, Taif University, Al Hawiyah 21995, Saudi Arabia; mmsoliman@tu.edu.sa

[16]   Department of Pharmacology and Therapeutics, Faculty of Veterinary Medicine, Damanhour University, Damanhour 22511, Egypt; gaberbatiha@gmail.com

*   Correspondence: rambhatta2436@gmail.com (R.D.B.); jitendra.upadhyaya@mcgill.ca (J.U.)

Citation: Bhatta, R.D.; Paudel, M.; Ghimire, K.; Dahal, K.R.; Amgain, L.P.; Pokhrel, S.; Acharya, S.; Kandel, B.P.; Aryal, K.; K. C., B.; et al. Production and Profitability of Hybrid Rice Is Influenced by Different Nutrient Management Practices. *Agriculture* 2022, 12, 4. https://doi.org/10.3390/agriculture12010004

Academic Editors: Mubshar Hussain, Sami Ul-Allah and Shahid Farooq

Received: 20 October 2021
Accepted: 25 November 2021
Published: 21 December 2021

Publisher's Note: MDPI stays neutral with regard to jurisdictional claims in published maps and institutional affiliations.

Abstract: The government of Nepal has recommended blanket fertilizer application for rice cultivation, which results in lower nutrient use efficiency (NUE) particularly under rainfed conditions. With the aim of finding an appropriate nutrient management practices concerning rice production and profitability, a field experiment was conducted during rainy season of 2017 and 2018 at Kavrepalanchowk and Dang district of Nepal. Altogether, five treatments comprising various nutrient management practices viz. Nutrient Expert Model (NE), use of Leaf Color Chart (LCC), Government Recommended Fertilizer Dose (GON), Farm Yard Manure (FYM), and Farmers' Field Practice (FFP), were laid out in RCBD with four replications in farmers' fields. The analysis of variance showed significant difference between treatments for test weight and grain yield in Kavrepalanchowk whereas all traits except number of effective tillers were significant in Dang. The significantly higher grain yield and harvest index were obtained in NE, followed by LCC; and the overall straw yield was highest in LCC, followed by NE in both the locations. Also, yield gap analysis suggested the NE had 44.44% and 23.97% increase in yield as compared to FPP in Kavrepalanchowk and Dang, respectively. The combined analysis with Best Linear Unbiased Estimator revealed the interaction of nutrient management and location significantly effects the straw yield and harvest index across both the locations. The estimated mean straw yield and harvest index were 10.93 t/ha and 34.98%, respectively. Both correlation study and biplot of principal component analysis signaled grain yield had positive correlation with all other traits. Furthermore, the net revenue was maximum for NE, followed by LCC in both the locations. The benefit: cost ratio was highest for NE which was 1.55 in Kavrepalanchowk and 2.61 in Dang. On the basis of these findings, NE and LCC can be effectively used as nutrient management practice by the farmers to obtain maximum production and profitability in Rice.

Keywords: hybrid rice; nutrient management practices; production and profitability

1. Introduction

Rice (*Oryza sativa*) is important cereals fulfilling food necessities of more than half of the world's population [1]. Globally, rice is cultivated over 162.06 million hectares, producing 504.17 million metric tons of milled rice in 2019 [2]. Rice is mainly grown and consumed in Asian countries (China, India, Indonesia, Bangladesh, Vietnam) with 80% of global production [3]. In Nepal, rice is cultivated in 1,458,915-hectare, producing 5,550,878 metric tons of rice, with an average productivity of 3.8 tons/ha during fiscal year 2019–2020 [4]. Nepal's Gross Domestic Product (GDP) is largely sustained by agriculture, of which rice alone represents 20.75 percent [5]. Nepal's current rice demand (2512 tons currently) is predicted to double by 2030 (4518 tons) due to population pressure. Further, the study of [6] forecasted that household demand and production would fluctuate from 19% to 80% by 2030. It is necessary to increase rice production and productivity to close the gap in supply and demand with the limited resource available, owing to the unstable rice yield due to insect, pest, nematodes, declination of soil fertility, imbalanced fertilizer use and poor nutrient management practices [7,8].

In Nepal, rice is grown with urea as the primary nitrogen source, but plants utilize only 30% of the applied urea, whereas the rest 70% is lost due to NH3 volatilization, surface runoff and leaching in lowland rice therefore, rice farmers need to maximize nitrogen use efficiency (NUE) while maintaining low nitrogen inputs [9,10]. Concept of Site-Specific Nutrient Management (SSNM) was developed to apply nutrients at the optimal rates with maximum nutrient use efficiency [11]. SSNM is a component of precision agriculture as it combines the nutrient requirements of plants at various growth stages with the soil's ability to supplying them. Further, SSNM integrates with Nutrient Expert; computer-based tool to estimate fertilizer requirement considering nutrient use efficiency, estimated yield, along with nutrient balance and added nutrients effect on yield [12]. SSNM principles recommend fertilizer based on the 4'R'-right dose, right method, right source, and right timing [13].

There have been different trails deploying Nutrient Expert in wheat and maize [14], in wheat [15] and in Rice [16]. These studies have demonstrated the significance of nutrient experts' recommendations over practices adapted by farmers concerning yield and profitability. The amount of fertilizer required for a field is calculated from the predicted yield in response to every fertilizer nutrient which is the contrast between the attainable yield and the nutrient-limited yield. Further, Nutrient osmosis trials are performed in farmer's fields to determine the attainable yield Attainable yield is the yield of particular without any nutrient limitation under best management practice in the farmer's location [17].

Government of Nepal's fertilizer dose recommendations are based on the fertility status of particular region in Nepal, rather than the soil fertility of the individual farmer. This leads to a need for SSNM techniques such as Leaf color chart (LCC) due to the lack of adequate dissemination of the developed approach. Thus, this research aims to evaluate the production and profitability of hybrid rice through different nutrient management practices at two locations; and to analyze the traits associated with higher yield.

2. Materials and Methods

2.1. Study Site and Weather Conditions

A field experiment was conducted in farmers' field of Mahadevsthan of Kavrepalanchok (Kavre) during rainy season of 2017 and in Lamahi municipality, Dang district during the rainy season of 2018. The site of Kavre is at 27°71′ North latitude and 85°61′ East longitudes which is 670 m above MSL. The site of Dang is at 27.8771° N, 82.5727° E, with 250 above meter sea level. An experimental site is located in a dry rain-fed region of Nepal with a subtropical climate consisting of wet summers and dry winters. The weather of the research area is presented in Figure 1.

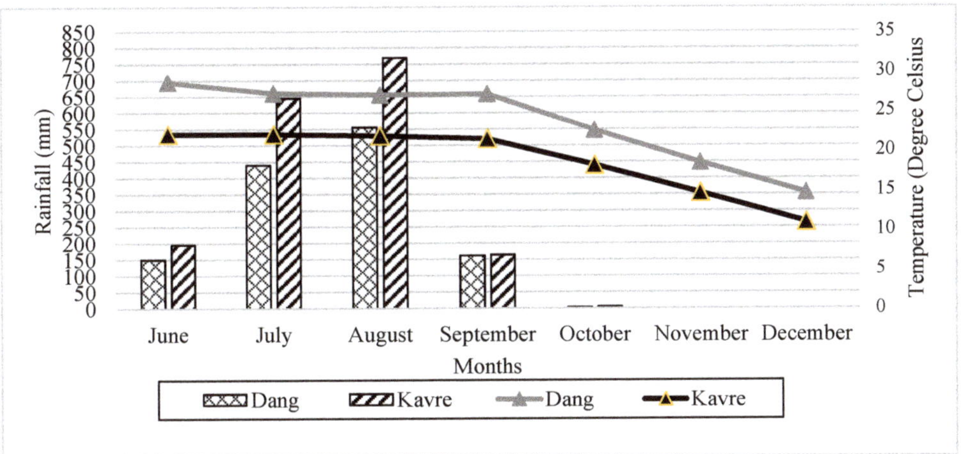

Figure 1. Climatic data of Research area.

2.2. Experimental Setup and Crop Management

A field experiment was conducted in Kavre and Dang districts of Nepal during the kharif seasons of 2017 and 2018 with hybrid rice transplanted to the main field after 22 days in the nursery bed as per farmer's practice. The details of soil chemical properties are given on Table 1. Experimental treatments were laid out on Randomized Complete Block Design with four replications in both locations. Seedlings were transplanted one per hill with a spacing of 25 cm × 25 cm i.e., the seed rate was 10 kg/ha. Each plot was fertilized according to the treatment (nutrient management practice) assigned to it in Table 2.

Table 1. Soil chemical properties of research area.

Properties	pH Value:	Organic Matter:	Total Nitrogen:	Available Phosphorus (P_2O_5):	Available Potassium (K_2O):	Zinc:	Boron:
Dang	7.091	2.71%	0.13%	73.128 kg/ha	396.539 kg/ha	0.612 ppm	1.283 ppm
Kavre	6.088	1.81%	0.07%	127.277 kg/ha	285.611 kg/ha	1.561 ppm	0.897 ppm

Three fertilizers named as Urea (46% N), Diammonium phosphate (18% N and 46% P_2O_5) and Muriate of Potash (60% K_2O) were used as a source of Nitrogen, Phosphorus and Potassium as per guidelines of Government of Nepal. In the Nutrient Expert (NE), fertilizer dose along with the application time as per the recommendation of NE model was applied. The model was run with the data of surveyed (interviewed farmer with questionnaire). In the Leaf Color Chart (LCC), P & K was applied as per the recommendation made by NE and N was applied as per need on the basis of Leaf Color Chart (LCC) score. In the Government of Nepal recommended fertilizer dose (GON), fertilizer in each plot was applied as per the recommendation by National Agriculture Research Council (NARC);

140:60:30 kg NPK ha^{-1} [18]. In the Farmyard Manure (FYM), FYM was applied to each field at 15-ton ha^{-1}.

Table 2. Treatment details for the experiment in both Kavre and Dang.

Treatment Number	Symbol	Treatment Combination
T$_1$	NE	Nutrient Expert recommended dose at Kavre -Farmer 1: 141:39:72 kg NPK/ha -Farmer 2: 132:38:61 kg NPK/ha -Farmer 3: 132:38:61 kg NPK/ha -Farmer 4–132:38:61 kg NPK/ha Nutrient Expert recommended dose at Dang -Farmer 1: 118:37:52 kg NPK/ha -Farmer 2: 93:23:29 kg NPK/ha -Farmer 3: 109:28:46 kg NPK/ha -Farmer 4: 109:28:46 kg NPK/ha
T$_2$	LCC	Nitrogen as per LCC and P & K calculated from Nutrient Expert
T$_3$	GON	Government of Nepal recommended dose -140:60:30 kg NPK/ha (Shah and Yadav, 2001)
T$_4$	FYM	-Farm Yard Manure (FYM) (15-ton ha^{-1})
T$_5$	FFP	Farmers' existing practices at Kavre -Farmer 1: 60.4:36.8:36 kg NPK/ha -Farmer 2: 71.2:64.4:0 kg NPK/ha -Farmer 3: 45.08:36.8:0 kg NPK/ha -Farmer 4: 31.8:46:0 kg NPK/ha Farmer's existing practice at Dang -Farmer 1, 2, 3, and 4: 16.5:15:0 kg NPK/ha

NE, Nutrient Expert; LCC, Leaf Color Chart, GON, Government of Nepal recommended dose of fertilizer; FFP, Farmer Field Practice.

2.3. Measurement and Data Collection

The agro-morphological traits such as plant height, number of effective tillers per meter square, length of panicle, test weight, straw and grain yield were recorded. The total cost and gross revenue of the rice production was calculated. The harvest index and fertility were taken and economic parameters (B:C ratio) were evaluated by using following formulae:

a. Fertility: Fertility was calculated as per [19] i.e.,

$$\text{Fertility\%} = (\text{number of fertile grains/total number of florets}) \times 100$$

b. Harvest Index: Harvest index (HI) was computed by

$$\text{H.I.\%} = (\text{grain yield/(grain yield} + \text{straw yield)}) \times 100$$

c. Benefit cost (B:C) ratio was calculated by using gross return and cost of cultivation [20]; Benefit: cost ratio = Net return/cost of cultivation

2.4. Data Analysis

MS Excel was used to enter and process all the data. An analysis of variance (ANOVA) was carried out using R version 3.6.0. R package-Agricoale to test for significant difference between treatment means at the 0.05 level of significance. The combined analysis of two sites was carried out by Meta-R with the Best Linear Unbiased Estimator (BLUE). MINITAB 14.0 was used to prepare the principal component biplot.

3. Results and Discussion

3.1. Result of Mean Performance of Rice Traits

3.1.1. Plant Height (PH)

An analysis of variance revealed a highly significant difference ($p < 0.01$) between the treatments for plants in Dang, but while in Kavre difference was not significant statistically (Figure 2). The overall mean of the plant height was 91.63 cm. The mean plant height was 87.47 cm and 95.78 cm in Kavre and Dang, respectively. Plant height was maximum in FYM (91.58 cm) and minimum in NE (83.76 cm) at Kavre while at Dang maximum plant height was observed in NE (99.89 cm), followed by GON (98.41 cm) and minimum plant height was observed in FFP (88.39), respectively). These findings are supported by report of [21].

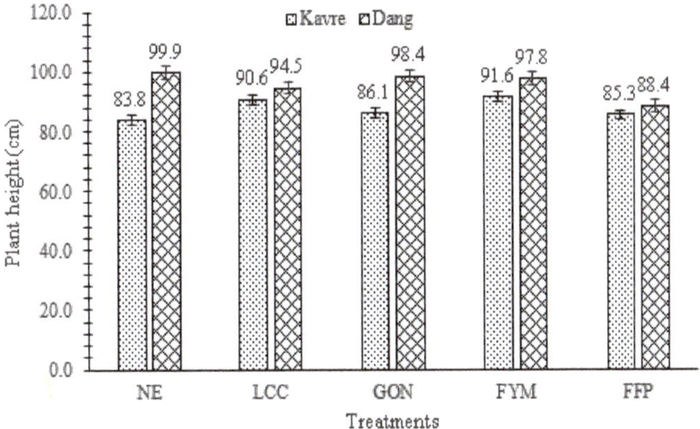

Figure 2. Effect of nutrient management practices on Plant Height of Rice at Kavrepalanchowk and Dang.

3.1.2. Number of Effective Tiller (NOET)

A significant difference ($p < 0.05$) was found for the number of effective tillers per m^2 at Kavre, while it was not significant at Dang (Figure 3). The mean number of tillers per m^2 was 258, 363, and 315.50 in Kavre, Dang and the overall field, respectively. The number of effective tillers per m^2 was highest in GON plots (278.40) followed by LCC (273.00) and minimum in FFP plots (224.80) in Kavre while LCC had the maximum number of effective tillers per m^2 (387.65) and NE had the lowest number of effective tillers per m^2 (355.20) at harvest in Dang.

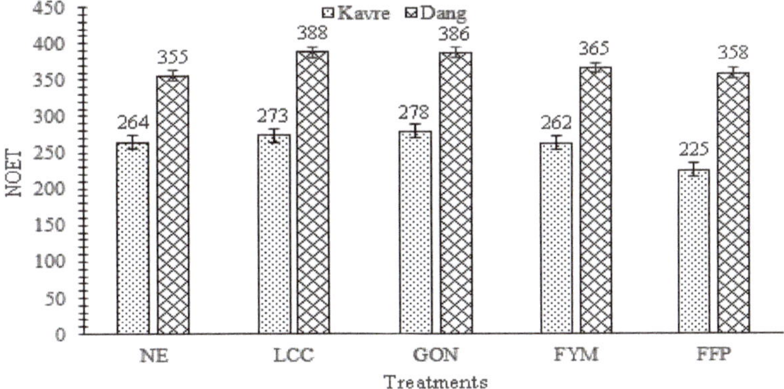

Figure 3. Effect of nutrient management practices on Number of Effective Tillers per Square Meter of Rice at Kavrepalanchowk and Dang.

3.2. Panicle Length (PL)

Panicle length was statistically significant ($p < 0.01$) in Dang while non-significant in Kavre (Figure 4). The overall mean of the panicle length was 24.76 cm. Kavre and Dang had mean panicle lengths of 26.92 cm and 22.59 cm, respectively.

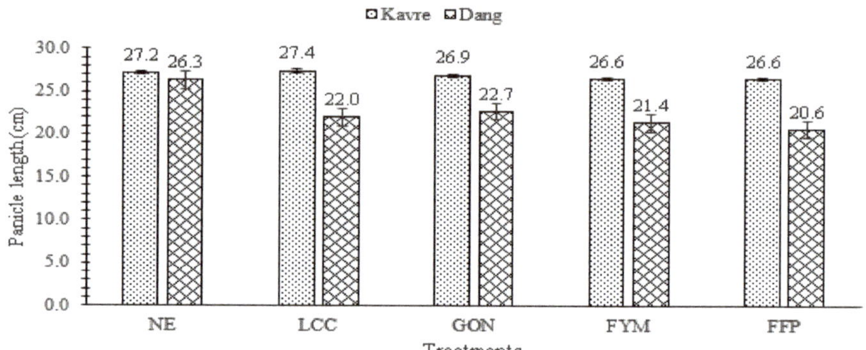

Figure 4. Effect of nutrient management practices on Panicle Length of rice at Kavrepalanchowk and Dang.

3.3. Kavre

LCC had the longest panicle length (27.41 cm), followed by NE (27.18 cm), and FYM and FPP both had the shortest panicles (26.56 cm). These result are in line with result obtained by [22].

3.4. Dang

The longest panicle length was found in NE (26.31 cm) which was statically superior to all other treatments. The shortest panicle was found in FPP (20.61 cm) which was statistically at par with LCC, GON, and FYM. Similar result was obtained by [23].

3.5. Fertility (FT)

In Kavre, there was no significant difference between treatments in regards to percentage fertility, while in Dang, there was significant difference (Figure 5). The overall mean of the fertility was 77.40%. The mean percentage of fertility in Kavre and Dang were 72.6% and 82.2%, respectively. The percentage of fertility ranged from 68.5% to 80.25% with highest fertility recorded in LCC and lowest fertility recorded in FYM in Kavre. In dang, the percentage of fertility ranged from 67.24% to 90.52%. The highest fertility recorded in FYM was statistically superior to other treatments. It was followed by LCC (88.12%) and NE (87.56%). The lowest fertility recorded in FPP.

3.6. Test Weight

The ANOVA showed significant effects ($p < 0.01$) of treatments on test weight of rice in both Kavre and Dang (Figure 6). The mean thousand grain weight was 15.7 g in Kavre and was 15.75 g in Dang. The overall mean was 15.73 g.

In Kavre, the maximum test weight recorded was 17.0 g in NE which was followed by LCC (16.5 g). While, the minimum test weight recorded was 14.0 g in FYM. Similarly, in Dang, the maximum test weight recorded was 17.00 g in FYM and LCC. While, the minimum test weight recorded was 14.0 g in FPP.

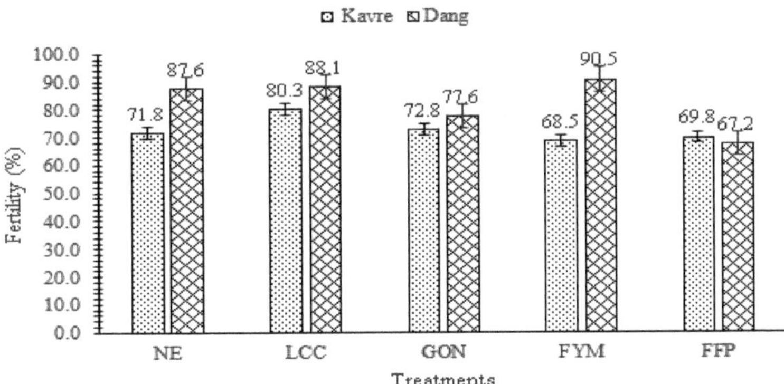

Figure 5. Effect of nutrient management practices on Fertility of rice at Kavrepalanchowk and Dang.

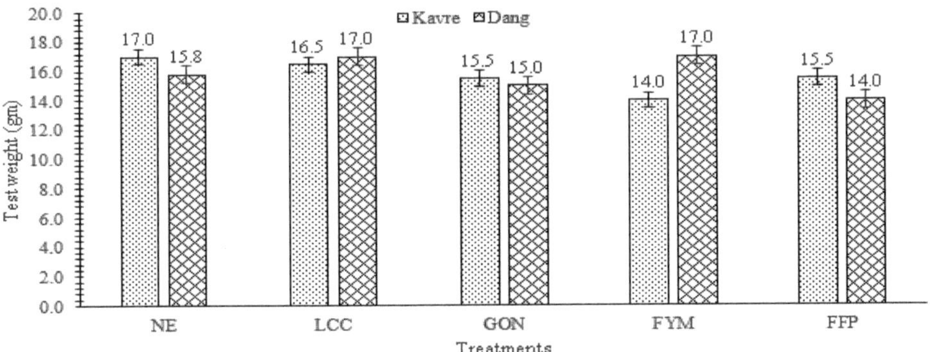

Figure 6. Effect of nutrient management practices on test weight of rice at Kavrepalanchowk and Dang.

3.7. Grain Yield

3.7.1. Kavrepalanchowk

The ANOVA elucidated significant effects of treatments on grain yield of rice at 0.05 level of significance (Figure 7). The highest grain yield (5.84 t/ha) was recorded in NE management followed by LCC (5.61 t/ha), GON (5.34 t/ha) and FYM (5.14 t/ha). The lowest yield (4.10 t/ha) was obtained in FYM.

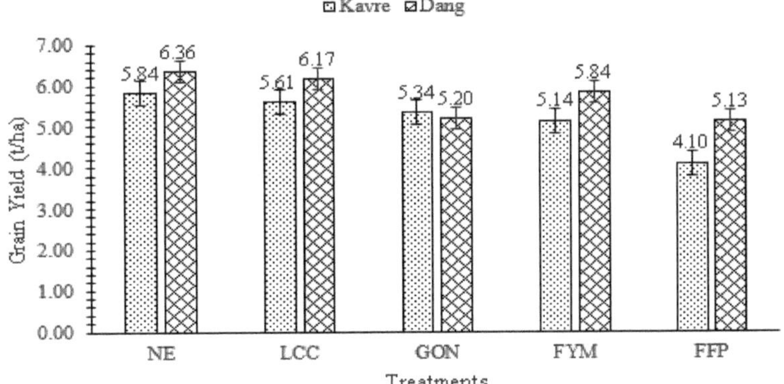

Figure 7. Effect of nutrient management practices on grain yield of rice at Kavrepalanchowk and Dang.

3.7.2. Dang

The ANOVA elucidated treatments caused significant variation on the grain yield of rice at 0.01 level of significance. The highest grain yield of rice (6.36 t/ha) was recorded in NE in NE management followed by LCC (6.17 t/ha), and FYM (5.8004 t/ha). The lowest yield (5.13 t/ha and 5.20 t/ha) was obtained in FPP and GON, statically at par with FYM.

3.8. Straw Yield

There was significant difference between nutrient management treatments for straw yield in Kavre and Dang (Figure 8). The mean straw yield in Kavre was 10.9 t/ha and in Dang was 10.97 t/ha. The overall mean was found to be 10.93 t/ha. The highest straw yield was 12.04 t/ha in FFP, followed by 11.50 t/ha in GON statistically at par with FFP while the lowest yield was found in LCC (9.50 t/ha) in Kavre. In Dang, the highest straw yield was found in LCC (12.87 t/ha) statistically at par with NE (12.62 t/ha). The lowest straw yield was found in FFP (12.04 t/ha).

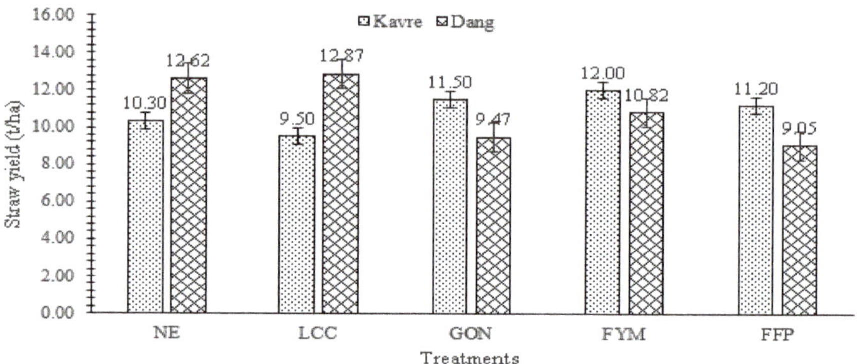

Figure 8. Effect of nutrient management practices on straw yield of rice at Kavrepalanchowk and Dang.

3.9. Harvest Index

The harvest index between treatments differed significantly in Dang while it did not differ significantly in Kavre (Figure 9). The mean harvest index was 33.92% and 35.76% in Kavre and Dang, respectively.

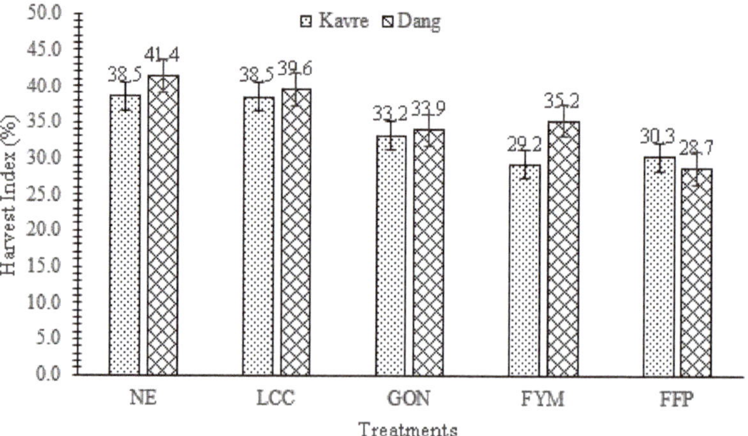

Figure 9. Effect of nutrient management practices on harvest index of rice at Kavrepalanchowk and Dang.

3.9.1. Kavrepalanchowk

In terms of harvest index, NE had the highest harvest index with 38.52%, followed by LCC with 38.47%. The lowest harvest index was found in FYM (29.20%). Ref. [24] reported NE in wheat produces 38% higher harvest index than FFP, which is consistent with our findings.

3.9.2. Dang

Significantly higher harvest index was obtained from NE fertilizer recommendation (41.38%) than other treatments, followed by LCC (39.57%) which are statistically at par. The lowest harvest index found was 28.73% in FYM.

4. Discussion on Mean Performance of Rice Traits

This research was similar to one by [25]. where balanced use of fertilizer boosted plant growth and produced maximum height, whereas poorly fertilized plots produced plants with minimal height. The application of macronutrients according to site requirements leads to higher plant heights in rice [26,27]. Further, this finding is in line with [28], who reported SSNM had more effective tillers than FFP. Optimal fertilizer application leads to higher tiller productivity per area when nutrient-balanced fertilizers are used [29–31] mention a similar pattern of straw yield under different nutrient management. The reason for high straw yield in LCC and NE because of balance nitrogen applied that influenced the vegetative growth, plant height and number of tillers leading to a high straw yield.

The nutritional management practices in Kavre and Dang differed significantly in terms of grain yield. A higher yield was recorded in NE, while a lower yield was recorded in FFP. This finding is in line with [32], they performed fifty six on-farm researches in rice and found out SSNM yielded 17% higher than FFP. In addition, they mention SSNM yield is higher than average yield for all other nutrient managements. The evidence of NE superior to over government recommendations or existing practices is reported by [13] using NE for maize and wheat. Hence, NE is far better nutrient management practice over GON and FFP [33]. Further, the authors claimed that rice managed under NE performs better than rice managed under FFP and GON due to the balanced and right source of fertilizer applied at the right time.

The grain yield is a complex trait attributed by different components. Ref. [34] found that higher yield in NE is the result of better panicle length, number of effective tillers per square meter, and test weight, which also result in higher harvest index and 30% less fertilizer use. Similar to this research, lowest HI in FFP and highest HI in NE is reported by [14,35].

The research of [15] also shows that under NE there is 38% more HI. Likewise, Ref. [32] found a trend in HI that is consistent with the results of this study.

4.1. Combined Analysis of Straw Yield and Harvest Index in Kavre and Dang

The Nutrient management and Location Interaction showed significant effect in the combined straw yield and harvest index across both the locations while treatment effect was found to be statistically non-significant as presented in Table 3. These findings suggest that combine performance of the nutrient management for biological yield is influenced by the environment of the location. This implies that combined straw yield and harvest index across both locations are determined by the interaction effect and are not influenced by nutrient management alone.

The mean of the overall straw yield estimated across both the environment was 10.93 t/ha. In terms of straw yield, the highest yield was given by NE (11.45 t/ha) and the lowest yield by GON (10.26 t/ha). The mean of overall straw yield estimated across each environment was found to be 34.98%. The maximum harvest index has been found in GON (37.67%) and the minimum harvest index has been found in FYM (31.14%).

Table 3. Combined Statistics (ANOVA) of Straw Yield and Harvest Index with Best Linear Unbaised Estimator (BLUE) across both the location.

Treatment	SY	HI
NE	11.45	34.65
LCC	11.20	34.50
GON	10.26	37.67
FYM	11.01	31.14
FPP	10.76	36.94
Residual Variance	2.808	16.673
Treatment significance	0.977	0.794
Trt x Loc significance	0.010	0.001
Grand Mean	10.93	34.98
LSD	5.31	15.53
CV	15.33	11.67

4.2. Yield Gap Analysis

4.2.1. Yield Gap at Kavre

In comparison with FFP, yields from NE increased by 42.44%, followed by LCC and GON by 36.82% and 30.24%, respectively (Table 4). Similar result of grain yield increased by 37.62% in NE-Rice recommendation over the FFP [36]. These results are in agreement with [15] in wheat reported NE produced 57% more grain yield than FFP. This implies that in contrast with FFP, different treatments showed increased grain yield.

Table 4. Increment on grain yield of hybrid rice over Farmer fertilizer practice through different nutrient management practices in Kavre and Dang.

Treatments	Kavre			Dang		
	Grain Yield (t/ha)	Yield Difference over FPP	Increase (%)	Grain Yield (t/ha)	Yield Difference over FPP	Increase (%)
NE	5.84	+1.74	42.44	6.36	+1.23	23.97
LCC	5.61	+1.51	36.82	6.17	+1.04	20.27
GON	5.34	+1.24	30.24	5.20	+0.07	1.36
FFP	4.10			5.13	-	-

4.2.2. Yield Gap at Dang

The study showed that NE based nutrient management can produce 1.23 tons ha^{-1} more grain yield than the existing FFP, which is 23.97% higher (Table 5). Additionally, LCC and GON yielded 20.27% and 1.36%, respectively. Shrestha et al. (2018) reported similar results for the Dang condition.

4.3. Correlation among Different Rice Trait across Both the Location

The correlation coefficient between different Rice trait and harvested index is presented in Table 5. The grain yield had significant strong positive correlation with Fertility (0.90), Test Weight (0.93), Straw Yield (0.89), and Harvest Index (0.94). There was significant strong positive correlation of Harvest Index with Test Weight (0.95) and Grain Yield. Further, the correlation between Test Weight, Fertility and Grain Yield was significant. [37] support the significant positive correlation between GY and NOET, PL, and SY. Positive correlation between GY and PL is in line with results of [38–40]; while positive correlation with NOET was supported by finding of by [41,42]. Moreover, [43] reported significant positive association between GY and NOET, TW. Similarly, positive correlation of GY with TW and PL was found in [44]. Ref. [45] also reported a positive relation of plant height and number of tillers per plant with grain yield of rice.

Table 5. Correlation between different traits of rice in both the locations.

	PH	NOET	PL	FT	TW	SY	HI
NOET	0.69						
PL	0.25	0.22					
FT	0.80	0.66	0.45				
TW	0.50	0.50	0.64	0.92 *			
SY	0.78	0.30	0.55	0.86	0.76		
HI	0.42	0.48	0.83	0.80	0.95 *	0.69	
GY	0.68	0.51	0.78	0.90 *	0.93 *	0.89*	0.94 *

* Significant at 0.05 level of significance.

4.4. Principal Component Analysis of Different Rice Traits

The correlation presented the relation among the tested variable. The PCA was performed with the view to clarify the relation and present the variation explained by each trait.

4.5. Eigen Value and Principal Component

The first four principal components explained 100% of the existing variation in which principal component 1 (PC1), PC2, PC3 and PC4 explained 72.2%, 15%, 0.81%, and 0.47% of the variation, respectively (Table 6).

Table 6. Eigen value and Principal Component of different Rice traits.

Traits	PC1	PC2	PC3	PC4
Plant Height	0.310	−0.531	−0.264	−0.406
Number of Effective Tillers	0.258	−0.507	0.679	−0.145
Panicle Length	0.292	0.533	0.167	−0.629
Fertility	0.395	−0.203	−0.076	0.351
Test Weight	0.387	0.151	0.093	0.518
Straw Yield	0.364	−0.024	−0.598	−0.057
Harvest Index	0.380	0.312	0.253	0.145
Grain Yield	0.411	0.126	−0.065	−0.063
Eigenvalue	5.776	1.1965	0.6478	0.3797
Proportion explained	0.722	0.15	0.081	0.047
Cumulative% explained	72.2	87.2	95.3	100

PC–Principal Component.

A positive correlation exists between all tested traits and PC1, and Grain Yield (0.411) is the main contributor to variability.PC2 was positively affected by Panicle Length, Test Weight, Harvest Index, and Grain Yield. Panicle length (0.533) was the major contributor toPC2. PC3 was positively influenced by Number of Effective Tillers, Panicle length, and Test Weight. The major contributing trait was the Number of Effective Tillers (0.679). PC4 was positive correlated with Fertility, Test Weight, and Harvest Index with Test weight (0.518) contributing the highest.

4.6. Biplot of PCA

The biplot of PCA was prepared to identify the trend of traits for main source of variation under both the location. The biplot component accounted 87.2%, PC1 (72.2%) and PC2 (15%), of the variability for the traits (Figure 10).

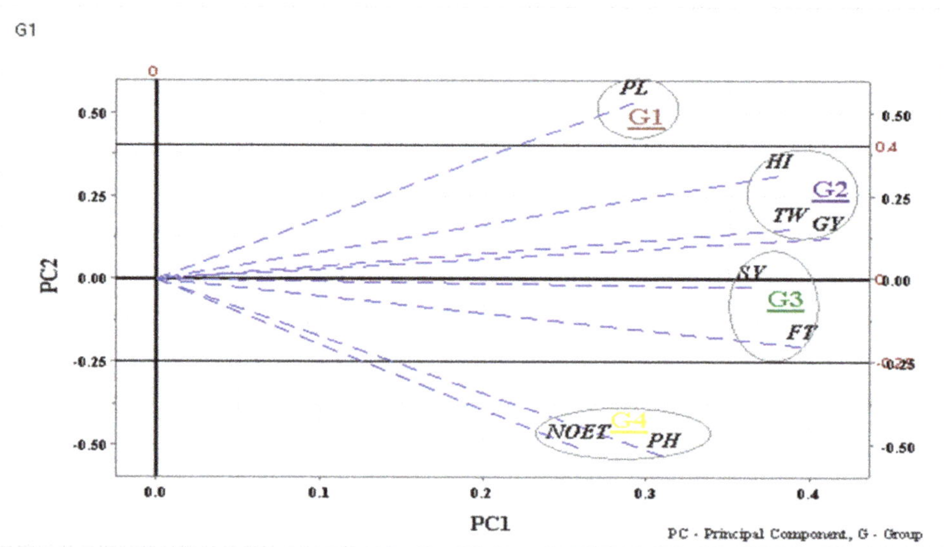

Figure 10. Loading biplot of different Rice traits in both the locations.

The biplot showed PC2 is able to separate the traits. The traits were grouped in four Groups (G). The Group 1 had low positive value of PC1 and high positive value of PC2; and consist of PL only. The Group 2 had high positive value of PC1 and low positive value of PC2 which comprises of TW, GY, and HI. The Group 3 had high positive value PC1 and low negative value of PC2; and consist of SY and FT. Finally, the Group 4 had low positive value of PC1 and high negative value of PC2 which comprises of NOET and PH.

The biplot shows that PC2 separates G1 from G2, G2 from G3 and G3 from G4. Also, PC1 separates G1 and G4 from G2 and G4. The traits in the G2 and G3 had the higher variance than the G1 and G4. Furthermore, GY had the highest variance and NOET had the lowest variance, among all the traits.

The vector angle of the traits has acute angle with each other revealed that each trait are positively correlated with another trait. Among the traits, within G2 (TW, GY and HI) have very small angles so, are strongly correlated to each other. Similarly, traits in G3 (SY and FT), traits in G4 (NOET and PH) are also strongly correlated. The traits in G1 (PL) have low positive correlation with traits in G4 (NOET and PH) because they have large acute angle between their vectors.

Correlation only shows strength of relationship between yield and other traits [46] while PCA is useful of identification of traits of high variability which helps in improvement of grain yield [47]. The similar analysis of different Rice trait with the help of PCA and biplot is observed in the report of [48] and [49]. Ref. [50] reported 65.4% of the total variation was because of first five principal components. This finding is in line with [51,52] and our finding.

4.7. Production Economics at Kavre and Dang

Significant result among the treatments was obtained for calculated economics of production viz. gross revenue, net revenue and benefit: cost ratio of rice production in Kavre and Dang (Table 7).

Table 7. Gross return, net return and B: C ratio of rice influenced by nutrient management practices in Kavrepalanchowk and Dang.

Treatment	Kavre			Dang		
	Gross Revenue (NRs ha^{-1})	Net Revenue (NRs ha^{-1})	B:C Ratio	Gross Revenue (NRs ha^{-1})	Net Revenue (NRs ha^{-1})	B:C Ratio
NE	182,710 [a]	64,516 [a]	1.55 [a]	206,211 [a]	75,925 [a]	2.61 [a]
LCC	161,936 [ab]	42,543 [a]	1.36 [a]	177,761 [b]	62,302 [ab]	2.26 [b]
GON	149,646 [ab]	32,338 [ab]	1.28 [ab]	173,289 [b]	58,380 [b]	2.02 [c]
FFP	151,859 [ab]	40,497 [a]	0.88 [a]	151,395 [c]	51,499 [b]	2.26 [b]
SEM	11,484.6	11,373.3	0.09	1.7328	3.842496	0.17
Grand mean	151,099	24,635	1.22	177,164	62,026.5	2.29
LSD (0.05)	34,618.4	34,282.8	0.25	19,699.41	13,618.16	0.05
CV (%)	15.2	92.3	14	6.72	14.15	1.41
F test	*	**	**	**	*	**

** Significant at 0.01 level of significance, * Significant at 0.05 level of significance, SEM—Standard Error Mean, CV—Coefficient of Variation, LSD—Least Significant Difference at 0.05 level of significance.

4.7.1. Kavre

The mean gross revenue and net revenue of all the nutrient management practices were 151,099 NRs/ha and 23,635 NRs/ha, respectively.

The highest gross revenue was obtained in NE (182,710 NRs/ha) which is followed by LCC (64,516 NRs/ha). The high gross revenue in NE can be attributed to its high grain and straw yield.

Similarly, the net revenue was found higher in NE (64,516 NRs/ha) followed by LCC (42,543 NRs ha^{-1}). The lowest net revenue was found in FYM (−12,869 NRs/ha). Ref. [53] mentioned in their reported that NE tools enables farmer to gain higher profits over the traditional fertilizer use practices.

Furthermore, NE (1.545) recorded the highest B:C ratio followed by LCC (1.356) and then FFP (0.879) with all three statistically at par. While lowest B:C ratio was recorded in FYM (0.879). This higher B:C ratio in NE treatments maybe because of N saving and increased yield of crop which ultimately increased gross and net return and reduced production cost. Ref. [54] reported the reduction of production cost with site specific nutrient management which agrees to findings of present research.

4.7.2. Dang

The highest gross revenue was obtained in NE (206,211 NRs/ha) which is followed by LCC (177,761 NRs/ha) and GON (173,289 NRs/ha). The lowest gross revenue is obtained in FFP (151,395 NRs/ha).

Similarly, the net revenue was found higher in NE (75,925 NRs/ha) which was statistically at par with LCC (62,302 NRs ha^{-1}).

Furthermore, NE (2.61) recorded the highest B:C ratio followed by LCC and FFP both with 1.356 B:C ratio and then GON (2.02) was the lowest B:C ratio recorded.

5. Conclusions

Nepal's rice production is much below its potential due to farmers' imbalanced fertilizer application practices. The proper nutrient management, i.e., the correct amount and timing of fertilizer application, minimizes the yield gap in crops.NE and LCC have both been proven effective in increasing grain and straw yield. Furthermore, the highest benefit-cost ratio for the Rice production was achieved in NE due to its high net revenue for farmers. According to BLUE estimates, the straw yield and harvest index were both affected by nutrient management and location interaction. Furthermore, all the tested traits were found positively correlated with the grain yield of Rice. This experiment concluded that nutrient management practices (Nutrient expert and Leaf Color Chart) are most appropriate for increasing the profitability of farms producing rice.

Author Contributions: Conceptualization, K.G., K.R.D. and L.P.A.; methodology, R.D.B., S.P. (Samiksha Pokhrel), M.P. (Mahendra Paudel), S.A. and B.P.K.; software, K.A., B.K.C., M.P. (Meena Pandey) and Y.K.; validation, S.P. (Samrat Paudel), M.S. and B.R.P.; formal analysis, T.R.B., N.K. and S.M.A.; investigation, S.P. (Samiksha Pokhrel) and M.S.; resources, M.M.S.; data curation, G.E.-S.B.; writing—original draft preparation, R.D.B. and J.U.; visualization, S.P. (Samiksha Paudel); supervision, J.U.; advisor, N.K.; funding acquisition, S.M.A. and M.M.S. All authors have read and agreed to the published version of the manuscript.

Funding: This study was supported by the Taif University Researchers Supporting Project (TURSP-2020/105), Taif University, Taif, Saudi Arabia.

Acknowledgments: We appreciate and thank Taif University for the financial support for Taif University Researchers Supporting Project (TURSP-2020/105), Taif University, Taif, Saudi Arabia.

Conflicts of Interest: The authors declare no conflict of interest.

References

1. Maiti, R.; Sarkar, N.C.; Rodriguez, H.G.; Kumari, A.; Begum, S.; Rajkumar, D. *Advances in Rice Science: Botany, Production, and Crop Improvement*; Apple Academic Press: Palm Bay, FL, USA, 2020.
2. Shahbandeh, M. World Production Volume of Milled Rice from 2008/2009 to 2020/2021. 2021. Available online: https://www.statista.com/statistics/271972/world-husked-rice-production-volume-since-2008/ (accessed on 9 June 2021).
3. Abdullah, A.; Kobayashi, H.; Matsumura, I.; Ito, S. World Rice Demand towards 2050: Impact of Decreasing Demand of Per Capita Rice Consumption for China and India. 2008. Available online: https://www.semanticscholar.org/paper/World-Rice-Demand-Towards-2050%3A-Impact-of-Demand-of-Ito/260f4d3b195718591a8143982ec4cffa66f8f338 (accessed on 9 June 2021).
4. MoALD. Selected Indicators of Nepalese Agriculture. Available online: https://nepalindata.com/ (accessed on 9 June 2021).
5. ABPSD. *Statistical Information on Nepalese Agriculture 2013/14 (2070/071), Agri Statistics Section, Agri Business Promotion and Statistics Division*; Government of Nepal; Ministry of Agriculture and Co-Operatives; ABPSD: Kathmandu, Nepal, 2014.
6. Prasad, K.; Singh, R.; Singh, S. Effect of seedlings age and number of seedlings per hill on the yield of rice in a sodic soil. *Curr. Agric.* **2011**, *16*, 67–70.
7. Katoch, P.; Katoch, A.; Podel, M.; Uperti, S. Bakanae of Rice: A serious disease in Punjab. *Int. J. Curr. Microbiol. Appl. Sci.* **2019**, *8*, 129–136. [CrossRef]
8. Baral, B.R.; Pande, K.R.; Gaihre, Y.K.; Baral, K.R.; Sah, S.K.; Thapa, Y.B. Farmers' fertilizer application gap in rice-based cropping system: A case study of Nepal. *SAARC J. Agric.* **2019**, *17*, 267–277. [CrossRef]
9. Matsushima, S. *High Yielding Rice Cultivation*; University of Tokyo Press: Tokyo, Japan, 1976.
10. Ladha, J.K.; Pathak, H.; Krupnik, T.J.J.; Six, J.; Van Kessel, C. Efficiency of fertilizer nitrogen in cereal production: Retrospects and prospects. *Adv. Agron.* **2005**, *87*, 85–156.
11. Joshy, D. Soil Fertility and Fertilizer Use in Nepal. 1997. Available online: https://agris.fao.org/agris-search/search.do?recordID=NP9700156 (accessed on 9 June 2021).
12. Witt, C.; Buresh, R.J.; Peng, S.; Balasubramanian, V.; Dobermann, A. Nutrient management. In *Rice A Practical Guide to Nutrient Management*; Fairhurst, T.H., Witt, C., Buresh, R., Dobermann, A., Eds.; International Rice Research Institute: Los Baños, Philippines, 2007.
13. Satyanarayana, T.; Majumdar, K.; Pampolino, M.; Johnston, A.M.; Jat, M.L.; Kuchanur, P.; Sreelatha, D.; Sekhar, J.C.; Kumar, Y.; Maheswaran, R.; et al. Nutrient Expert TM: A Tool to Optimize Nutrient Use and Improve Productivity of Maize. *Better Crop.* **2013**, *97*, 21–24.
14. Dahal, S.; Shrestha, A.; Dahal, S.; Amgain, L.P. Nutrient Expert Impact on Yield and Economic In Maize and Wheat. *Int. J. Appl. Sci. Biotechnol.* **2018**, *6*, 45–52. [CrossRef]
15. Bhatta, R.D.; Amgain, L.P.; Subedi, R.; Kandel, B.P. Assessment of productivity and profitabilty of wheat using Nutrient Expert®-Wheat model in Jhapa district of Nepal. *Heliyon* **2020**, *6*, e04144. [CrossRef]
16. Amgain, L.P.; Timsina, J.; Dutta, S.; Majumdar, K. Nutrient expert® rice—An alternative fertilizer recommendation strategy to improve productivity, profitability and nutrient use efficiency of rice in Nepal. *J. Plant Nutr.* **2021**, *44*, 2258–2273. [CrossRef]
17. Sapkota, T.B.; Majumdar, K.; Jat, M.L.; Kumar, A.; Bishnoi, D.K.; McDonald, A.J.; Pampolino, M. Precision nutrient management in conservation agriculture based wheat production of Northwest India: Profitability, nutrient use efficiency and environmental footprint. *Field Crop. Res.* **2014**, *155*, 233–244. [CrossRef]
18. Shah, M.L.; Yadav, R. Response of Rice Varieties to Age of Seedlings and Transplanting Dates. *Nepal Agric. Res. J.* **2001**, *4–5*, 14–17. [CrossRef]

19. Prasad, R.; Shivay, Y.S.; Kumar, D.; Sharma, S.N. *Learning by Doing Exercises in Soil Fertility (A Practical Manual for Soil Fertility)*; Division of Agronomy, IARI: New Delhi, Indian, 2006.

20. Pal, S.; Sana, M.; Banerjee, H.; Lhungdim, L. Effectiveness of nitrogen and bio-fertilizer choice for improving growth and yield of hybrid rice (*Oryza sativa* L.) during dry season. *Oryza-An Int. J. Rice* **2020**, *57*, 302–309. [CrossRef]

21. Meena, S.L.; Singh, S.; Shivay, Y.S. Response of hybrid rice (*Oryza sativa*) to nitrogen and potassium application in sandy clay-loam soils. *Indian J. Agric. Sci.* **2003**, *73*, 8–11.

22. Aslam, M.M.; Zeeshan, M.; Irum, A.; Hassan, M.U.; Ali, S.; Hussain, R.; Ramzani, P.M.A.; Rashid, M.F. Influence of Seedling Age and Nitrogen Rates on Productivity of Rice (*Oryza sativa* L.): A Review. *Am. J. Plant Sci.* **2015**, *6*, 1361–1369. [CrossRef]

23. Subedi, R. Nursery Management Influences Yield and Yield Attributes of Rainfed Lowland Rice. *J. Sustain. Soc.* **2013**, *2*, 86–91. [CrossRef]

24. Dutta, S.K.; Majumdar, K.; Satyanarayana, T. India, Nutrient Expert: A precisionnutrient management tool for smallholder production systems of India. *Crop Soils Mag.* **2014**, *47*, 23–25. [CrossRef]

25. Malghani, A.L.; Malik, A.U.; Sattar, A.; Hussain, F.; Abbas, G.; Hussain, J. Response of Growth and Yield of Wheat to NPK fertilizer. *Sci. Int.* **2010**, *24*, 185–189.

26. Budhathoki, S.; Amgain, L.P.; Subedi, S.; Iqbal, M.; Shrestha, N.; Aryal, S. Assessing Growth, Productivity and Profitability of Drought Tolerant Rice Using Nutrient Expert—Rice and Other Precision Fertilizer Management Practices in Lamjung, Nepal. *Acta Sci. Agric.* **2018**, *2*, 153–158. [CrossRef]

27. Chaturvedi, I. Effect of Nitrogen Fertilizers on Growth, Yield and Quality of Hybrid Rice (*Oryza sativa*). *J. Cent. Eur. Agric.* **2005**, *6*, 611–618. [CrossRef]

28. Marahatta, S. Increasing productivity of an intensive rice based system through site specific nutrient management in Western Terai of Nepal. *J. Agric. Environ.* **2017**, *18*, 140–150. [CrossRef]

29. Haq, M.T.; Sattar, M.A.; Hossain, M.M.; Hasan, M.M. Effects of Fertilizers and Pesticides on Growth and Yield of Rice. *J. Biol.* **2002**, *2*, 84–88. [CrossRef]

30. Hussain, N.; Khan, M.A.; Javed, M.A. Effect of foliar application of plant micronutrient mixture on growth and yield of wheat (*Triticum aestivum*). *Pakitstan J. Biol. Sci.* **2005**, *8*, 1096–1099. [CrossRef]

31. Virmani, S.S.; Mao, C.X.; Hardy, B. *Hybrid Rice for Food Security, Poverty Alleviation, and Environmental Protection*; International Rice Research Institute: Metro Manila, Philippines, 2003.

32. Dobermann, A.; Witt, C.; Dawe, D.; Abdulrachman, S.; Gines, H.; Nagarajan, R.; Satawathananont, S.; Son, T.; Tan, P.; Wang, G.; et al. Site-specific nutrient management for intensive rice cropping systems in Asia. *Field Crop. Res.* **2002**, *74*, 37–66. [CrossRef]

33. Gaire, A.; Koirala, S.; Shrestha, R.K.; Amgain, L.P. Growth and Productivity of Different Cultivars of Rice Under Nutrient Expert© and Other Fertilizer Management Practices at Lamjung. *Int. J. Appl. Sci. Biotechnol.* **2016**, *4*, 178–182. [CrossRef]

34. Wang, G.; Zhang, Q.C.; Witt, C.; Buresh, R.J. Opportunities for yield increases and environmental benefits through site-specific nutrient management in rice systems of Zhejiang province, China. *Agric. Syst.* **2007**, *94*, 801–806. [CrossRef]

35. Biradar, D.; Aladakatti, Y.; Rao, T.; Tiwari, K. Site-specific nutrient management for maximization of crop yields in Northern Karnataka. *Better Crop.* **2006**, *90*, 33–35.

36. Shrestha, S.; Amgain, L.P.; Subedi, R.; Shrestha, P.; Shahi, S. Productivity and Profitability Analysis of Old Aged Hybrid Rice Seedling Using Nutrient-Rice Expert and Other Precision Nutrient Management Practices at Lamjung, Nepal. *Int. J. Appl. Sci. Biotechnol.* **2018**, *6*, 232–237. [CrossRef]

37. Adhikari, B.N.; Joshi, B.P.; Shrestha, J.; Bhatta, N.R. Genetic variability, heritability, genetic advance and correlation among yield and yield components of rice (*Oryza sativa* L.). *J. Agric. Nat. Resour.* **2018**, *1*, 149–160. [CrossRef]

38. Aditya, J.P.; Bhartiya, A. Genetic variability, correlation and path analysis for quantitative characters in rainfed upland rice of Uttarakhand Hills. *J. Rice Res.* **2013**, *6*, 24–34.

39. Gyawali, S.; Poudel, A.; Poudel, S. Genetic variability and association analysis in different rice genotypes in mid hill of western Nepal. *Acta Sci. Agric.* **2018**, *2*, 69–76.

40. Ogunbayo, S.A.; Sie, M.; Ojo, D.K.; Sanni, K.A.; Akinwale, M.G.; Toulou, B.; Shittu, A.; Idehen, E.O.; Popoola, A.R.; Daniel, I.; et al. Genetic variation and heritability of yield and related traits in promising rice genotypes (*Oryza sativa* L.). *J. Plant Breed. Crop Sci.* **2014**, *6*, 153–159. [CrossRef]

41. Madhavilatha, L.; Sekhar, M.R.; Suneetha, Y.; Srinivas, T. Genetic variability, correlation and path analysis for yield and quality traits in rice (*Oryza sativa* L.). *Res. Crop.* **2005**, *6*, 527.

42. Qamar, Z.U.; Cheema, A.A.; Ashraf, M.; Rashid, M.; Tahir, G.R. Association analysis of some yield influencing traits in aromatic and non aromatic rice. *Pakistan J. Bot.* **2005**, *37*, 613–627.

43. Rashid, M.; Cheema, A.A.; Ashraf, M. Numerical analysis of variation among Basmati rice Mutants. *Pakistan J. Bot.* **2008**, *40*, 2413–2417.

44. Iftekharuddaula, K.M.; Akter, K.; Hassan, M.S.; Fatema, K.; Badshah, A. Genetic Divergence, Character Association and Selection Criteria in Irrigated Rice. *J. Biol. Sci.* **2002**, *2*, 243–246. [CrossRef]

45. Khan, A.S.; Imran, M.; Ashfaq, M. Estimation of genetic variability and correlation for grain yield components in Rice (*Oryza sativa* L.). *Am. J. Agric. Environ. Sci.* **2009**, *6*, 585–590.

46. Ravikumar, B.N.V.S.R.; Kumari, P.N.; Rao, P.V.R.; Rani, M.G.; Satyanarayana, P.V.; Chamundeswari, N.; Vishnuvardhan, K.M.; Suryanarayana, Y.; Bharatalakdhmi, M.; Vishnuvardhan, R. Principal component analysis and character association for yield components in rice (*Oryza sativa* L.) cultivars suitable for irrigated ecosystem. *Curr. Biot.* **2015**, *9*, 25–35.

47. Babu, V.R.; Shreya, K.; Dangi, K.S.; Usharani, G.; Shankar, A.S. Correlation and Path Analysis Studies in Popular Rice Hybrids of India. *Int. J. Sci. Res. Publ.* **2012**, *2*, 1–5.

48. Kashyap, A.; Yadav, V.K. Principal Component Analysis and Character Association for Yield Components in Rice (*Oryza sativa* L.) Genotypes of Salt Tolerance under Alkaline Condition. *Int. J. Curr. Microbiol. Appl. Sci.* **2020**, *9*, 481–495. [CrossRef]

49. Anandan, A.; Pradhan, S.K.; Das, S.K.; Behera, L.; Sangeetha, G. Differential responses of rice genotypes and physiological mechanism under prolonged deepwater flooding. *Field Crop. Res.* **2015**, *172*, 153–163. [CrossRef]

50. Gana, A.S.; Shaba, S.Z.; Tsado, E.K. Principal component analysis of morphological traits in thirty-nine accessions of rice (*Oryza sativa* L.) grown in a rainfed lowland ecology of Nigeria. *J. Plant Breed. Crop Sci.* **2013**, *4*, 120–126. [CrossRef]

51. Aliyu, B.; Akoroda, M.; Padulosi, S. Variation with Vigna Reticulata Hooker FII. *Niger. J. Genet.* **2000**, *15*, 1–8. [CrossRef]

52. Gana, A.S. Variability Studies of the Response of Rice Varieties to Biotic and Abiotic Stresses. Ph.D. Thesis, University of Ilorin, Ilorin, Nigeria, 2006.

53. Kumar, A.; Majumdar, K.; Jat, M.L.; Pampolino, M.; Kamboj, B.R.; Bishnoi, D.K.; Kumar, V.; Johnston, A.M. Evaluation of Nutrient Expert TM for Wheat. *Better Crop.* **2012**, *6*, 27–29.

54. Attanandana, T.; Kongton, S.; Boonsompopphan, B.; Polwatana, A.; Verapatananirund, P.; Yost, R. Site-specific nutrient management of irrigated rice in the central plain of Thailand. *J. Sustain. Agric.* **2010**, *34*, 258–269. [CrossRef]

Article

Response of Indian Dwarf Wheat and Persian Wheat to Sowing Density and Hydrothermal Conditions of the Growing Seasons

Małgorzata Szczepanek [1],*, Grzegorz Lemańczyk [2], Rafał Nowak [1] and Radomir Graczyk [3]

[1] Department of Agronomy, Bydgoszcz University of Science and Technology, 85-796 Bydgoszcz, Poland; rafnow003@pbs.edu.pl

[2] Department of Biology and Plant Protection, Bydgoszcz University of Science and Technology, 85-796 Bydgoszcz, Poland; Grzegorz.Lemanczyk@pbs.edu.pl

[3] Department of Biology and Animal Environment, Bydgoszcz University of Science and Technology, 85-039 Bydgoszcz, Poland; Radomir.Graczyk@pbs.edu.pl

* Correspondence: Malgorzata.Szczepanek@pbs.edu.pl

Abstract: The need for foods with high nutritional value has led to the rediscovery of ancient wheat species *Triticum sphaerococcum* and *T. persicum* as raw materials with valuable consumption properties, but their reintroduction requires assessment of their productivity under different agricultural practices. The field experiments were carried out for three years (2018–2020) to test the hypothesis that the sowing density of *T. sphaerococcum* and *T. persicum* (400, 500 and 600 no m^{-2}) will affect their agronomic traits, yield and occurrence of diseases, but the response will depend on the hydrothermal conditions of the growing seasons. In this study, a significant correlation of the grain yield with the amount of precipitation in tillering, and from booting to the beginning of fruit development was demonstrated. The sowing density of *T. sphaerococcum* had an impact on the grain yield only under moderate drought stress during the growing season (2019), when the highest yield was obtained at a sowing density of 600 m^{-2}. In 2019, the most favorable sowing density was also the highest for *T. persicum*. In the year with the lowest amount of rainfall during the growing season (2018), the yield of *T. persicum* was the highest in the lowest sowing density. At the shooting stage, a greater intensity of powdery mildew was observed on *T. persicum*, especially with higher sowing densities. Increasing the sowing density also increased the occurrence of root rot symptoms in both wheat species in the year that favored the occurrence of this disease (2018). It can be concluded that in the integrated low-input cultivation of *T. sphaerococcum* and *T. persicum*, it is justified to use a sowing density of 600 pcs. m^{-2}, in an agroclimatic zone with moderate droughts during the growing season.

Keywords: Fusarium foot rot; Indian dwarf wheat; tillering; powdery mildew; Persian wheat; root rot; yield components

Citation: Szczepanek, M.; Lemańczyk, G.; Nowak, R.; Graczyk, R. Response of Indian Dwarf Wheat and Persian Wheat to Sowing Density and Hydrothermal Conditions of the Growing Seasons. *Agriculture* **2022**, *12*, 205. https://doi.org/10.3390/agriculture12020205

Academic Editors: Mubshar Hussain, Sami Ul-Allah and Shahid Farooq

Received: 15 January 2022
Accepted: 31 January 2022
Published: 1 February 2022

Publisher's Note: MDPI stays neutral with regard to jurisdictional claims in published maps and institutional affiliations.

1. Introduction

One of the key objectives of long-term EU policy is to increase the production of foods with high nutritional value using low-input cultivation technologies [1]. Ancient wheat species characterized by an increased content of macro and micronutrients [2] and much higher (as compared to other cultivated wheat species) contents of phenolic acids and alkylresorcinols, with proven pro-health effects [2], may be useful for these purposes. Examples of ancient species of cereals with increased nutritional value are Indian dwarf wheat (*Triticum sphaerococcum* Perc. = *T. aestivum* L. ssp. *sphaerococcum* (Perc.) Mac Key) and Persian wheat (*Triticum persicum* Vav. = *T. carthlicum* Nevski) [3,4]. The grains of both genotypes are a rich source of phenolic acids and alkylresorcinols. Moreover, in these wheats, an enhanced share of ferulic acid was found, accompanied by higher quantitative and qualitative variability of homologues within sterols, tocols and carotenoids [3]. Indian dwarf wheat is an endemic plant in Pakistan and India. It is known to have been one of the main crops grown by ancient Indian cultures. However, its cultivation disappeared during the early twentieth century [5]. Indian dwarf wheat is a hexaploid species and has naked grains [6]. Persian wheat is a tetraploid, early-maturing spring wheat, and originates from the southern slopes of the Greater Caucasus in Georgia [6,7]. Reintroduction of ancient wheat species such as Indian dwarf wheat or Persian wheat requires the assessment of productivity under different agricultural practices. Among them, one basic measure is the sowing density, which influences the inter- and intra-plant competition for nutrients, light and water, which may significantly affect the yield components and grain yield [8]. The researchers' opinions on the effect of sowing density on the yield of wheat grain are ambiguous. Poutala et al. [9] did not demonstrate the effect of sowing density (600 and 900 pcs. M^{-2}) on the grain yield of nine common wheat cultivars in organic cultivation. Similarly, Auškalnienė and Auškalnis [10] found no effect on the grain yield of six spring wheat sowing densities (from 200 to 700 pcs. M^{-2}) in conventional cultivation. Fisher et al. [11], summarizing a 30-year study of modern irrigated short wheat cultivars under favorable climatic conditions, found that grain yield was remarkably insensitive to planting density within the range of 80–400 plants m^{-2}. In contrast, Silva et al. [12] showed an increase in the yield of common wheat grain with an increase in sowing density (75–375 pcs. m^{-2}). Similar conclusions are presented by Hussain et al. [13], who compared the wheat response to seed rates of 50, 100, 150 and 200 kg ha^{-1}. The increase in yield shown in this study resulted from the decreased weed pressure along with an increase in the sowing density of wheat. In the study by Olsen et al. [14], the weed biomass was 46% lower and the grain yield was 25% higher in the high-density uniform sowing pattern (721 pcs. m^{-2}; treatment with the greatest weed suppression) than in the medium-density row sowing pattern (449 pcs. m^{-2}). Cardoso et al. [15] report that an increase in sowing density favors the seed yield; however, it reduces the vigor of the seeds produced. Sowing density affects the morphological traits of plants, including those directly related to the yield (yield components). The number of generative shoots usually increases with an increase in sowing density [12,16], but the number of grains per spike may decrease [8,17], as well as 1000-grain weight [8,9]. However, another study has shown no effect of density on the number of grains per spike or 1000-grain weight [13]. The relationship between plant density and yield is not directly proportional due to the plasticity of the wheat plant (the ability of the crop to sustain yield by changing yield components) [16].

Morphological traits and yield of wheat are severely affected by different weather stresses, such as drought, cold, and heat [18]. Wheat yield can be reduced by 65% after heat stress [19]. For every 1 °C increase in average temperature from 23 °C, the wheat yield can decrease by about 10% [20]. Heat stress increases evapotranspiration, leading to drought stress in plants. A combined effect of drought and heat stress is more detrimental than individual stresses [21]. Wheat shows the highest susceptibility under water stress from tillering to maturity [22]. Zhang et al. [23] report that the reductions in common wheat yield under mild, moderate, and severe stress were 21.0%, 25.8%, and 32.0%, respectively. Filled grain percentage or seed setting rate, grain number per spike, spike number per plant or m^2, and 1000-grain weight show significantly larger reductions under severe stress than under mild stress [23]. In study of primary wheat (*Triticum beoticum*) lack of water significantly affected morphological parameters such as the number of fertile spikes per plant, number of spikelets per spike, 1000-grain weight, leaf area index, relative water content, main root length and volume, as well as root dry weight and area [24]. A comparison of the stress resistance of di-, tetra- and hexaploid wheat shows that hexaploid wheat held the productive and adaptive advantage as it combined higher yield and stability when compared to the tetraploid and diploid groups [25]. Moreover, the number of spikes per m^2 was found to be more important than grain number per spike in determining grain yield under heat stress in hexaploid and tetraploid wheat species, while the grain number per spike was more important in diploids.

The incidence of fungal diseases is also a known factor significantly affecting the size of wheat crops [26–28]. Among them, the ones most often observed on the leaves of wheat plants are powdery mildew (caused by *Blumeria graminis*), Septoria leaf spot (caused by *Zymoseptoria tritici*, synonym *Septoria tritici*) and wheat leaf rust (caused by *Puccinia recondita*). Foot and root rot diseases, in properly cultivated spring forms of wheat, do not pose a great threat. However, the symptoms of root rot (caused by *Gaeumannomyces graminis, Fusarium* spp., *Rhizoctonia* spp.), Fusarium foot rot (caused by *Fusarium* spp.) and eyespot (caused by *Oculimacula acuformis*, and *Oculimacula yallundae*) can be observed during the growing season [29,30]. Diseases such as powdery mildew, Fusarium foot rot and root rot are of particular importance in the cultivation of Persian wheat and Indian dwarf wheat [31]. The incidence of disease symptoms in wheat significantly depends on weather conditions. Rainfall, not only the total, but also the frequency, distribution in particular stages of plant development, and moisture in the canopy play a special role here [32]. For pathogens whose spores are easily spread by raindrops splashing from the lower infected parts of the plant to the upper leaves, the closer the leaves are to each other, the easier the spores are spread to the upper parts [33]. The incidence of disease symptoms in wheat crops also depends on various agrotechnical practices, including the date and density of sowing [28,34]. This effect may be direct, but often the sowing method indirectly influences the incidence of diseases on plants [35].

The aim of the research was to evaluate the response of yield, biometric traits and incidence of diseases on the ancient wheat species, Indian dwarf and Persian wheat, using integrated, low-input (limited N rate and pesticide consumption) technology depending on the sowing density under varied hydrothermal conditions of the growing seasons. It was assumed that (1) increasing the sowing density would increase the number of production shoots per area unit and increase the grain yield, but at the same time would contribute to greater disease pressure; (2) the response of yield, plant biometric traits and the incidence of diseases will depend on the hydrothermal conditions during the growing season.

2. Materials and Methods

2.1. Site Description and Crop Management

Field experiments were located in Mochełek, Kuyavian-Pomeranian voivodeship (53°13′ N; 17°51′ E), Poland. The soils at experimental sites were characterized as Alfisol (USDA). The abundance of available macronutrients, pH and C-org content in the site of the research is presented in Table 1. The precipitation and thermal conditions in the subsequent years of field experiments were very diverse (Table 2). The year 2018 was warm (except March) with a large amount of rainfall in April and July. The hottest month in 2019 was June, with heavy rainfall only in May. 2020 was cool, with average rainfall in May and heavy rainfall in June and July.

Table 1. Results of soil analysis before sowing.

N-NO$_3$	N-NH^4	pH (KCl)	P	K	Mg	C-org.
g kg^{-1} of Soil				g kg^{-1} of Soil		
0.0042	0.0044	7.0	0.112	0.153	0.056	11.0

Table 2. Mean air temperature and precipitation in the growing seasons at experimental site.

Year	Days	March	April	May	June	July	March	April	May	June	July
		Temperature °C					Precipitation mm				
2018	1–10	3.5	9.2	14.8	19.8	18.7	5.7	18.2	7.6	7.2	2.0
	11–20	1.2	13.7	16.3	18.9	19.5	8.2	12.8	4.5	0.0	74.9
	20–30/31	1.7	13.2	19.5	16.6	23.0	2.7	9.4	2.1	19.2	9.1
	Mean/Sum	2.1	12.0	16.9	18.4	20.5	16.6	40.4	14.2	26.4	86.0
2019	1–10	4.6	7.6	8.8	21.4	16.0	12.1	0.0	9.3	0.0	14.7
	11–20	4.5	6.5	12.3	22.8	18.0	11.7	0.0	56.4	16.2	2.0
	21–30/31	6.8	13.7	15.0	21.6	21.6	5.0	1.5	23.5	1.5	5.7
	Mean/Sum	5.4	9.3	12.1	21.9	18.6	28.8	1.5	89.2	17.7	22.4
2020	1–10	4.2	7.3	11.2	14.9	17.9	13.1	0.0	16.4	63.0	32.5
	11–20	5.5	7.0	9.6	19.5	17.8	13.0	0.0	11.0	33.5	47.1
	21–30/31	2.1	10.4	12.0	19.4	18.3	0.0	0.7	7.2	57.4	5.5
	Mean/Sum	3.9	8.2	10.9	17.9	18.0	26.1	0.7	34.6	153.9	85.1

At the beginning of August, immediately after harvesting of the previous crop (triticale), the catch crop of a pea was sown. This catch crop was plowed at the end of November, after the above-ground part had frozen over. At the beginning of March, in each year of the research, pre-sowing mineral fertilization was performed at the following rates: 30 P_2O_5 kg ha^{-1} (superphosphate), 50 K_2O kg ha^{-1} (potassium salt) and 30 N kg ha^{-1} (ammonium sulphate) and additionally, 30 N kg ha^{-1} at the beginning of the shooting stage. In early spring (middle of march), the soil was cultivated with a tillage unit. Sowing of the wheat was performed 9 April 2018, 1 April 2019 and 23 March 2020 using OYORD plot seeder, in a row spacing of 16 cm. In determining the seeding rate, the 1000-grain weight and germination capacity as well as plant establishment (the ratio of seedlings vs. seeds), a value of 90% was set. Persian wheat was harvested at the end of July, and Indian dwarf wheat 10 days later, using a Wintersteiger plot harvester.

In our study, the cultivar 'Trispa' of Indian dwarf wheat was used [36]. This cultivar has long (65–80 cm), stiff shoots. The stems are covered with wax coating. The spike is short (5 cm) and awned. The spikelets are one- or two-grained. The grain of Indian dwarf wheat is rounded and relatively small (average 1000-grain weight is 30 g). For Persian wheat, the cultivar 'Persa' was used [36]. This cultivar develops thin and long stems (65–80 cm), which are susceptible to lodging. There is no wax coating on the leaves. The medium long spike (7–8 cm) is brownish, loose, and awned. The spikelets are two- or three-flowered, and one- or two-grained. The grain of Persian wheat is elongated and small (average 1000-grain weight is 28 g). Both Persian wheat and Indian dwarf wheat grains are naked.

Indian dwarf wheat (*Triticum sphaerococcum* Perc.), cv 'Trispa' and Persian wheat (*Triticum persicum* Vav.) cv 'Persa' were sown in three sowing densities: 400, 500, 600 grains m^{-2}. The experiment was established in a split-plot design, in four replications. The area of the plot for harvest was 21 m^2.

Pesticide treatments were applied using the same technology for both wheat species, adjusting the selection of preparations to the current threats from pests. Weed control was performed using florasulam 5 g a.i. ha^{-1}, aminopyralid 10 g a.i. ha^{-1} +2,4-dichlorophenoxyacetic acid 180 g a.i. ha^{-1} (Mustang Forte 1 dm^3 ha^{-1}) at the end of the tillering stage (BBCH 29) and, after 7 days, fenoxaprop-P-ethyl 77 g a.i. ha^{-1} (Fenoxinn 0.7 dm^3 ha^{-1}). In order to reduce the pressure of disease, fenpropimorph 250 g a.i. ha^{-1} and epoxiconazole 84 g a.i. ha^{-1} (Tango Star 1 dm^3 ha^{-1}) were applied once, at the full-fledged flag leaf stage (BBCH 39). Plant growth regulators were applied: chlormequat chloride 720 g a.i. ha^{-1} (CCC 1 dm^3 ha^{-1}) with ethephon 225 g a.i. ha^{-1} (Kobra 0.5 dm^3 ha^{-1}), at the first node stage (BBCH 31). Pests were controlled with the use of chlorpyrifos 228 g a.i. ha^{-1} (Insodex 0.6 dm^3 ha^{-1}) and deltamethrin 5 g a.i. ha^{-1} (Delcaps 0.1 dm^3 ha^{-1}), to prevent the intensity of occurrence of cereal leaf beetles and aphids from exceeding the thresholds of economic harmfulness.

2.2. Measurements of Morphological and Agronomic Traits

At the heading stage, the number of generative and vegetative tillers per plant was determined on subsequent plants collected from 1 linear meter from each plot. At full maturity, the number of generative fertile tillers (with grains) and sterile generative tillers (with spikes without grains) was determined on test plots of 1 m^2 on each plot. Moreover, 50 spikes were randomly collected from each plot, on which the number of grains per spike were measured. During harvest, the grain and straw yields as well as their moisture content were determined. Then, yields were converted into a constant humidity of 15%. The 1000-grain weight was determined based on 500 pcs, in four replications of each treatment.

2.3. Diseases Occurrence Assessment

The plant health assessment included an analysis of the severity of the most dangerous leaf diseases, the stem base and spring wheat roots, i.e., powdery mildew, Fusarium foot rot and root rot. The severity of disease symptoms caused by mildew was determined at stem elongation (Se) and at fruit development (Df). On the first assessment date, carried out prior to the fungicide application, the health of the lower leaves (L3 and L4) was assessed, and on the second date, the health of the flag leaf (L1), which was described as a percentage of the leaf area that was covered with *Blumeria graminis* mycelium or disease symptoms caused by this pathogen, was assessed. On the second assessment date, the severity of disease symptoms was also determined based on damage to wheat stalks caused by *Fusarium* fungi (Fusarium foot rot) and the severity of disease symptoms on the roots caused by a complex of pathogens, i.e., *Gaeumannomyces graminis*, *Fusarium* spp., *Rhizoctonia* spp. (root rot). The assessment of the severity of powdery mildew was performed directly on the experimental plots and the severity of Fusarium foot rot and root rot was determined in the laboratory. One hundred generative tillers of each combination were assessed for health status and the occurrence or severity of disease symptoms. In laboratory conditions, after separating of the tillers and removal of the leaf sheaths, the degree of infection of the stem base and roots was assessed on a scale of 0-4, where 0 meant no symptoms of a given disease, and 4 meant severe infection, when the symptoms of Fusarium foot rot covered the stem base around followed by its rot or symptoms of root rot were observed on more than 60% of the surface of the assessed roots. Infection rates were transferred to the disease index (DI) according to the Townsend and Heuberger formula and expressed as a percentage.

2.4. Statistical Analyses

The basic statistical descriptors included arithmetic mean ($\bar{x}$) and standard deviation (SD). Analysis of the results was performed using statistical inference methods [37]. Normality was examined by the Kolmogorov-Smirnov test and homogeneity of variance with the Levene's test. The data were analyzed by two-way analysis of variance (ANOVA), in which the sowing density and study years (temperature and precipitation were included as passive variables) were the main factors, and then to determine significant differences between the means, the Tukey test was used. Based on the same data set, the analysis of the structure of agronomic traits and disease occurrence, depending on the years of study and sowing density, was performed via Canonical Correspondence Analysis (CCA). Pearson's correlation coefficient was used to measure the correlation between grain yield, biometric features of *T. sphaerococcum* and *T. persicum* and Sielianinov's hydrothermal coefficient. Quadratic polynomial regression was used to relate the grain yield and sowing density of *T. sphaerococcum* and *T. persicum* in the study years 2018–2020. The analyses were performed using PAST 3.2 software, CANOCO 4.5, MS EXCEL 365 and STATISTICA 13.3 (Hammer

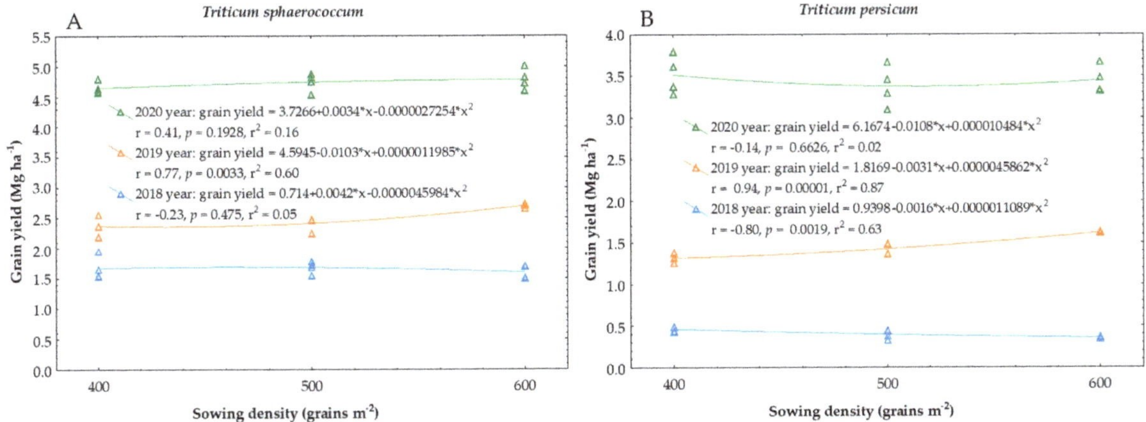

Figure 1. Relationships between sowing density and grain yield of *Triticum sphaerococcum* (**A**) and *Triticum persicum* (**B**).

In 2018, Indian dwarf wheat developed the least generative tillers and the most vegetative tillers per plant (Table 4). In this unfavorable year for plant growth and development, the number of generative tillers per plant was higher at lower sowing densities (400 and 500 pcs. m^{-2}). Only in 2019, at a sowing density of 600 pcs. m^{-2}, the number of vegetative tillers per plant was lower compared to the number of such tillers at the sowing density of 400 pcs. m^{-2}. Moreover, in 2019, the number of generative tillers per m^2 and the number of total generative tillers were higher at the sowing density of 600 pcs. m^{-2} compared to that recorded in the sowing density of 400 pcs. M^{-2}. In 2019, the number of sterile generative tillers per m^{-2} and their share in total generative tillers was the highest. Among the analyzed biometric features of Indian dwarf wheat, the number of sterile tillers per m^2 was characterized by the highest variation coefficient.

Table 4. Number of generative and vegetative tillers of *Triticum sphaerococcum* at different sowing densities in study years 2018–2020.

Year	Sowing Density (Grains m^{-2})	Generative Tillers (No Plant^{-1})	Vegetative Tillers (No Plant^{-1})	Fertile Generative Tillers (No m^{-2})	Sterile Generative Tillers (No m^{-2})	Total Generative Tillers (No m^{-2})
2018	400	0.96 ± 0.07 a [1]	1.56 ± 0.24 a	273 ± 20 a	11.3 ± 0.9 a	285 ± 19 a
	500	0.88 ± 0.08 a	1.44 ± 0.16 a	275 ± 20 a	14.7 ± 9.6 a	589 ± 21 a
	600	0.72 ± 0.03 b	1.56 ± 0.28 a	280 ± 10 a	20.0 ± 11.8 a	300 ± 4 a
	CV%	13.96	14.39	5.76	57.20	5.63
2019	400	1.04 ± 0.03 a	1.43 ± 0.26 a	313 ± 21 b	18.5 ± 4.1 a	331 ± 20 b
	500	1.05 ± 0.06 a	1.06 ± 0.19 ab	351 ± 12 ab	23.0 ± 5.3 a	374 ± 14 ab
	600	1.05 ± 0.06 a	0.69 ± 0.06 b	378 ± 34 a	38.0 ± 18.7 a	416 ± 35 a
	CV%	4.39	33.71	10.19	51.10	11.36
2020	400	1.07 ± 0.17 a	1.47 ± 0.38 a	601 ± 67 a	3.5 ± 1.9 a	605 ± 68 a
	500	1.06 ± 0.07 a	1.23 ± 0.28 a	590 ± 82 a	5.0 ± 4.8 a	595 ± 84 a
	600	1.09 ± 0.08 a	1.27 ± 0.35 a	596 ± 14 a	3.3 ± 1.0 a	599 ± 13 a
	CV%	9.82	24.46	9.38	72.59	6.51

[1] Mean values ± standard deviation (SD) in column followed by different letters indicate significant differences between treatments at $p \leq 0.05$.

3.2. Yield and Biometric Features of Triticum persicum

The highest grain yield of Persian wheat was obtained in 2020 (3.45 Mg ha^{-1}) (Table 5). It was 2.4 times higher than the grain yield in 2019. In 2018, the most unfavorable for plant growth, the average grain yield of Persian wheat was only 0.41 Mg ha^{-1}. In 2018, the grain yield was higher at the sowing density of 400 pcs. M^{-2} compared to that obtained in the density of 600 pcs. M^{-2}. In contrast, in 2019, the highest grain yield was obtained at a density of 600 pcs. M^{-2}, the yield was significantly lower at the sowing density of 500 pcs. M^{-2} and the lowest at a sowing density of 400 pcs. M^{-2} (Table 5, Figure 1B). The number of grains per spike was the highest in 2020 and the lowest in 2018. In 2019, increasing the sowing density from 500 to 600 pcs. M^{-2} had a positive effect on the number of grains per spike of Persian wheat. The Persian wheat grain obtained in 2018 had the highest 1000-grain weight. There was no significant effect of Persian wheat sowing density on 1000-grain weight or straw weight found in any of the study years.

Table 5. Grain and straw yield, number of grains per spike, and 1000-grain weight of *Triticum persicum* at different sowing density in study years 2018–2020.

Year	Sowing Density (Grains m^{-2})	Grain Yield (Mg ha^{-1})	Number of Grain per Spike	1000-Grain Weight (g)	Straw Yield (Mg ha^{-1})
2018	400	0.46 ± 0.03 a [1]	10.7 ± 2.8 a	32.4 ± 0.3 a	1.02 ± 0.12 a
	500	0.40 ± 0.06 ab	9.3 ± 1.9 a	33.3 ± 2.8 a	1.05 ± 0.16 a
	600	0.36 ± 0.01 b	9.8 ± 2.0 a	34.5 ± 4.1 a	1.00 ± 0.08 a
	CV%	13.97	21.58	8.31	10.91
2019	400	1.32 ± 0.05 c	17.9 ± 2.0 ab	25.2 ± 0.8 a	3.27 ± 0.66 a
	500	1.42 ± 0.07 b	15.6 ± 1.5 b	24.8 ± 1.7 a	3.44 ± 0.07 a
	600	1.62 ± 0.01 a	19.1 ± 1.5 a	26.6 ± 1.3 a	3.55 ± 0.17 a
	CV%	9.47	12.26	5.55	10.99
2020	400	3.52 ± 0.23 a	25.9 ± 2.6 a	27.7 ± 1.8 a	6.77 ± 1.35 a
	500	3.38 ± 0.24 a	25.4 ± 2.3 a	25.9 ± 2.1 a	6.36 ± 1.17 a
	600	3.45 ± 0.16 a	29.7 ± 2.6 a	26.7 ± 0.93 a	6.56 ± 0.43 a
	CV%	5.87	11.22	6.33	14.87

[1] Mean values ± standard deviation (SD) in column followed by different letters indicate significant differences between treatments at $p \leq 0.05$.

In 2018, Persian wheat developed the least generative tillers and the most vegetative tillers per plant (Table 6). In the same year, the number of sterile generative tillers per m^2 was higher at the sowing density of 600 pcs. m^{-2} compared with the number of such tillers at the sowing density of 400 pcs. m^{-2}. In 2019, it was shown that the number of generative tillers per plant, as well as the number of such tillers per m^2, was the highest at the highest sowing density of 600 pcs. m^{-2}. The coefficients of variation of the grain yield and most of the analyzed biometric features were higher in 2018 compared to 2019 and 2020. Only the coefficients of variation of the number of vegetative tillers per plant and sterile generative tillers per m were the highest in 2020.

Table 6. Number of generative and vegetative tillers of *Triticum persicum* at different sowing density in study years 2018–2020.

Year	Sowing Density (Grains m^{-2})	Generative Tillers (No Plant^{-1})	Vegetative Tillers (No Plant^{-1})	Fertile Generative Tillers (No m^{-2})	Sterile Generative Tillers (No m^{-2})	Total Generative Tillers (No m^{-2})
2018	400	0.42 ± 0.08 a [1]	2.08 ± 0.24 a	168 ± 36 a	22.7 ± 4.1 b	191 ± 37 a
	500	0.57 ± 0.16 a	1.76 ± 0.09 a	120 ± 12 a	30.7 ± 5.0 ab	151 ± 14 a
	600	0.34 ± 0.11 a	1.86 ± 0.14 a	146 ± 29 a	38.0 ± 10.2 a	184 ± 22 a
	CV%	33.82	10.94	22.19	29.84	17.02
2019	400	1.09 ± 0.11 b	1.40 ± 0.30 a	273 ± 13 b	20.0 ± 2.8 a	293 ± 10 b
	500	1.19 ± 0.16 b	1.37 ± 0.23 a	298 ± 31 b	21.5 ± 1.0 a	319 ± 32 b
	600	1.48 ± 0.11 a	1.61 ± 0.28 a	401 ± 54 a	23.0 ± 5.3 a	424 ± 54 a
	CV%	16.46	18.50	20.63	15.93	19.65
2020	400	1.14 ± 0.10 a	1.48 ± 0.42 a	551 ± 86 a	14.0 ± 10.8 a	565 ± 96 a
	500	1.15 ± 0.06 a	1.43 ± 0.21 a	667 ± 83 a	12.5 ± 6.6 a	679 ± 86 a
	600	1.20 ± 0.14 a	1.56 ± 0.37 a	602 ± 68 a	13.5 ± 4.7 a	615 ± 71 a
	CV%	10.17	21.17	14.39	53.26	14.74

[1] Mean values ± standard deviation (SD) in column followed by different letters indicate significant differences between treatments at $p \leq 0.05$.

3.3. Occurrence of Diseases

On Indian dwarf wheat, the severity of powdery mildew varied over the years of the study and observation dates (Table 7). On the first assessment date, most symptoms of the disease on the lower leaves (L3 and L4) were found in 2019 (45.58–56.67% of the area of assessed leaves with symptoms of the disease), where the high total rainfall in May was conducive to the development of this disease (Table 2). However, the least were found in 2018 (1.0–1.75% of the leaf area), where rainfall in May was the lowest. On the second assessment date, the severity of powdery mildew symptoms on the flag leaf (L1) was small and was observed only in 2019 and 2020. The occurrence of powdery mildew did not depend on the sowing density. The large amount of rainfall in May, during the period of intensive growth and development of plants, limited the occurrence of roots and stem base diseases. Hence, the fewest symptoms of root rot were observed in 2019 and the most in 2018 (Table 7). The severity of these disease symptoms, with the exception of 2019, significantly depended on the sowing density. In the year with the greatest number of symptoms of root rot (2018), the severity of the disease increased with an increase in sowing density. In 2020, the lowest number of disease symptoms was recorded at the sowing density of 500 pcs. m^{-2}, while at the sowing densities of 400 and 600 pcs. m^{-2} there was a significant increase in infection. Relatively few symptoms of Fusarium foot rot were recorded on Indian dwarf wheat, and in 2019 they were observed only sporadically. Only in 2020, a significant effect of the sowing density on the severity of the symptoms of Fusarium foot rot was demonstrated. Increasing the sowing density of Indian dwarf wheat from 400 or 500 pcs. m^{-2} to 600 pcs. m^{-2} resulted in a significant increase in the infestation of the stem base by fungi of the genus *Fusarium*.

Table 7. Occurrence of diseases on *Triticum sphaerococcum* at different sowing density in study years 2018–2020.

Year	Sowing Density (Grains m^{-2})	Powdery Mildew Se	Powdery Mildew Df	Root Rot	Fusarium Foot Rot
		Leaf or Spike Area with Disease Symptoms (%)			
2018	400	1.00 ± 0.00 a [1]	0.00 ± 0.00 a	34.25 ± 5.12 b	10.25 ± 2.87 a
	500	1.50 ± 0.58 a	0.00 ± 0.00 a	46.00 ± 3.5 ab	16.25 ± 3.77 a
	600	1.75 ± 0.96 a	0.00 ± 0.00 a	51.75 ± 9.91 a	12.25 ± 2.50 a
	CV%	47.19	0.00	22.18	29.61
2019	400	45.58 ± 6.23 a	0.38 ± 0.47 a	5.50 ± 2.08 a	0.50 ± 0.58 a
	500	56.67 ± 0.20 a	0.20 ± 0.15 a	6.50 ± 1.75 a	0.00 ± 0.00 a
	600	49.58 ± 5.99 a	0.07 ± 0.06 a	4.75 ± 2.22 a	0.00 ± 0.00 a
	CV%	15.47	133.13	35.38	180.91
2020	400	7.75 ± 1.59 a	0.26 ± 0.09 a	22.71 ± 1.72 a	9.38 ± 1.58 b
	500	5.92 ± 1.66 a	0.25 ± 0.14 a	13.54 ± 2.67 b	11.04 ± 1.85 b
	600	5.25 ± 1.45 a	0.14 ± 0.06 a	22.92 ± 6.89 a	17.29 ± 2.29 a
	CV%	28.55	20.01	30.65	31.54

[1] Mean values ± standard deviation (SD) in column followed by different letters indicate significant differences between treatments at $p \leq 0.05$.

Many symptoms of powdery mildew were observed on Persian wheat, especially before the fungicide was applied. Their intensity depended on the year of the study, the observation date and the sowing density (Table 8). At the first assessment date, on average, the greatest number of disease symptoms caused by *B. graminis* was found in 2019 (34.92–58.75%), which was characterized by the highest total rainfall in May. There were slightly fewer of them in 2020 (41.26–49.42%), and clearly the fewest in 2018 (2.75–7.50%). In all years of observation, the most symptoms of powdery mildew were found at the highest sowing density (600 pcs. m^{-2}). In 2019, many symptoms of the disease were also found at the sowing density of 500 pcs. m^{-2}. At the second assessment date, the severity of powdery mildew symptoms on Persian wheat plants was low. It was observed only in 2019 and 2020 and was not dependent on the sowing density. The large amount of rainfall in May 2019, during the period of intensive plant growth and development, limited the occurrence of symptoms of foot and root rot diseases on Persian wheat plants. Hence, the fewest symptoms of Fusarium foot rot were found in 2019, when they were observed only occasionally, and the most were observed in 2018. The incidence of this disease significantly depended on the sowing density. In 2019, the most symptoms of Fusarium foot rot were observed at the sowing density of 500 pcs. m^{-2}, and in 2020, at the sowing density of 400 pcs. m^{-2}. In general, the highest sowing density was not conducive to the occurrence of this disease. Furthermore, the symptoms of root rot were least observed in 2019 and the most in 2018, when the disease index value was high and ranged from 50.25 to 63.25% (Table 7). Despite such a high intensity of disease symptoms, there was no significant effect of sowing density on the severity of symptoms of root rot.

Table 8. Occurrence of diseases on *Triticum persicum* at different sowing densities in study years 2018–2020.

Year	Sowing Density (Grains m^{-2})	Powdery Mildew Se	Powdery Mildew Df	Root Rot	Fusarium Foot Rot
		Leaf or Spike Area with Disease Symptoms (%)			
2018	400	2.75 ± 0.96 b [1]	0.00 ± 0.00 a	50.25 ± 4.57 a	10.25 ± 1.89 b
	500	4.50 ± 0.58 b	0.00 ± 0.00 a	56.25 ± 9.03 a	35.25 ± 5.12 a
	600	7.50 ± 1.73 a	0.00 ± 0.00 a	63.25 ± 6.45 a	9.00 ± 1.83 b
	CV%	47.07	0.00	14.80	71.46
2019	400	34.92 ± 11.18 b	0.11 ± 0.07 a	6.25 ± 1.50 a	0.50 ± 1.00 a
	500	52.67 ± 4.12 a	0.07 ± 0.00 a	7.00 ± 2.83 a	0.00 ± 0.00 a
	600	58.75 ± 3.70 a	0.06 ± 0.01 a	6.50 ± 0.58 a	0.00 ± 0.00 a
	CV%	25.44	54.69	26.28	346.41
2020	400	45.00 ± 3.04 ab	0.18 ± 0.07 a	23.96 ± 1.72 a	7.50 ± 1.18 a
	500	41.26 ± 2.84 b	0.13 ± 0.00 a	23.33 ± 2.45 a	2.92 ± 1.08 b
	600	49.42 ± 2.77 a	0.13 ± 0.01 a	24.38 ± 2.75 a	2.92 ± 0.48 b
	CV%	9.63	29.55	9.09	54.42

[1] Mean values ± standard deviation (SD) in column followed by different letters indicate significant differences between treatments at $p \leq 0.05$.

3.4. Variable Dependency Analysis

Among the grain yield structure components of Indian dwarf wheat and Persian wheat, biometric features such as the number of grains per spike and the number of fertile generative tillers per m^2 were significantly positively correlated with the grain yield (Table 9). These yield components and the grain yield were significantly positively correlated with the Sielianinov's hydrothermal coefficient in the first ten days of May, and with the values of this coefficient in June (Tables S1 and S2). The number of sterile generative tillers per m^2 was negatively correlated with the grain yield and Sielianinov's hydrothermal coefficient in June. Both studied wheat species showed a significant positive correlation between grain and straw yield. Canonical correspondence analysis (CCA) of variables from individual research years, including grain and straw yield, number of grains per spike, 1000-grain weight, number of generative tillers per plant, fertile generative tillers per m^2 and the incidence of diseases (powdery mildew, root rot and Fusarium foot rot), clearly indicates a very strong relationship with the research year (weather conditions in the growing season) and, to a lesser extent, with the sowing density (Figure 2). Biometric features, yield of wheat and incidence of diseases are distributed horizontally, according to vectors: (longer) to the left—Year and (shorter) to the right—Sowing densities (the vectors—green lines—almost coincide with the line of the first axis). Axis 1 explains the system at 100% and is statistically significant, while the second axis is statistically insignificant. According to the conducted analysis, the significance of weather conditions (years of research) on shaping the analyzed features of the studied wheat species was highest in 2019, mainly at the lowest sowing density (400 pcs. m^{-2}). A separate group is formed by 2018, where the effect of sowing density on the analyzed features is clearly visible, while the species is also important (*Triticum sphaerococcum* and *T. persicum* are grouped separately). In 2020, *T. sphaerococcum* also forms a separate group.

Table 9. Significant correlation coefficients ($p \leq 0.05$) between grain yield and biometric features.

Biometric Features	*Triticum sphaerococcum*	*Triticum persicum*
Number of grains per spike	0.86	0.94
1000-grain weight	n.s.	−0.58
Straw yield	0.99	0.98
Generative tillers (plants $^{-1}$)	0.59	0.62
Vegetative tillers (plants^{-1})	n.s.	−0.44
Fertile generative tillers (m^{-2})	0.96	0.95
Sterile generative tillers (m^{-2})	-0.47	−0.69

n.s.—not significant.

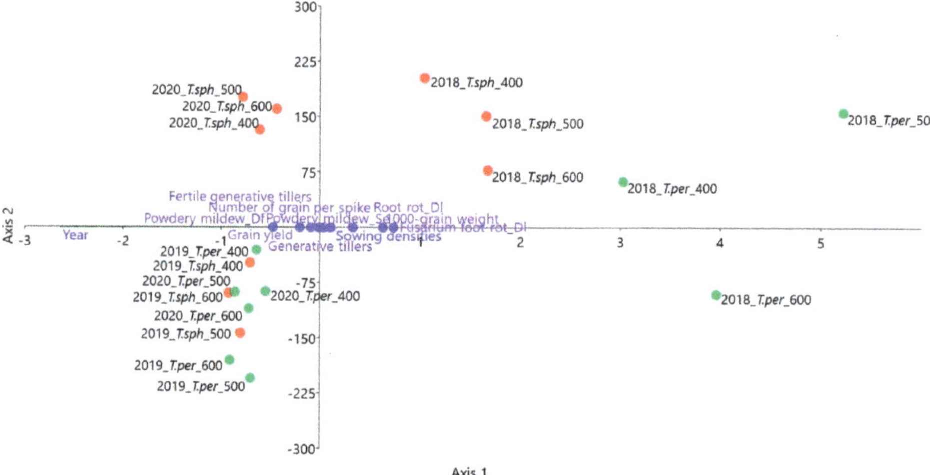

Figure 2. Canonical correspondence analysis (CCA) for grain and straw yield, biometric features (number of grains per spike, 1000-grain weight, number of generative tillers per plant, number of fertile generative tillers per m^2) and occurrence of diseases (powdery mildew at stem elongation Se and at fruit development Df, root rot and Fusarium foot rot) on *Triticum sphaerococcum* and *T. persicum* at different sowing densities. Variables from three consecutive study years (2018, 2019, 2020) were analyzed. Eigenvalues for axis 1 = 0.04 (100%), for axis 2 = <0.001 (<0.001%), significance of first axis: Monte Carlo-permutation test: (F-ratio = 6.30, p = 0.001).

Canonical correspondence analysis of variables, on the average of three years of research, indicates that powdery mildew in flag leaf development significantly affects the formation of biometric features, more in the case of *T. sphaerococcum*, especially at the sowing densities of 400 and 500 pcs. m^{-2} (Figure 3). Powdery mildew in steam elongation, in turn, as well as root rot and Fusarium foot rot affect the biometric features of *T. persicum* to a greater extent, especially at the densities of 600 and 500 pcs. m^{-2}. Axis 1 of the CCA analysis statistically significantly explains the system to a degree over 90%, while the second axis is statistically insignificant.

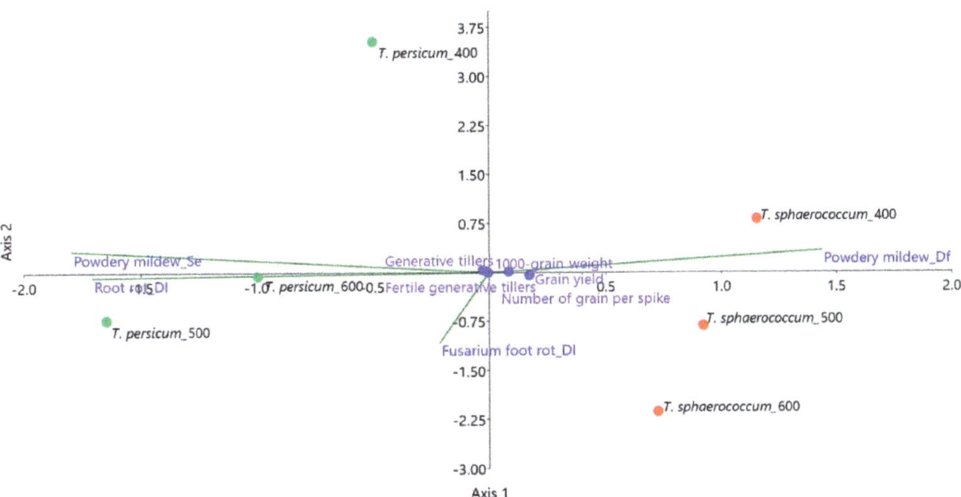

Figure 3. Canonical correspondence analysis (CCA) for grain and straw yield, generative tiller number, and leaf and foot and root rot diseases of *Triticum sphaerococcum* and *T. persicum* at different sowing densities. Mean values of variables from three years of research were considered. Eigenvalues for axis 1 = 0.02 (94.41%), for axis 2 = 0.007 (3.71%), significance of first axis: Monte Carlo-permutation test: (F-ratio = 4.46, p = 0.031), significance of second axis: Monte Carlo-permutation test: (F-ratio = 0.94, p = 0.614).

4. Discussion

4.1. Morphological and Agronomic Traits

Canonical correspondence analysis of grain yield and agronomical traits (number of grains per spike, 1000-grain weight, number of generative tillers per plant and fertile generative tillers per m^2) of Indian dwarf wheat and Persian wheat indicates a significant relationship with the year of the study (weather conditions in the growing season) and, to a lesser extent, with the sowing density. Auškalnienė and Auškalnis [10] indicate the greater importance of weather conditions compared to the sowing density of common wheat. Tokatlidis [38] concludes that the optimal wheat population is primarily affected by the water regime, while sowing date, heat, terminal drought, frost and type of soil are also indicators.

The least favorable year for the growth and yield of Indian dwarf wheat and Persian wheat was 2018, with low rainfall at the end of April, in May and June (from the stem elongation stage to the beginning of the grain development stage). Under severe water deficit, both wheat species exhibited reduced fertile generative tillers per m^2, number of grains per spike, and the number of generative tillers per plant, morphological traits significantly correlated with the grain yield. Moreover, the water deficit and high air temperatures in May limited tillering, which resulted in a reduction in the number of generative tillers per plant.

According to Zhang et al. [23], drought decreases wheat grain yield the most during the complete growth cycle, which is achieved from a combination of decreases during the tillering stage (−27.4%), the stem elongation stage (−21.4%), the heading stage (−16.8%), the flowering stage (−17.7%), and the development of grain stage (−16.3%). In the study of *Triticum aestivum* it was proven that water stress affects morpho-physiological traits such as plant height, number of tillers per plant, number of spikelets per spike, number of grains per spike, 1000-grain weight, flag leaf area, relative water content, and chlorophyll content (d). Moreover, drought or heat stress enhances enzymatic antioxidant activity (catalase, dismutase, peroxidase), and proline content. These chemicals help to stabilize membranes, sub-cellular structures and cellular redox potential by destroying the free radicals [21]. In a study of the ancient wheat *Triticum boeoticum,* some proteins (MICP, in the group of lipid metabolism proteins) were found [24]. These proteins may play an important role in the metabolic pathways of wheat and its drought tolerance [24].

In our study, the number of generative tillers per plant of Indian dwarf wheat in 2018 was significantly higher at lower sowing densities (400 and 500 pcs. M^{-2}). Similarly, Silva et al. [12], in the study of the response of common wheat cultivars to sowing density (75, 150, 225, 300, 375 pcs. m^{-2}), showed an increase in the number of generative tillers per plant along with a decrease in sowing density. In the study by Beavers at al. [16] the number of spikes per plant was significantly larger at a sowing density of 330 pcs. m^{-2} compared to the number found at densities of 412 and 495, and the smallest at a sowing density of 660 pcs. m^{-2}. In 2018, the grain yield of Persian wheat was higher at the lowest sowing density (400 pcs. m^{-2}) compared to that obtained in the sowing density of 600 pcs. m^{-2}. At the lowest sowing density, the number of sterile generative tillers per m^{-2} was significantly lower and the number of fertile generative tillers per m^2 was slightly higher, compared to the number of such tillers at a sowing density of 600 pcs. m^{-2}. It is worth noting that, at the lowest sowing density, the share of sterile generative tillers in the total of generative tillers was lower (12%) compared to the share of such tillers at sowing densities of 500 and 600 pcs. m^{-2} (20 and 21%, respectively). The reason for the increase in the number of sterile spikes at higher sowing densities was the increased competition for water under conditions of heavy shortage, which started in May and deepened in June. The study by Pinheiro et al. [39] clearly shows an increase in the number of non-productive tillers as the sowing density of wheat increases. Marinho et al. [40] prove that competition between plants and tillers can increase the incidence of infertile tillers, especially at high sowing densities.

The year 2019 was characterized by a moderate water deficit. Heavy rainfall occurred in the second half of May (at the stem elongation stage), while a rainfall deficit and high air temperatures were recorded in June (from the booting to the beginning of the grain development stage). According to Sattar et al. [21], heat stress along with low water availability at the reproductive stage are major contributing factors towards lower wheat production. In 2019, increasing the sowing density from 400 or 500 pcs. m^{-2} to 600 pcs. m^{-2} resulted in a significant increase in the grain yield of Indian dwarf wheat. Similarly, in Persian wheat, the highest grain yield was obtained at the highest sowing density (600 pcs. m^{-2}), while it was significantly lower at a density of 500 pcs. m^{-2} and was lowest at the density of 400 pcs. m^{-2}. The increase in grain yield at the highest sowing density was mostly due to the increased number of fertile generative tillers per m^2, the most important yield component. Moreover, there was an increase in generative tillers per plant in Persian wheat and a reduction in number of vegetative tillers per plant in Indian dwarf wheat. In the study by Hussain et al. [13] among the yield components, only the number of generative tillers per m^{-2} increased in proportion to the increase in the grain yield at higher seed rates. According to Wu et al. [41], high sowing density as a purposeful agricultural practice increases yields, reduces weed infestation and increases sustainability by increasing biomass and soil cover during the growing season of the plants, which reduces nutrient loss and erosion. In 2019, there was a significant increase in the straw weight of Indian dwarf wheat at the sowing density of 600 pcs. m^{-2}. This weight consists mainly of generative tillers, hence the significant correlation of the straw yield with the grain yield. Furthermore, we have noted that the 1000-grain weight was higher at the lowest sowing density of 400 pcs. m^{-2} compared to the size of this parameter at a sowing density of 500 pcs. M^{-2}. Poutala et al. [9] showed a reduction in the 1000-grain weight of common wheat due to an increase in sowing density. In the study by Tahir et al. [8], a higher grain weight in common wheat was found at seed rates of 100 kg ha^{-1} and lower at 175 kg ha^{-1}.

In Persian wheat grown in 2019, despite the increase in the number of fertile generative tillers per m^2 at the sowing density of 600 pcs. m^{-2}, the number of grains per spike in this treatment was higher compared to wheat in a sowing density of 500 pcs. m^{-2}. Opposite relationships are presented by Tahir et al. [8], who showed a reduction in the number of grains per spike as a result of increasing the density of common wheat sowing from 125 to 175 kg ha^{-1}. The different response of the compared wheat species could result from a different structure of the spike. Common wheat in the studies by Tahir et al. [8] had a number of spikelets per spike twice as high as Persian wheat in 2019, which was significantly influenced by drought stress during flowering, heading and the beginning of grain development stages.

In the present study, the most favorable year for the growth of Indian dwarf wheat and Persian wheat was 2020, where heavy rainfalls in May and June favored the yield. The distribution of rainfall favored the development of numerous fertile generative tillers per m^2 and grains per spike—yield components significantly correlated with the grain yield. It is noteworthy that under these weather conditions, almost all generative tillers of Indian dwarf wheat and Persian wheat remained fertile (the shares of fertile generative tillers as a percentage of total generative tillers per m^2 were 99% and 98%, respectively). In that year, no significant effect of the sowing density on any of yield components or the grain yield of both studied wheat species was demonstrated. The obtained results confirm the conclusion of Fisher et al. [11] on the lack of response of common wheat grain yield to the sowing density under favorable weather conditions.

Our study did not show a significant effect of sowing density on the number of grains per spike of Indian dwarf wheat in any of the study years. However, a correlation was observed between the number of generative tillers per m^2 and the number of grains per spike. If, in a given year, the number of generative tillers per m^2 was low, the number of grains per spike was also low. Therefore, there was no compensation of the yield components. The opposite conclusions were formulated by Lloveras et al. [42] who claimed that common wheat can compensate for low populations by modifying the number of grains per spike. However, in our research, the number of grains per spike in 2018 and 2019 was mostly limited by drought from the booting stage to the beginning of the grain development stage, which made it impossible to compensate for the limited density of generative shoots in the canopy.

4.2. Occurrence of Diseases

The analyzed Indian dwarf wheat and Persian wheat plants showed a similar intensity of disease symptoms as in previous studies on those wheat species conducted in organic farming crops, in which powdery mildew, Fusarium foot rot and root rot were also mainly observed [31]. On the first observation date, a high intensity of powdery mildew symptoms was observed on the assessed lower leaves of both wheat species, and its incidence was significantly correlated with the total rainfall. According to Te Beest et al. [43], the disease is favored by warm and humid conditions, but even under favorable conditions for the development of the disease, appropriate chemical protection can significantly reduce the disease's severity. Hence, on the first observation date, the fewest symptoms were observed in 2018, with low rainfall at the end of April and in May (Table 2). The lesser severity of powdery mildew symptoms on the second observation date, found on the analyzed flag leaves of wheat, was influenced by the use of the Tango Star fungicide at the fully developed flag leaf stage. According to Semaškienė et al. [44], a mixture of fenpropimorph and epoxiconazole, active substances contained in this preparation, are highly effective in limiting the development of powdery mildew on spring wheat.

In the present authors' research, the occurrence of powdery mildew on Persian wheat also depended on the sowing density, which was observed only at the stem elongation stage. Generally, with increasing sowing density, the severity of the symptoms of this disease increased. This is consistent with the observations made on the form of winter common wheat, as an increase in plant density strongly contributed to the increased occurrence of powdery mildew symptoms [34,45]. It is known that dense sowing brings the leaves closer to each other, which facilitates the spread of pathogen spores. In addition, the sowing density may also indirectly affect the occurrence of pathogens and severity of diseases by changing the microclimate inside the canopy. In dense crops, the temperature tends to be more uniform, the air humidity is higher, and water remains on the leaves for longer, favoring infection by *B. graminis* [32,35]. Moreover, the distance between the infected and healthy tissues is also of great importance. The closer the leaves are to each other, the easier it is for the fungus spores to splash, and thus for pathogens to spread through the canopy [33]. On the other hand, Finckh et al. [46] found that, in the case of susceptible wheat cultivars, the severity of powdery mildew symptoms increased with the decrease in sowing density. They did not observe such an effect for resistant wheat genotypes. Such relationships were also not observed in previous studies conducted in organic crops, where in both Persian wheat and Indian dwarf wheat the sowing density had no effect on the severity of powdery mildew [31].

In the present authors' research, the severity of the symptoms of foot and root rot diseases significantly depended on the sowing density. On Indian dwarf wheat, the most symptoms of root rot and Fusarium foot rot were found in the highest sowing density of 600 pcs. m^{-2}. The sowing density of Persian wheat, however, had an ambiguous effect on the occurrence of Fusarium foot rot, and usually the fewest disease symptoms were found at the highest sowing density. Moreover, in this wheat, there was no significant effect of sowing density on the severity of root rot disease. Other authors, studying the winter forms of common wheat, also observed an increase in the severity of the symptoms of diseases observed on the roots and stem base at a higher sowing density [47,48], which was especially noticeable at the early development stages of the analyzed plants. They found that in the case of foot and root rot diseases, an increased plant density means a shorter distance between the host plant and the inoculum source, which increases the probability of the pathogen reaching the adjacent plant through the spores or mycelium and infecting it. In previous studies, when Indian dwarf wheat or Persian wheat were grown on organic farms, no significant influence of their sowing density on the severity of foot and root rot diseases, including the severity of Fusarium foot rot and root rot, was found [31]. In addition, Eken et al. [49] found that there was no significant effect of the sowing density on the severity of root and crown rot caused by Fusarium fungi in common wheat. In organic crops, due to the lower intensity of cultivation, the growth and development of plants is less intense, the canopy architecture is different, and thus the influence of factors on the development of pathogens is different as well [26,32].

5. Conclusions

Our research presents, for the first time, the characteristics of morphological and agronomic features and the intensity of disease symptoms in Indian dwarf wheat and Persian ancient wheat species depending on the sowing density under cultivation using low-input technology. The three-year field experiments demonstrated that the grain yield and occurrence of diseases depended primarily on the weather conditions in the growing season and to a lesser extent on sowing density. Increasing the sowing density to 600 pcs. m^{-2} increased the number of fertile generative tillers per m^{-2} and the grain yield in the year in which moderate drought stress occurred during the growing season. However, the increased sowing density also contributed to greater disease pressure. In low-input integrated cultivation, Indian dwarf wheat and Persian wheat sowing should be performed at a sowing density of 600 pcs. m^{-2}, in an agroclimatic zone with moderate droughts during the growing season. Further research is needed on the physiological and biochemical responses of ancient species of cereals such as Indian dwarf wheat and Persian wheat on sowing density under various agroclimatic conditions.

Supplementary Materials: The following are available online at https://www.mdpi.com/article/10.3390/agriculture12020205/s1, Table S1: Significant correlation coefficients ($p \leq 0.05$) between Sielianinov's hydrothermal coefficient and grain yield, and biometric features of *Triticum sphaerococcum*, Table S2: Significant correlation coefficients ($p \leq 0.05$) between Sielianinov's hydrothermal coefficient and grain yield, and biometric features of *Triticum persicum*.

Author Contributions: Conceptualization, M.S.; methodology, M.S., G.L. and R.G.; software, R.G.; validation, M.S., G.L. and R.G.; formal analysis, M.S.; investigation, M.S. and G.L.; resources, M.S.; data curation, M.S. and G.L.; writing—original draft preparation, M.S., G.L., R.G. and R.N.; writing—review and editing, M.S., G.L., R.G. and R.N.; visualization, M.S. and R.G.; supervision, M.S.; project administration, M.S.; funding acquisition, M.S. All authors have read and agreed to the published version of the manuscript.

Funding: "European Agricultural Fund for Rural Development: Europe investing in rural areas". The publication co-financed from the European Union funds under the COOPERATION of the Rural Development Programme for 2014–2020. The Managing Authority of the Rural Development Programme for 2014–2020—the Minister of Agriculture and Rural Development.

Institutional Review Board Statement: Not applicable.

Informed Consent Statement: Not applicable.

Data Availability Statement: Data sharing not applicable.

Conflicts of Interest: The authors declare no conflict of interest.

References

1. European Commission. Available online: https://ec.europa.eu/ (accessed on 1 September 2021).
2. Suchowilska, E.; Wiwart, M.; Kandler, W.; Krska, R.A. Comparison of macro- and microelement concentrations in the whole grain of four *Triticum* species. *Plant Soil Environ.* **2012**, *58*, 141–147. [CrossRef]
3. Skrajda-Brdak, M.; Konopka, I.; Tańska, M.; Szczepanek, M.; Sadowski, S.; Rychcik, B. Low molecular phytochemicals of Indian dwarf (*Triticum sphaerococcum* Percival) and Persian wheat (*T. carthlicum* Nevski) grain. *J. Cereal Sci.* **2020**, *91*, 102887. [CrossRef]
4. Josekutty, P.C. Defining the Genetic and Physiological Basis of *Triticum sphaerococcum* Perc. Master's Thesis, University of Canterbury, Christchurch, New Zealand, 2008.
5. Mori, N.; Ohta, S.; Chiba, H.; Takagi, T.; Niimi, Y.; Shinde, V.; Kajale, M.D.; Osada, T. Rediscovery of Indian dwarf wheat (*Triticum aestivum* L. ssp. *sphaerococcum* (Perc.) MK.) an ancient crop of the Indian subcontinent. *Genet. Resour. Crop. Evol.* **2013**, *60*, 1771–1775. [CrossRef]
6. Matsuoka, Y. Evolution of polyploid *Triticum* wheats under cultivation: The role of domestication, natural hybridization and allopolyploid speciation in their diversification. *Plant Cell Physiol.* **2011**, *52*, 750–764. [CrossRef]
7. Mosulishvili, M.; Bedoshvili, D.; Maisaia, I. A consolidated list of *Triticum* species and varieties of Georgia to promote repatriation of local diversity from foreign genebanks. *Ann. Agrar. Sci.* **2017**, *5*, 61–70. [CrossRef]
8. Tahir, S.A.; Khaliq, A.T.; Cheema, M.J.M. Evaluating the impact of seed rate and sowing dates on wheat productivity in semi-arid environment. *Int. Agric. Biol.* **2019**, *22*, 57–64. [CrossRef]
9. Poutala, R.; Korva, J.; Varis, E. Spring wheat cultivar performance in ecological and conventional cropping systems. *J. Sustain. Agric.* **1993**, *3*, 63–84. [CrossRef]
10. Auškalnienė, O.; Auškalnis, A. The influence of spring wheat plant density on weed suppression and grain yield. *Zemdirbyste* **2008**, *95*, 5–12.
11. Fischer, R.A.; Moreno-Ramos, O.H.; Monasterio-Ortiz, I.; Sayre, K.D. Yield response to plant density, row spacing and raised beds in low latitude spring wheat with ample soil resources. *Field Crops Res.* **2019**, *22*, 95–105. [CrossRef]
12. Silva, F.L.; Carvalho, I.R.; Barbosa, M.H.; Conte, G.G.; Hutra, D.; Moura, N.B.; Souza, V.Q. Sowing density and clipping management: Effects on the architecture and yield of dual-purpose wheat. *Biosci. J.* **2020**, *36*, 2060–2067.
13. Hussain, I.; Khan, E.A.; Hassan, G.; Gul, J.; Ozturk, M.; Alharby, H.; Hakeem, K.R.; Alamri, S. Integration of high seeding densities and criss cross row planting pattern suppresses weeds and increases grain yield of spring wheat. *J. Environ. Biol.* **2017**, *38*, 1139–1145. [CrossRef]
14. Olsen, J.; Kristensen, L.; Weiner, J. Influence of sowing density and spatial pattern of spring wheat (*Triticum aestivum*) on the suppression of different weed species. *Weed Biol. Manag.* **2006**, *6*, 165–173. [CrossRef]
15. Cardoso, C.P.; Bazzo, J.H.; Marinho, J.L.; Zucareli, C. Effect of seed vigor and sowing densities on the yield and physiological potential of wheat seeds. *J. Seed Sci.* **2021**, *43*, 1–11. [CrossRef]
16. Beavers, R.L.; Hammermeister, A.M.; Frick, B.; Astatkie, T.; Martin, R.C. Spring wheat yield response to variable seeding rates in organic farming systems at different fertility regimes. *Can. J. Plant. Sci.* **2008**, *88*, 43–52. [CrossRef]
17. Turner, N.C.; Prasertsak, P.; Setter, T.L. Plant spacing, density, and yield of wheat subjected to postanthesis water deficits. *Crop Sci.* **1994**, *34*, 741–748. [CrossRef]
18. Jeyasri, R.; Muthuramalingam, P.; Satish, L.; Pandian, S.K.; Chen, J.-T.; Ahmar, S.; Wang, X.; Mora-Poblete, F.; Ramesh, M. An overview of abiotic stress in cereal crops: Negative impacts, regulation, biotechnology and integrated omics. *Plants* **2021**, *10*, 1472. [CrossRef]

UiO, 2018, Microcomputer Power, USA, Microsoft 2020, Statsoft 2019). The significance level for all analyses was assumed to be minimal, $\alpha = 0.05$.

3. Results

3.1. Yield and Biometric Features of Triticum sphaerococcum

The grain yield of Indian dwarf wheat varied over the years of the research and ranged from 1.60 to 4.79 Mg ha^{-1} (Table 3). The least favorable year for the growth and development of wheat was 2018, where low rainfall in May and June resulted in a strong reduction of the yield (Table 2). The most favorable, in turn, was 2020, where heavy rainfall in May and June was beneficial for the yield. In 2019, there was heavy rainfall only in May. Under these conditions, the grain yield was average (2.50 Mg ha^{-1}). A significant effect of the sowing density on the grain yield of Indian dwarf wheat was demonstrated only in 2019. Increasing the sowing density from 400 or 500 pcs. m^{-2} to 600 resulted in a significant increase in the yield (Table 3, Figure 1A). There was no significant effect of sowing density on the number of grains per spike in any of the study years. In contrast, 1000-grain weight in 2019 was higher at the sowing density of 400 pcs. m^{-2} compared to the size of this parameter at the sowing density of 500 pcs. m^{-2}. Indian dwarf wheat developed the lowest straw weight in 2018 and the highest in 2020. The straw weight was higher at the sowing density of 600 pcs. m^{-2} compared to the straw weight at other sowing densities only in 2019. The variation coefficients of the grain and straw yield and the number of grains per spike of Indian dwarf wheat were highest in 2018.

Table 3. Grain and straw yield, number of grains per spike, and 1000-grain weight of *Triticum sphaerococcum* at different sowing density in study years 2018–2020.

Year	Sowing Density (Grains m^{-2})	Grain Yield (Mg ha^{-1})	Number of Grain Per Spike	1000-Grain Weight (g)	Straw Yield (Mg ha^{-1})
2018	400	1.67 ± 0.20 a [1]	19.0 ± 3.3 a	32.1 ± 1.4 a	1.40 ± 0.18 a
	500	1.68 ± 0.10 a	19.7 ± 4.0 a	32.6 ± 0.8 a	1.64 ± 0.25 a
	600	1.60 ± 0.11 a	18.9 ± 2.3 a	31.1 ± 0.7 a	1.52 ± 0.11 a
	CV%	8.14	15.46	3.57	13.27
2019	400	2.37 ± 0.15 b	25.2 ± 1.9 a	28.3 ± 0.5 a	4.39 ± 0.08 b
	500	2.42 ± 0.11 b	23.6 ± 1.9 a	26.7 ± 0.4 b	4.26 ± 0.11 b
	600	2.70 ± 0.03 a	23.4 ± 3.1 a	27.0 ± 1.0 ab	4.74 ± 0.11 a
	CV%	7.25	9.51	3.61	5.10
2020	400	4.65 ± 0.10 a	28.9 ± 1.6 a	32.1 ± 0.3 a	9.25 ± 0.32 a
	500	4.75 ± 0.15 a	28.8 ± 1.2 a	31.3 ± 0.9 a	9.26 ± 0.95 a
	600	4.79 ± 0.17 a	29.8 ± 2.1 a	32.2 ± 0.8 a	9.52 ± 0.56 a
	CV%	3.03	5.37	2.46	6.58

[1] Mean values ± standard deviation (SD) in column followed by different letters indicate significant differences between treatments at $p \leq 0.05$.

19. Altaf, A.; Zhu, X.; Zhu, M.; Quan, M.; Irshad, S.; Xu, D.; Aleem, M.; Zhang, X.; Gull, S.; Li, F.; et al. Effects of environmental stresses (heat, salt, waterlogging) on grain yield and associated traits of wheat under application of sulfur-coated urea. *Agronomy* **2021**, *11*, 2340. [CrossRef]

20. Prasad, P.V.; Pisipati, S.; Ristic, Z.; Bukovnik, U.; Fritz, A. Impact of nighttime temperature on physiology and growth of spring wheat. *Crop. Sci.* **2008**, *48*, 2372–2380. [CrossRef]

21. Sattar, A.; Sher, A.; Ijaz, M.; Ul-Allah, S.; Rizwan, M.S.; Hussain, M.; Jabran, K.; Cheem, M.A. Terminal drought and heat stress alter physiological and biochemical attributes in flag leaf of bread wheat. *PLoS ONE* **2020**, *15*, e0232974. [CrossRef]

22. Panhwar, N.A.; Mierzwa-Hersztek, M.; Baloch, G.M.; Soomro, Z.A.; Sial, M.A.; Demiraj, E.; Panhwar, S.A.; Afzal, A.; Lahori, A.H. Water stress affects the some morpho-physiological traits of twenty wheat (*Triticum aestivum* L.) genotypes under field condition. *Sustainability* **2021**, *13*, 13736. [CrossRef]

23. Zhang, J.; Zhang, S.; Cheng, M.; Jiang, H.; Zhang, X.; Peng, C.; Lu, X.; Zhang, M.; Jin, J. Effect of drought on agronomic traits of rice and wheat: A meta-analysis. *Int. J. Environ. Res. Public Health* **2018**, *15*, 839. [CrossRef]

24. Moosavi, S.S.; Abdi, F.; Abdollahi, M.R.; Enferadi, S.T.; Maleki, M. Phenological, morpho-physiological and proteomic responses of Triticum boeoticum to drought stress. *Plant Physiol. Biochem.* **2020**, *156*, 95–104. [CrossRef]

25. Khanna-Chopra, R.; Viswanathan, C. Evaluation of heat stress tolerance in irrigated environment of *T. aestivum* and related species. I. Stability in yield and yield components. *Euphytica* **1999**, *106*, 169–180. [CrossRef]

26. Brandsæter, L.O.; Mangerud, K.; Andersson, L.; Børresen, T.; Brodal, G.; Melander, B. Influence of mechanical weeding and fertilisation on perennial weeds, fungal diseases, soil structure and crop yield in organic spring cereals. *Acta Agric Scand* **2020**, *B*, 1–15. [CrossRef]

27. Simón, M.R.; Börner, A.; Struik, P.C. Editorial: Fungal wheat diseases: Etiology, breeding, and integrated management. *Front. Plant Sci.* **2021**, *12*, 671060. [CrossRef]

28. Turner, J.A.; Chantry, T.; Taylor, M.C.; Kennedy, M.C. Changes in agronomic practices and incidence and severity of diseases in winter wheat in England and Wales between 1999 and 2019. *Plant Pathol.* **2021**, *70*, 1759–1778. [CrossRef]

29. Figueroa, M.; Hammond-Kosack, K.E.; Solomon, P.S. A review of wheat diseases—A field perspective. *Mol. Plant Pathol.* **2018**, *19*, 1523–1536. [CrossRef]

30. Khan, M.S.; Ullah, M.; Ahmad, W.; Shah, S.U.A. The use of modern technologies to combat stripe rust in wheat. *Rom. Biotech. Lett.* **2020**, *25*, 1281–1288. [CrossRef]

31. Szczepanek, M.; Lemańczyk, G.; Lamparski, R.; Wilczewski, E.; Graczyk, R.; Nowak, R.; Prus, P. Ancient wheat species (*Triticum sphaerococcum* Perc. and *T. persicum* Vav.) in organic farming: Influence of sowing density on agronomic traits, pests and diseases occurrence, and weed infestation. *Agriculture* **2020**, *10*, 556. [CrossRef]

32. Baccar, R.; Fournier, C.; Dornbusch, T.; Andrieu, B.; Gouache, D.; Robert, C. Modelling the effect of wheat canopy architecture as affected by sowing density on *Septoria tritici* epidemics using a coupled epidemic–virtual plant model. *Ann. Bot.* **2011**, *108*, 1179–1194. [CrossRef]

33. Robert, C.; Garin, G.; Abichou, M.; Houlès, V.; Pradal, C.; Fournier, C. Plant architecture and foliar senescence impact the race between wheat growth and *Zymoseptoria tritici* epidemics. *Ann. Bot.* **2018**, *121*, 975–989. [CrossRef] [PubMed]

34. Beres, B.L.; Turkington, T.K.; Kutcher, H.R.; Irvine, B.; Johnson, E.N.; O'Donovan, J.T.; Harker, K.N.; Holzapfel, C.B.; Mohr, R.; Peng, G.; et al. Winter wheat cropping system response to seed treatments, seed size, and sowing density. *J. Agron.* **2016**, *108*, 1101–1111. [CrossRef]

35. Walters, D. *Disease Control in Crops: Biological and Environmentally Friendly Approaches*; John Wiley & Sons: Edinburgh, UK, 2009; pp. 1–226.

36. Pradawne Ziarno. Available online: http://www.pradawneziarno.pl/ (accessed on 1 September 2021).

37. McDonald, J.H. *Handbook of Biological Statistics*, 2nd ed.; Sparky House Publishing: Baltimore, MD, USA, 2009; pp. 1–319.

38. Tokatlidis, I.S. Addressing the yield by density interaction is prerequisite to bridge the yield gap of rain-fed wheat. *Ann. Appl. Biol.* **2014**, *65*, 27–42. [CrossRef]

39. Pinheiro, M.G.; Souza, C.A.; Fioreze, S.L.; Sangoi, L.; Carneiro, J.F.; Bisato, M.M. Cultivar, sowing density, or time of emission: What influences mortality and performance of wheat tillers the most? *Rev. Ciênc. Agrovet.* **2021**, *20*, 19–31. [CrossRef]

40. Marinho, J.L.; Silva, S.R.; Nascimento-Souza, D.; Fonseca-Batista, I.C.; Bazzo-Bizzarri, J.H. Wheat yield and seed physiological quality affected by initial seed vigor, sowing density, and environmental conditions. *Semin. Ciênc. Agrá.* **2021**, *42*, 1595–1614. [CrossRef]

41. Wu, Y.; Xi, N.; Weiner, J.; Zhang, D.Y. Differences in weed suppression between two modern and two old wheat cultivars at different sowing densities. *Agronomy* **2021**, *11*, 253. [CrossRef]

42. Lloveras, J.; Manent, J.; Viudas, J.; Lopez, A.; Santiveri, P. Seeding rate influence on yield and yield components of irrigated winter wheat in a Mediterranean climate. *Agron. J.* **2004**, *96*, 1258–1265. [CrossRef]

43. Te Beest, D.E.; Paveley, N.D.; Shaw, M.W.; Van Den Bosch, F. Disease–weather relationships for powdery mildew and yellow rust on winter wheat. *Phytopathology* **2008**, *98*, 609–617. [CrossRef]

44. Semaškienė, R.; Tamošiūnas, K.; Dabkevičius, Z. Control of powdery mildew, *Blumeria graminis* (DC), in spring and winter wheat with decision support system based on assessments and weather data. *Plant Prot. Sci.* **2002**, *38*, 667–669.

45. Lemańczyk, G.; Piekarczyk, M. Effect of fertilization, chemical crop protection and sowing density on health status of winter wheat grown in short-time monoculture on light soil. *Prog. Plant Prot.* **2013**, *53*, 487–493.

46. Finckh, M.R.; Gacek, E.S.; Czembor, H.J.; Wolfe, M.S. Host frequency and density effects on powdery mildew and yield in mixtures of barley cultivars. *Plant Pathol.* **1999**, *48*, 807–816. [CrossRef]
47. Colbach, N.; Lucas, P.; Meynard, J.M. Influence of crop management on take-all development and disease cycles on winter wheat. *Phytopathology* **1997**, *7*, 26–32. [CrossRef]
48. Lemańczyk, G. Effects of farming system, chemical control, fertilizer and sowing density on sharp eyespot and *Rhizoctonia* spp. in winter wheat. *J. Plant Prot. Res.* **2012**, *52*, 381–396. [CrossRef]
49. Eken, C.; Bulut, S.; Genc, T.; Öztürk, A. Effects of different fertilizer sources and sowing density on root and crown rot disease agents of organic wheat. *Gaziosmanpaşa Üniversitesi Ziraat Fakültesi Derg.* **2014**, *31*, 12–19.

Article

Manure Maturation with Biochar: Effects on Plant Biomass, Manure Quality and Soil Microbiological Characteristics

Tereza Hammerschmiedt [1], Jiri Holatko [1], Jiri Kucerik [2], Adnan Mustafa [1,2,3,*], Maja Radziemska [1,4], Antonin Kintl [1,5], Ondrej Malicek [1], Tivadar Baltazar [1], Oldrich Latal [1] and Martin Brtnicky [1,2,6,*]

[1] Department of Agrochemistry, Soil Science, Microbiology and Plant Nutrition, Faculty of AgriSciences, Mendel University in Brno, Zemedelska 3, 613 00 Brno, Czech Republic; tereza.hammerschmiedt@mendelu.cz (T.H.); jiri.holatko@mendelu.cz (J.H.); maja_radziemska@sggw.edu.pl (M.R.); kintl@vupt.cz (A.K.); ondrej.malicek@mendelu.cz (O.M.); tivadar.baltazar@mendelu.cz (T.B.); oldrich.latal@mendelu.cz (O.L.)

[2] Faculty of Chemistry, Institute of Chemistry and Technology of Environmental Protection, Brno University of Technology, Purkynova 118, 612 00 Brno, Czech Republic; kucerik@fch.vut.cz

[3] Faculty of Science, Institute for Environmental Studies, Charles University in Prague, Benatska 2, 128 00 Praha, Czech Republic

[4] Institute of Environmental Engineering, Warsaw University of Life Sciences, 159 Nowoursynowska, 02-776 Warsaw, Poland

[5] Agricultural Research, Ltd., 664 41 Troubsko, Czech Republic

[6] Department of Geology and Soil Science, Faculty of Forestry and Wood Technology, Mendel University in Brno, Zemedelska 3, 613 00 Brno, Czech Republic

* Correspondence: adnanmustafa780@gmail.com (A.M.); martin.brtnicky@seznam.cz (M.B.)

Citation: Hammerschmiedt, T.; Holatko, J.; Kucerik, J.; Mustafa, A.; Radziemska, M.; Kintl, A.; Malicek, O.; Baltazar, T.; Latal, O.; Brtnicky, M. Manure Maturation with Biochar: Effects on Plant Biomass, Manure Quality and Soil Microbiological Characteristics. *Agriculture* **2022**, *12*, 314. https://doi.org/10.3390/agriculture12030314

Academic Editors: Mubshar Hussain, Sami Ul-Allah, Shahid Farooq and Sven Marhan

Received: 29 December 2021
Accepted: 18 February 2022
Published: 22 February 2022

Publisher's Note: MDPI stays neutral with regard to jurisdictional claims in published maps and institutional affiliations.

Abstract: Application of biochar and composts prepared from organic wastes as soil amendments has been recognized as a beneficial strategy to enhance soil fertility and crop production. However, the modification of manures with applied organic amendments such as biochar has not been well explained. Therefore, the preliminary study was designed to evaluate the impact of two doses of biochar (low 0.4 kg + 10 kg of manure and high 4 kg + 10 kg of manure) on the modification of resulting co-composted manure properties, and subsequently to evaluate the effect of matured manure amendment on the soil chemical and biological properties and plant yield in the pot experiment with barley (*Hordeum vulgare* L.). The following variants were tested: control, manure (M), manure + low biochar dose (M + LB), manure + high biochar dose (M + HB). Results revealed that, the M + HB significantly improved the co-composted manure properties as compared to control and M + LB, respectively. The most pronounced effects of M + HB treatment were observed on pH, NH_4-N and humic acid to fulvic acid ratio (used as an index for manure maturity) relative to other treatments. Similarly, significant variations were observed between AOB (ammonium oxidizing bacteria) and *nirs* genes under M + HB which lowered the AOB and increased the *nirs* abundance as compared to other treatments. Moreover, when applied to soil, M + HB increased the observed soil chemical parameters with the exception of TN contents as compared to M and M + LB treatments. Similarly, plant biomass was significantly enhanced under the applied M + HB treatment. However, statistically insignificant differences were observed regarding soil enzyme activities and soil respiration values under the applied amendments. Thus, it was concluded that the co-composted manure with high biochar dose can have the potential to enhance the manure properties, soil fertilization value and plant biomass. However, its effects on soil microbiological and enzyme activities were intended be explored under long-term field experiments.

Keywords: manure; crop production; soil enzymes; plant nutrients; respiration; microbial biomass

1. Introduction

Biochar is a heterogeneous material produced by the pyrolysis of various biomass such as wood, agriculture wastes, sewage sludge, etc., under controlled conditions in the absence or limited access of oxygen. It has several specific properties such as stability, high carbon (C) content, large specific surface area, microporosity and sorption capacity [1]. Biochar is resistant to decomposition, which predetermines its potential year-long persistence in nature [2] and positive contribution to soil C sequestration [3]. Its surface and sorption properties enable reduction of greenhouse gas emissions [4], removal of diverse organo-mineral pollutants, e.g., agrochemicals, antibiotics, polycyclic aromatic hydrocarbons, polychlorinated biphenyls [5,6], heavy metals [7], ammonium [8], and hydrogen sulfide [9]. Yet, the foremost benefit of biochar was considered a 2000 year-long effect in the Amazonian "Terra Preta" soils, due to joint amendment with biochar and other organic materials (e.g., bones, plant tissues, animal feces) exerting higher pH, nutrient content and more abundant and diverse microbial populations than unamended Oxisols [10,11]. Taking these findings into account, biochar represents a carbonaceous material, which provides agronomic benefits in the improvement of soil fertility (carbon and other nutrients sequestration [12]) and quality, such as mitigation of soil acidity and salinity [13–15].

Application of biochar, manure or compost mixtures and co-composted biochar as soil amendment is the most advantageous agriculture usage of biochar [16,17]. Biochar as a supplement for composting and fermentation of different types of manure [18] showed a positive influence on mitigation of greenhouse gases [19,20] and ammonia [21] emissions, prevention of nutrient loss [22], process of manure maturation (thermodynamics, heat generation [23]), abundance and functional diversity of manure microflora [24] and microorganism-dependent biochemical processes of nutrient metabolism and mineralization [25]. The physicochemical properties of different biochar types, depending on the biochar feedstock and pyrolysis conditions [26,27], determine the variation in the process of composting and the final traits of blended product [28]. It is known that biochar prepared at low temperatures ($\leq$400 °C) has a much higher ratio of potentially mineralizable C to stable C as well as a higher content of low-molecular-weight acids [29]. This feature enables greater incorporation of ammonium ions (NH_4^+) into organic compounds during the microbial utilization of labile, soluble organic C and subsequent lower NH_3 emissions [29]. Therefore, maximum ammonia (NH_3) sorption occurs in biochar prepared by low-temperature pyrolysis at almost neutral pH (7.0–7.5).

The effect of application rate on the biochar behavior in either manure or solid waste has also been a focus of various scientists [21,30]. Depending on the increasing application rate in the composted material, biochar can exert either positive [31] or negative effects [32] on microbial diversity, composting efficiency and quality, and C mineralization in soil and amendment priming effect. There are many studies regarding biochar, manure or compost mixtures and composting process modification by biochar [33–35]. Yet, the knowledge of co-composted biochar-manure importance for soil fertility and quality improvement via affecting, e.g., biological traits and the microbe-controlled nutrient transformation, still calls for further improvement [36]. Therefore, the objectives of our pilot research were to evaluate the impact of different doses of biochar on the properties of resulting co-composted manure product. Subsequently, we investigated the effect of non-enriched and dose-dependent enriched manure amendment on the soil chemical and biological properties and plant biomass yield in the pot experiment using barley (*Hordeum vulgare* L.) as a test crop. We hypothesized that: (I) co-composted manure would positively affect the maturity of final product, mainly nutrient (C and plant-available nitrogen (N)) content, microbial abundance and fertilization value; (II) subsequently, the soil amended with the biochar-enriched manure would stimulate the microbial activity in soil, enhance the nutrient transformation and mineralization, increase their (C, N, S) content and C:N ratio, in addition to increased plant biomass.

2. Materials and Methods

2.1. Manure Modification

This pot experiment was intended as a preliminary study for further field trial. Experimental manures were prepared by mixing of unmatured manure with biochar, which was (according to manufacturer) pyrolyzed from agricultural waste (85% spelled husks and sunflower husks, 10% wood chips and 5% fruit pulp) at 600 °C for 30 min (extinguished with water). Properties of the commercial biochar (Sonnenerde GmbH, Riedlingsdorf, Austria) were the same as referred [37].

Unmatured manure (from cattle breeding without marketable milk production) and biochar were dosed into the 50-litre tightly closeable barrels (Table 1). Each manure variant was prepared in 3 replicates. In biochar-amended variants, the manure and biochar were thoroughly mixed.

Table 1. Manure variants and additives dosage.

Variant	Abbrev.	Manure (M) per Barrel	Biochar (B) per Barrel	Dry Matter Ratio M:B
Manure	M	10 kg	0	0
Manure + biochar low dose	M + LB	10 kg	0.4 kg	12.5:1
Manure + biochar high dose	M + HB	10 kg	4.0 kg	1.25:1

The barrels were tightly covered, and fermentation process was carried out for 8 weeks under laboratory temperature 22.2–25.2 °C and relative humidity 60–78%. The flow of air was not controlled but was limited due to the covering. At the end of process, the mixed sample of matured manure was taken from each variant and was analyzed. The determined properties, methods and relevant references are described in Table 2.

Table 2. Determined manure properties, methods used for measurement, relevant references.

Abbrev.	Property, Method	Unit	Reference
pH	pH determined in $CaCl_2$	-	[38]
DM	dry matter, gravimetry	%	[39]
P	available phosphorus, extraction	$g \cdot kg^{-1}$	[40]
Ca	calcium, extractable (Mehlich III)	$g \cdot kg^{-1}$	[41]
TN	total Kjeldahl nitrogen	%	[42]
TC	total carbon, dry combustion	%	[43]
TOC	total organic carbon, dry combustion		
HA:FA	humic acid:fulvic acid ratio	-	[44]
N-min	mineral nitrogen	%	
$N-NH_4$	ammonium nitrogen	$mg \cdot kg^{-1}$	[45]
$N-NO_3$	nitrogen in nitrate form	$mg \cdot kg^{-1}$	[45]
AOB	ammonium-oxidizing bacteria, qPCR (gene *amoA*)	$copies \cdot g^{-1}$	[46]
nirS	denitrifying bacteria, qPCR (gene *nirS*)	$copies \cdot g^{-1}$	[47]

2.2. Pot Experiment

This pot experiment was intended as a preliminary study for further field trials. All three types of experimental manures were used as organic fertilizers in a pot experiment with barley (*Hordeum vulgare* L.). Each experimental pot of volume 5 dm^3 was filled up with soil substrate prepared by mixing of fine quartz sand (0.1–1.0 mm) with topsoil (0–15 cm) from the rural area near the town Troubsko, Czech Republic (49°10′28″ N 16°29′32″ E) sieved through 2.0 mm in ratio 1:1 (*w*/*w*). The soil was a silty clay loam (USDA Textural Triangle), Haplic Luvisol (WRB soil classification), the soil properties were the same as previously reported [37].

The tested variants were made by thorough mixing of 5 kg of experimental soil with 200 g of particular manure type per pot (equal to field manure dose of 50 $t \cdot ha^{-1}$). Un-amended control contained only 5 kg of experimental soil without any fertilizer. Each soil variant was prepared in 4 replications, in a minimal design sufficient for a preliminary study. The variants were marked equally to the used manure type: (1) (unamended) control, (2) manure (M), (3) manure + low dose of biochar (M + LB), (4) manure + high dose of biochar (M + HB). Each pot was sown with 16 barley seeds 2 cm under soil surface and was watered with distilled water to achieve 65% water holding capacity (WHC). The moisture level was maintained at $65 \pm 5\%$ WHC throughout the experiment. All pots were placed randomly into the grow chamber (CLF Plant Climatics GmbH, Wertingen, Germany) and rotated every other day to ensure homogeneity of conditions for the treatments. Controlled conditions were set as follows: 12 h long photoperiod, temperature (day/night) 20/12 °C, relative air humidity (day/night) 45/70%. The number of plants was reduced to 12 in each pot after 14 days.

2.3. Plant Biomass

The barley seedlings were grown for 12 weeks. After that, they were cut at the ground level and dried at 60 °C to constant weight. The dry above ground biomass (AGB) was determined gravimetrically using the analytical scales.

2.4. Soil Sampling, Chemical and Biological Analyses

A mixed soil sample was taken from each pot after the harvesting of barley. The samples were homogenized by sieving through a 2 mm mesh. Air-dried samples were used for total soil carbon (TC), nitrogen (TN), and sulphur (S) content determination by analyzer TruSpec (LECO Corporation, St. Joseph, MO, USA). The freeze-dried samples were prepared for the enzyme activity assays according to (ISO 20130: 2018) [48], namely: β-glucosidase (GLU), arylsulfatase (ARS), phosphatase (Phos), *N*-acetyl-β-D-glucosaminidase (NAG), and urease (Ure). The samples stored at 4 °C were used for determination of dehydrogenase activity (DHA) according to [49], and for determination of soil basal (BR) and substrate-induced respiration (expressed in µg $CO_2 \cdot g^{-1} \cdot h^{-1}$) via MicroResp method according to [50], namely: D-glucose (Glc-SIR), *N*-acetyl-β-D-glucosamine (NAG-SIR), L-lysine (Lys-SIR), L-arginine (Arg-SIR).

2.5. Statistical Analysis

Data obtained from the performed measurements were statistically analyzed using the methods of principal component analysis (PCA), one-way analysis of variance (ANOVA), Tukey HSD post-hoc test (at significance level $p = 0.05$), and Pearson correlation analysis (Program R, version 3.6.1).

3. Results

3.1. Effect of Amendments on Physical and Chemical Properties of Manure after Maturation

The pH value of the un-amended manure (M) was significantly higher in comparison to the M + HB variant, but there was no difference compared to M + LB variant (Table 3). We found a very highly positive correlation of pH with TN (r = 0.96), AOB (r = 0.96) and N-NO₃ (r = 0.82), whereas the correlation with N-NH₄ (r = −0.85) was negative (Figure A1a).

Content of DM was affected by the added biochar and therefore the un-amended manure (M) showed a significantly lower DM compared to M + LB and M + HB. The significantly highest TC was detected in the M + HB variant. However, M + LB did not differ from the un-amended manure variants, which was unexpected (Table 3). Further, the very high negative correlation (Figure A1a) with TOC (r = −0.93) and mutual antagonism (PCA biplot—Figure A2a) was found. Contrarily, TOC content in M + HB was significantly decreased as compared to both M + LB and M, and TOC value of M + LB was lower in comparison to the unamended manure (Table 3). The TOC property positively highly correlated with AOB (r = 0.85), and N-NO₃ (r = 0.91).

TN value of M + HB was lowered significantly in comparison to both other variants. However, M + LB variant did not differ from the control manure M in the TN content (Table 3). The positive correlation of TN with AOB (r = 0.91), TOC (r = 0.87), and N-NO₃ (r = 0.77) was observed. The M variant showed a significantly higher N-min content in comparison to both biochar-amended variants (Table 3). N-min correlated highly positively with N-NO₃ (r = 0.92), P (r = 0.78), and TOC (r = 0.74). N-NH₄ was significantly the highest in the M + HB variant as compared to the other two variants, and the M variant was also surprisingly higher compared to M + LB (Table 3). The high negative correlation and antagonism (Figures A1a and A2a) of N-NH$_4$ with TN (r = −0.87) and AOB (r = −0.78) was found. The content of N-NO$_3$ was significantly the highest in the control manure M in comparison to both biochar-amended variants, and M + HB showed the significantly lowest value (Table 3). The correlation analysis (Figures 1 and A1) revealed high relation between N-NO$_3$ and P (r = 0.80), AOB (r = 0.82), *nirS* (r = −0.86).

Table 3. Properties of matured manures—pH, dry matter (DM), total carbon (TC), total organic carbon (TOC), total nitrogen (TN), mineral nitrogen (N-min), ammonium nitrogen (N-NH$_4$), nitrogen in nitrate form (N-NO$_3$), humic acid:fulvic acid ratio (HA:FA), available phosphorus (P), available calcium (Ca).

Property [Unit]	M	M + LB	M + HB
	Mean ± SD *	Mean ± SD *	Mean ± SD *
pH [-]	9.04 ± 0.01 a	9.05 ± 0.01 a	8.71 ± 0.01 b
DM [%]	30.01 ± 0.02 c	31.48 ± 0.02 b	36.01 ± 0.02 a
TC [%]	9.10 ± 0.12 b	9.13 ± 0.18 b	21.01 ± 0.34 a
TOC [%]	13.50 ± 0.24 a	13.01 ± 0.12 b	11.89 ± 0.09 c
TN [%]	2.48 ± 0.05 a	2.54 ± 0.02 a	1.99 ± 0.07 b
N-min [%]	16.70 ± 0.03 a	14.58 ± 0.09 b	14.33 ± 0.55 b
N-NH$_4$ [mg·kg^{-1}]	2.06 ± 0.01 b	1.44 ± 0.07 c	2.58 ± 0.07 a
N-NO$_3$ [mg·kg^{-1}]	14.64 ± 0.02 a	13.14 ± 0.03 b	11.75 ± 0.50 c
HA:FA [-]	0.79 ± 0.01 b	0.69 ± 0.02 b	2.04 ± 0.18 a
P [g·kg^{-1}]	4.22 ± 0.31 a	3.45 ± 0.55 ab	3.04 ± 0.30 b
Ca [g·kg^{-1}]	20.93 ± 0.76 c	32.95 ± 1.69 b	159.34 ± 8.03 a

Mean values calculated as average from independent replicates (n = 4) ± SD (standard deviation). * Different letters express the statistical differences at the significance level $p \leq 0.05$.

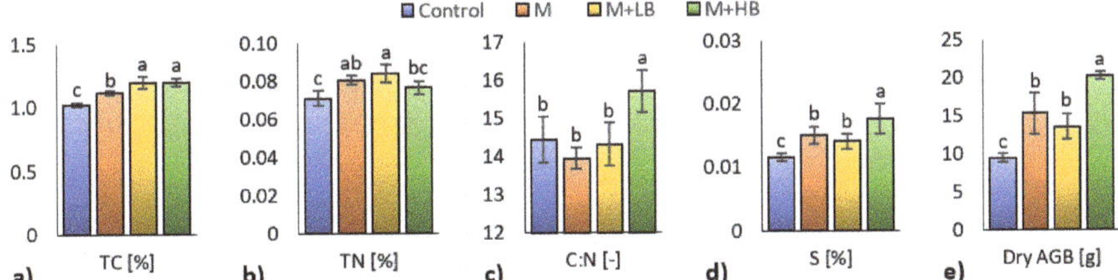

Figure 1. Mean values of chemical properties of soil and above ground biomass production. (**a**) TC = total soil carbon, (**b**) TN = total soil nitrogen, (**c**) C:N = total soil carbon/nitrogen ratio, (**d**) S = total soil sulphur, (**e**) Dry AGB = dry plant aboveground biomass. Average values of independent replicates (n = 3), error bars = standard deviation; the different letters express the statistical differences at the significance level $p \leq 0.05$.

The P content was significantly lowered in M + HB in comparison to the control manure (Table 3). A high positive correlation of P with N-NO$_3$ (r = 0.80) and N-min (r = 0.78) and synergism in PCA biplot (Figure A2a) was revealed. On the contrary, Ca was significantly higher in the M + HB variant as compared to the other variants (Table 3), and M + LB was higher compared to the control manure. The significantly highest ratio of HA:FA was observed in M + HB variant (Table 3). However, we observed decreased HA:FA ratio in the M + LB variant (in average) as compared to the un-amended manure (M).

3.2. Effect of Amendments on Microbial Properties of Manure after Maturation

A significantly decreased amount of ammonium oxidizing bacteria (AOB, indicated by gene marker *amoA*) was observed in M + HB, compared to the variants M and M + LB, which did not differ (Table 4). High correlation and synergism (Figures A1a and A2a) of AOB with TOC (r = 0.85), TN (r = 0.91) and pH (r = 0.96) was already mentioned.

Table 4. Microbial properties of matured manures—ammonium-oxidizing bacteria (AOB), denitrifying bacteria (*nirS*).

Property [Unit]	M	M + LB	M + HB
	Mean ± SD *	Mean ± SD *	Mean ± SD *
AOB [copies·g^{-1}]	$2.11 \times 10^8 \pm 2.48 \times 10^7$ a	$2.09 \times 10^8 \pm 3.43 \times 10^6$ a	$7.46 \times 10^7 \pm 5.26 \times 10^6$ b
nirS [copies·g^{-1}]	$1.07 \times 10^9 \pm 1.50 \times 10^8$ c	$1.40 \times 10^9 \pm 1.59 \times 10^8$ b	$1.85 \times 10^9 \pm 2.06 \times 10^8$ a

Mean values calculated as average from independent replicates (n = 4) ± SD (standard deviation). * Different letters express the statistical differences at the significance level $p \leq 0.05$.

M + HB was significantly more abundant in *nirS* (determinant of microorganisms reducing nitrates) in comparison to both other variants (Table 4). *nirS* of the M + LB variant was significantly increased compared to the control manure M.

3.3. Effect of Matured Manure Types on Soil Chemical Properties and Plant Biomass

The soil TC was significantly increased in the variants M + LB and M + HB, as compared to the variants M and control, and the un-amended control soil showed the significantly lowest TC value (Figure 1a). The moderate correlation (Figure A1b) with dry AGB (r = 0.67), DHA (r = 0.70), and Ure activity (r = 0.74) was shown by the Pearson's correlation and synergism was found in the PCA biplot (Figure A2b).

The soil TN was significantly increased as compared to the control in the variants M and M + LB (Figure 1b). The Pearson's analysis (Figure A1b) revealed moderate positive correlation with TC (r = 0.70), DHA (r = 0.60), Ure (r = 0.59). The significantly highest soil C:N ratio (Figure 1c) was received in the M + HB variant. However, neither strong negative correlation (Figure A1b) nor any antagonism (Figure A2b) was observed in the relation to the dry AGB. The soil S was significantly the highest in the M + HB variant as well (Figure 1d). Both other manure-amended soil variants M and M + LB showed also significantly higher S compared to the control.

The significantly highest dry AGB was found in the M + HB variant (Figure 1e). Both other manure-amended soil variants M and M + LB exerted also significantly higher dry AGB as compared to the control. Dry AGB correlated positively (Figure A1b) with BR (r = 0.71), various types of SIRs (0.46 < r < 0.82), Ure (r = 0.58).

3.4. Effect of Matured Manure Types on Soil Microbial Properties

The BR and SIR were used as the indicators of aerobic C and N mineralization rate in soil. BR was significantly the highest in M + HB (Figure 2a), and the control soil showed the significantly lowest BR. The Glc-SIR was significantly higher (compared to the control and M) in both M + LB and M + HB, BR and Glc-SIR correlated positively with dry AGB (r = 0.71 and 0.48, respectively), with NAG-SIR (r = 0.80 and 0.81, respectively), with Lys-SIR (r = 0.81 and 0.78, respectively) and each other (r = 0.67). The significance of highest and lowest values was identical for BR and NAG-SIR, Arg-SIR (Figure 2a,c,e), and similar significant differences among variants were found for Glc-SIR, NAG-SIR and Lys-SIR (Figure 2b–d).

All manure-amended variants showed the significant increase in the DHA (dehydrogenase activity) in comparison to the control (Figure 3a). The ARS (arylsulfatase) was significantly decreased in all manure-amended variants in comparison to the control (Figure 3b). The activity of *N*-acetyl-b-D-glucosaminidase (NAG) was significantly increased only in M + LB variant (Figure 3c) compared to the control. Significantly increased Ure was revealed in all manure-based variants as compared to the control (Figure 3d). The significantly highest Phos value (compared to the control) was observed in M + HB (Figure 3f).

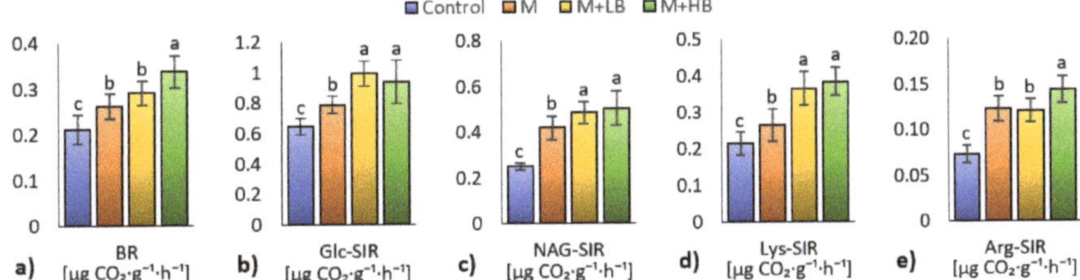

Figure 2. Mean values of soil basal (BR) and substrate induced (SIR) respirations. (**a**) BR = basal respiration, (**b**) Glc-SIR = respiration induced by D-glucose, (**c**) NAG-SIR = respiration induced by *N*-acetyl-β-D-glucosamine, (**d**) Lys-SIR = respiration induced by L-lysine, (**e**) Arg-SIR = respiration induced by L-arginine. Average values of independent replicates ($n = 3$), error bars = standard deviation; the different letters express the statistical differences at significance level $p \leq 0.05$.

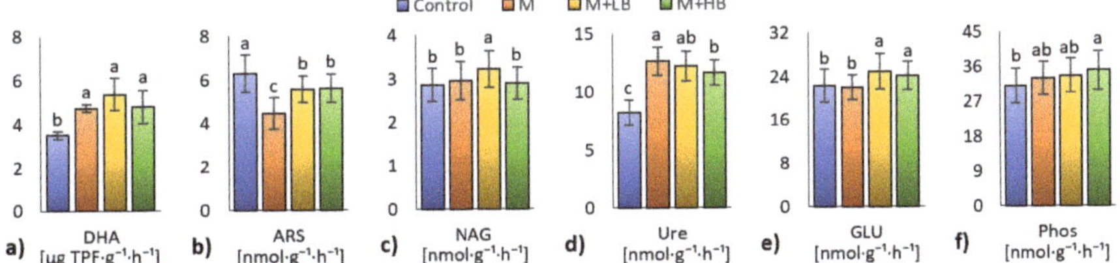

Figure 3. Mean values of soil enzyme activities. (**a**) DHA = dehydrogenase activity, (**b**) ARS = arylsulfatase activity, (**c**) NAG = *N*-acetyl-β-D-glucosaminidase activity, (**d**) Ure = urease activity, (**e**) GLU = β-glucosidase activity, (**f**) Phos = phosphatase activity. Average values of independent replicates ($n = 3$), error bars = standard deviation; the different letters express the statistical differences at significance level $p \leq 0.05$.

4. Discussion

4.1. Effect of Amendments on Physical and Chemical Properties of Manure after Maturation

The M + HB variant showed significantly lower pH compared to un-amended manure (M) and M + LB variant (Table 3). We assume that, similarly to the previously observed results by Hammerschmiedt et al. [51], the high biochar dose caused a negative priming effect on the mineralization of N, which was coupled with increased content of NH_4^+, which has a slight acidifying effect [52]. We ascribed this presumption from the PCA biplot (Figure A2a) and the calculated positive correlation with both AOB and $N\text{-}NO_3$ as well as negative correlation with $N\text{-}NH_4$ (Figure A1a).

The M + HB had significantly increased DM compared to M + LB, which was simply explainable by the higher dose of added biochar with lower moisture than the amended fresh manure had (Table 3).

TC content was closely related to the DM values as well and the significantly highest TC was detected in the M + HB variant, in comparison to the M and M + LB variants. TC of M + LB was insignificantly higher compared to the un-amended manure value (Table 3). We ascribed from the very high negative correlation (Figure A1a) of TC and TOC (and antagonism apparent in the PCA biplot—Figure A2a) that the high portion of biochar-derived C was recalcitrant to enter the metabolism of manure microorganisms. The study by Jien et al. [53] referred to biochar-mediated increase in C mineralization at the beginning of application after 70 days incubation. Although this feature was observed for compost-amended soil, we presumed this feature occurs in manure as well. Thus, we assume that primarily enhanced utilization of organic C led to significantly lower TOC content in M + HB as compared to both M + LB and M, and M + LB in comparison to the unamended manure (Table 3). The positive high correlation of TOC with AOB and $N\text{-}NO_3$ implied that the higher TOC access stimulated the nitrification process and N mineralization.

N content of pyrolyzed matter was 0.3% and the manure N content was assumed to be eight-fold higher. The amendments were mixed in the dry matter ratio 1.25:1 in the M + HB variant, therefore, the N concentration in the final mixture was lowered significantly in comparison to both other variants. Less clear was the result received for the M + LB variant, which did not differ from the control manure M (Table 3). We presumed that early fermentation sorption of volatile forms of N (NH_3, N_2O), mediated by low biochar dose, caused the difference between M + LB and M variant. There was a report that biochar reduced the losses of N in the matured composted product [18]. The positive correlation of TN with AOB and $N-NO_3$ corroborated the presumption that mainly higher access of mineralizable organic N and C in manure organic matter (OM) enhanced nitrification during the maturation.

Nevertheless, the mineralization of N seemed to be repressed by the biochar adsorbing and stabilizing impact of the organic N-compounds in the treated manure: M revealed a significantly higher N-min content in comparison to both biochar-amended variants (Table 3). The study by Lentz et al. [54] showed that use of wood biochar led to 33% less cumulative net mineral N compared to manure. N-min correlated highly positively with $N-NO_3$ (r = 0.92), P (r = 0.78), and TOC (r = 0.74), which corroborated the relation between nitrification and the availability of nutrients for manure microorganisms.

$N-NH_4$ was significantly the highest in the M + HB variant as compared to the other two variants, and the M variant was also surprisingly higher compared to M + LB (Table 3). We explain these results obtained for M + HB with the biochar adsorption of NH_4^+ reported by Kizito et al. [55], whereas in case of M + LB, we presumed higher losses of N during nitrification due to denitrification indicated by higher *nirS* number, which is also proven by low N-min. The high negative correlation of $N-NH_4$ with TN and AOB implied that a putative retardation in early nitrification phase (organic N deamination) could be involved. The content of $N-NO_3$ was significantly the highest in the control manure M in comparison to both biochar-amended variants. On the other hand, M + HB showed the significantly lowest value (Table 3). Overall restriction from further N oxidation via adsorption, availability of P, and putative losses of NO_3^- due to the denitrification are the presumed causes of the observed differences. These presumptions were corroborated by the high positive correlation and synergism between $N-NO_3$ and P, AOB, *nirS* (Figures A1a and A2a).

Similarly to the amount of TN in the manure types, the other nutrients (P, Ca) were determined by their content in the un-manured manure and biochar. Therefore, the P content was significantly lowered in M + HB in comparison to the control manure (Table 3), as the biochar exerted significantly lower P content (2.45% in DM). A high positive correlation of P with $N-NO_3$ and N-min which indicated the positive effect of P access on the nitrification in manure could be seen in PCA (Figure A2a). This finding is in line with the study by [56] reporting that P addition to composted manure positively affects N fixation and restrict denitrification. On the contrary, Ca was significantly higher in the M + HB variant as compared to the other variants (Table 3), and M + LB was higher compared to the control manure. This remarkable difference was presumably caused by the high cation sorption capacity of biochar, which was reported by several studies [57,58].

The recalcitrant nature of biochar may affect the character of humic substances formed during the manure composting; it was referred that biochar increases the polycyclic aromatic C in humified matter of biochar [59]. Further, it was reported that humic acids were adsorbed and co-flocculated on the biochar surface [60], a feature which might have decreased the HA:FA ratio in the M + LB variant, together with weaker contribution of low dose of biochar to the formation of humic acids. On the other hand, the significantly highest ratio of HA:FA was observed in the M + HB variant (Table 3). The HA:FA value, determining the degree of maturity of composted organic matter [61], indicated that M + HB may represent the most qualitatively fermented manure variant.

4.2. Effect of Amendments on Microbial Properties of Manure after Maturation

The gene marker *amoA* [46] was used as a determinant of bacteria capable to oxidize ammonium nitrogen (AOB). The variants M and M + LB exerted significantly increased AOB compared to M + HB, although AOB values of both M and M + LB did not differ (Table 4). This finding corresponded to the availability of N-NH$_4$ substrate to oxidation and to the overall nutrient availability and soil conditions as well. It was proven by high correlation and synergism of AOB with TOC, TN, and pH (Figures A1a and A2a).

The gene marker *nirS* [47] was used as a determinant of microorganisms which mediated reduction of nitrate. We received inverted proportions of AOB and *nirS* and thus, M + HB was significantly more abundant in *nirS* in comparison to both other variants (Table 4). And *nirS* of the M + LB variant was significantly increased compared to the control manure M. Putatively, the decreased pH, the lowest abundance of counteracting AOB and lowest humification intensity determined by low HA:FA ratio resulted in the enhanced denitrification coupled with the lower N-NO$_3$ content in the final manure M + LB. Other studies referred to increased nitrate reduction to nitrite in low pH [62] and increase in denitrification which was counteracted by the release of ammonium from OM [63]. These features (lower pH and N-NO$_3$, higher N-NH$_4$) were observed in the M + HB variant and may explain a detected high *nirS* value which represented abundance of denitrifiers in this manure.

4.3. Effect of Matured Manure Types on Soil Chemical Properties and Plant Biomass

Significantly increased TC was detected in M + LB and M + HB compared to the variants M and control, with the significantly lowest value in the un-amended control soil (Figure 1a). The increased access of C from the high dose biochar-enriched manure may explain the result in M + HB, however surprisingly high TC in M + LB implied that a C-sequestration role of biochar in the soil may be involved. The less easily degradable biochar-derived C putatively lowered the mineralization rate in early phase of the experiment, which was described similarly in previous study [64], and this short priming effect preserved more C to be degraded at the end of experiment. We evidenced moderate correlation of TC with dry AGB, DHA, and Ure activity (Figure A1b) which proved higher TC-mediated decomposition activity in soil, joint increased N mineralization and concurrent plant biomass yield. The PCA biplot data supported these findings in the mutual relation of traits (Figure A2b).

The soil TN of the variants M and M + LB was significantly increased as compared to the control (Figure 1b). These results clearly correspond to the values of N which were determined in the respective manure types, which were added to the soil variants. The Pearson's analysis (Figure A1b) revealed moderate positive correlation of TN with TC, DHA, Ure—these relations presumed the positive effect of the biochar-enriched manure on the soil OM decomposition.

The M + HB variant exerted the significantly highest soil C:N ratio (Figure 1c). However, neither strong negative correlation (Figure A1b) nor any antagonism (Figure A2b) was observed in the relation to the dry AGB, which was in contrast to the previously observed relation between C:N and plant biomass values in soil treated with a biochar-based amendment [17].

The soil S was the significantly highest in M + HB variant as well (Figure 1d). Both other manure-amended soil variants M and M + LB showed also significantly higher S compared to the control. We explain the results by the reported positive effect of the biochar on sorption of S forms in manure and other waste OM during composting [65]. However, partial immobilization of S in some soil types was observed too [66], which corresponded with higher S content in the M variant.

Further, the M + HB variant showed the highest dry AGB (Figure 1e). Both other manure-amended soil variants M and M + LB exerted also significantly higher dry AGB as compared to the control, which clearly corresponded to the values of TC and S in soil, and to the content of Ca and NH$_4^+$ in the prepared manures, used for the fertilization. From this reason we consider role of biochar in the soil manure amendment as crucial for crop yield, similar results were already observed [67]. The used manures impacted the microbial properties of soil which exerted a certain effect on the plant growth, probably due to the enhanced mineralization activities, monitored as respiration and enzyme activities. This was corroborated by the positive correlation (Figure A1b) of dry AGB with BR, various types of SIRs, and Ure.

4.4. Effect of Matured Manure Types on Soil Microbial Properties

The BR and SIR were used as the indicators of aerobic C and N mineralization rate in soil. BR was significantly the highest in M + HB (Figure 2a), and the control soil showed the significantly lowest BR. The findings for Glc-SIR, which was significantly higher (compared to the control and M) in both M + LB and M + HB, were similar (Figure 2b). These findings corresponded to the overall content of TC and S. Moreover, BR and Glc-SIR correlated positively with dry AGB, with NAG-SIR, with Lys-SIR, and each other. These relations indicate that the soil respiration was coupled also with nitrification and N mineralization, as the substrates NAG, Lys, and Arg are important organic N sources. The significance of highest and lowest values and their comparison was identical for BR and NAG-SIR, Arg-SIR (Figure 2), and similar statistical significance and differences among variants were found for Glc-SIR, Tre-SIR and Lys-SIR (Figure 2). The study by [68] referred to biochar–manure interaction for increased CO_2 emissions representing higher BR, but the concurrent sequestration of manure-derived C. We proved similar features of biochar-enriched manure amendment in relation to the soil respiration and observed that the respective soil variant significantly more increased both, BR and SIR, without any significant effect on the soil functional diversity; we ascribe this from the similarities among the various SIR types.

The DHA monitors overall decomposition rate of soil OM and it is considered as one of the best indicators of microbiological redox systems [69]. All manure-amended variants showed the significant increase in the DHA, which was in relation to the increased access of degradable OM (Figure A1a). The study by [70] reported that soil amended with biochar and animal manure increased DHA. The activities of enzymes involved in the nutrient mineralization corresponded to the particular nutrient contents in the soil variants. The ARS is the enzyme that catalyzes desulphurization of organosulphates [71], and its activity was significantly decreased in all manure-amended variants in comparison to the control (Figure 3b), probably because of the higher access of mineralized S and lower demand for the organic S mineralization. The respective manure-amended soil variants showed both significantly higher total S and dry AGB and lower ARS activity, thus we presume a higher S content in the utilizable form for both plants and microbes in these soil variants.

The NAG is an enzyme involved in the N mineralization, which hydrolyzes the intermediate of chitin degradation [72]. Its activity was significantly increased only in the M + LB variant (Figure 3c). The enhanced NAG corresponded with the highest soil TN, its activity may also indicate higher fungi content in the M + HB soil. The study by [35] referred to increased fungi abundance, but lowered NAG activity due to co-composted biochar-chicken manure addition to soil.

The Ure activity is the most ubiquitous enzyme among the soil microbiota, which catalyzes the deamination of urea [73]. Significantly increased Ure (as compared to the control) was revealed in all manure-based variants (Figure 3d). Consist with our findings, it was reported that biochar and animal manure increased Ure activity in soil [70]. Ure is one of the key enzymes in the early nitrification pathway and we explain the results of its activity with the chemical and biological properties of the added manure types. The non-enriched manure showed the highest AOB abundance and N-NO_3, whereas the manure M + HB exerted the lowest values of all traits. These data of manures anticipated significantly enhanced nitrification in the M variant as compared to M + HB.

The slightly different aspects were the results of GLU estimation. GLU is a part of cellulase complex and is involved in the C mineralization [74]. Hussain et al. referred to maximum conversion of feed OC to biochar recalcitrant OC at 500 °C [75]. The biochar serves as a source of both recalcitrant and labile C, and despite relatively high pyrolysis temperature (600 °C), we consider significant enrichment of soil with available carbon. Therefore, the significantly increased GLU was detected in M + LB and M + HB variants in comparison to the control and M variants (Figure 3e). The last evaluated soil enzyme was Phos, which catalyzes dephosphorylation of organophosphates in P mineralization [76]. Minhas et al. reported that biochar significantly promoted the availability and acquisition of nitrogen and phosphorus nutrients, as 2 t·ha^{-1} of amended biochar with half-recommended dose of NP fertilizer improved the fertilization performance of full dose of NP fertilizer [77]. The significantly highest Phos value was observed in M + HB (Figure 3f). We assumed that phosphatase activity corresponded to the amount of P added with respective amendment to soil, thus we put into context the highest Phos value and the lowest P content in the M + HB manure. In other variants, the manure-derived P was high enough to mitigate the demand for P mineralization. Our findings agreed with reported enhancement of Phos activity in soil amended with biochar-blended compost [78].

5. Conclusions

In this preliminary study, it was concluded that the co-composted manure with high biochar dose can lower the pH of the resultant matured amendment and hence has strong effect on nutrient availability. This was further supported by the increased NH_4-N content which further suggests the acidifying effects of co-composted amendment. Moreover, co-composted manure with high biochar dose caused higher maturation of the final product as indicated by increased HA:FA ratio. Further, biochar in manure caused a retardation in nitrification, whereas denitrification was promoted as revealed by reduced AOB and higher *nirS* abundance. The plant biomass was also enhanced, which suggests the ability of co-composted manures to supply essential plant nutrients in higher amounts. However, this is only a preliminary study with a small sized experiment, so it is necessary to investigate this issue on a larger scale to draw clear conclusions regarding soil enzyme activity and microbial respiration.

Author Contributions: Conceptualization, T.H., J.H., A.M. and M.B.; methodology, T.H., J.H. and M.B.; software, T.H. and T.B.; validation, M.B., T.H. and J.K.; formal analysis, O.L. and T.B.; investigation, J.H., J.K., A.M. and M.B.; resources, O.L., O.M. and A.K.; data curation, T.H., M.R., O.M. and O.L.; writing—original draft preparation, T.H. and J.H.; writing—review and editing, T.H., J.H., J.K., A.M., M.R. and M.B.; visualization, T.H. and A.K.; supervision, M.R. and M.B.; project administration, M.B. and A.K.; funding acquisition, M.B. and A.K. All authors have read and agreed to the published version of the manuscript.

Funding: The work was supported by the project of Technology Agency of the Czech Republic TH03030319, and by Ministry of Education, Youth and Sports of the Czech Republic, grant number FCH-S-21-7398.

Institutional Review Board Statement: Not applicable.

Informed Consent Statement: Not applicable.

Data Availability Statement: The data presented in this study are available on request from the corresponding author.

Conflicts of Interest: The authors declare no conflict of interest.

Appendix A

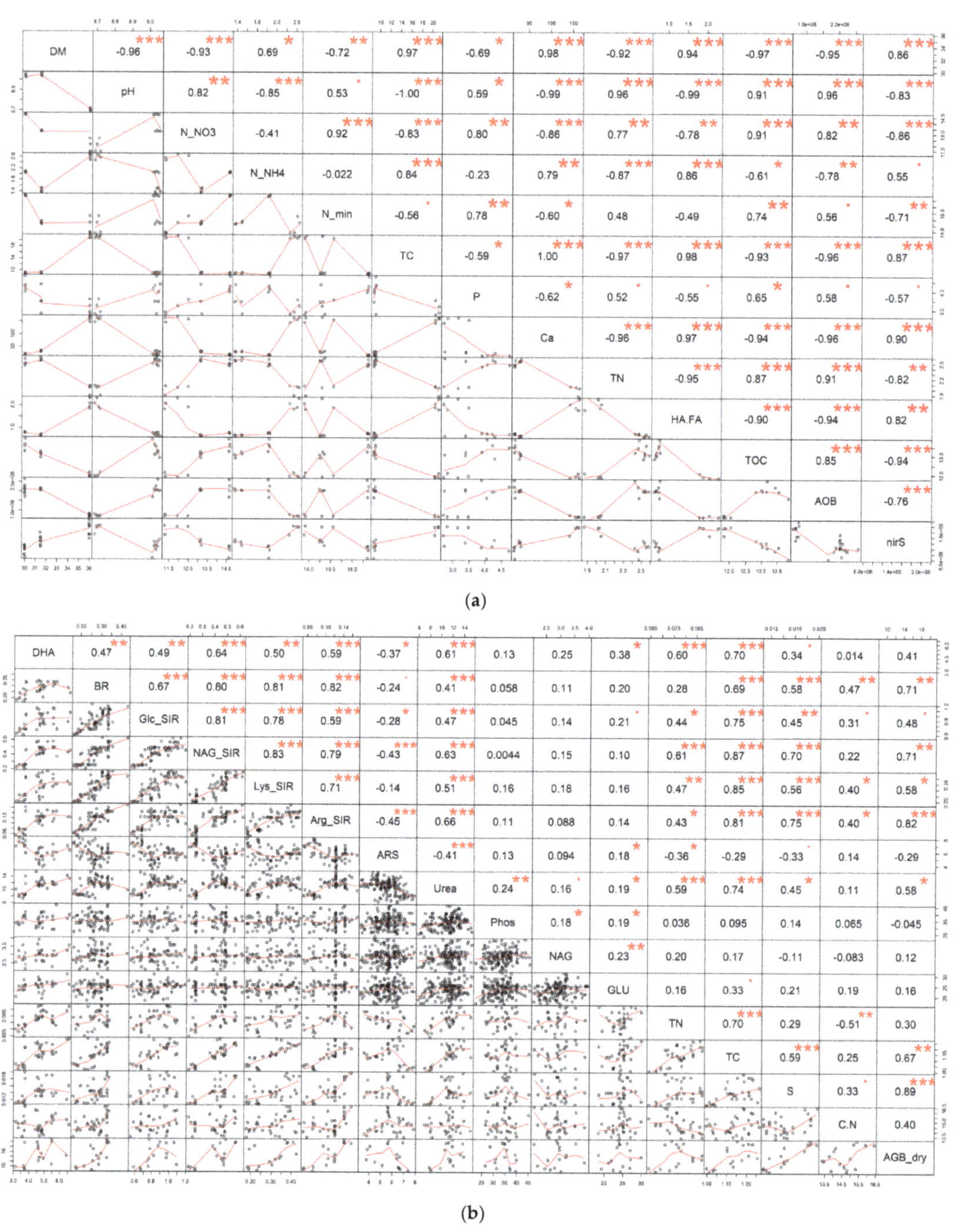

(a)

(b)

Figure A1. The Pearson's correlation matrix of (**a**) matured manure properties and (**b**) soil and plant properties; values indicate a correlation coefficient (r). Explanation: · Significant at 0.10 level; * Significant at 0.05 level; ** Significant at 0.01 level; *** Significant at 0.001 level.

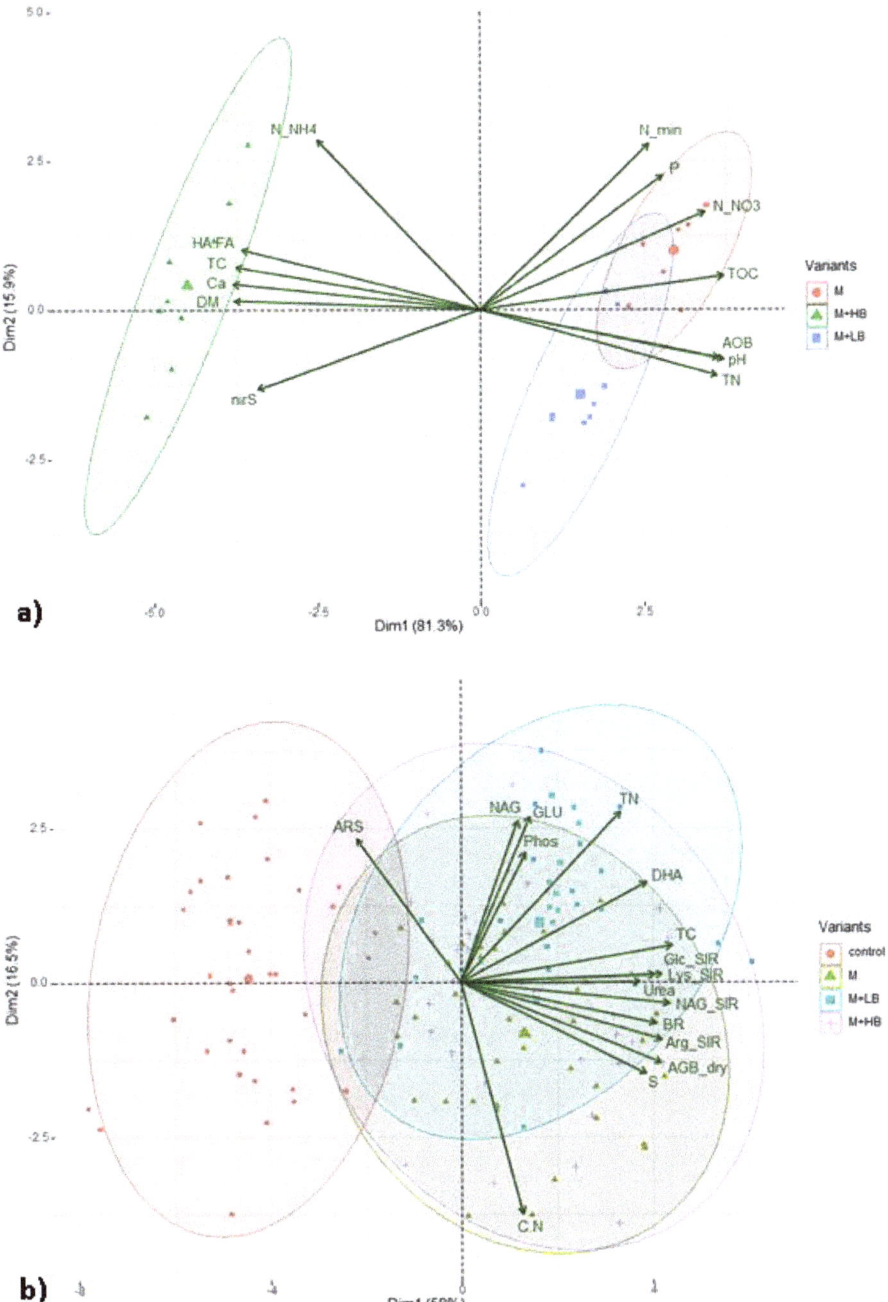

Figure A2. The Rohlf's PCA biplot of (**a**) matured manure properties and (**b**) soil and plant properties. (**a**) M = manure, M + LB = manure + biochar low dose, M + HB = manure + biochar high dose; (**b**) control = unamended soil, M = soil amended with manure, M + LB = soil amended with manure + biochar low dose, M + HB = soil amended with manure + biochar high dose; position of points / arrows is placed on the basis of their role in the first and second component (Dim1 and Dim2); arrow length equals the rate of property effect.

References

1. Li, S.; Harris, S.; Anandhi, A.; Chen, G. Predicting biochar properties and functions based on feedstock and pyrolysis temperature: A review and data syntheses. *J. Clean. Prod.* **2019**, *215*, 890–902. [CrossRef]
2. Verheijen, F.G.A.; Jeffery, S.; Bastos, A.C.; van der Velde, M.; Diafas, I. *Biochar Application to Soils—A Critical Scientific Review of Effects on Soil Properties, Processes and Functions*; Office for the Official Publications of the European Communities: Luxembourg, 2009; p. 149.
3. Ennis, C.J.; Evans, A.G.; Islam, M.; Ralebitso-Senior, T.K.; Senior, E. Biochar: Carbon sequestration, land remediation, and impacts on soil microbiology. *Crit. Rev. Environ. Sci. Technol.* **2012**, *42*, 2311–2364. [CrossRef]
4. Stewart, C.E.; Zheng, J.; Botte, J.; Cotrufo, M.F. Co-generated fast pyrolysis biochar mitigates green-house gas emissions and increases carbon sequestration in temperate soils. *GCB Bioenergy* **2013**, *5*, 153–164. [CrossRef]
5. Zheng, W.; Guo, M.; Chow, T.; Bennett, D.N.; Rajagopalan, N. Sorption properties of greenwaste biochar for two triazine pesticides. *J. Hazard. Mater.* **2010**, *181*, 121–126. [CrossRef]
6. Teixidó, M.; Pignatello, J.J.; Beltrán, J.L.; Granados, M.; Peccia, J. Speciation of the Ionizable Antibiotic Sulfamethazine on Black Carbon (Biochar). *Environ. Sci. Technol.* **2011**, *45*, 10020–10027. [CrossRef]
7. Xu, X.; Cao, X.; Zhao, L. Comparison of rice husk- and dairy manure-derived biochars for simultaneously removing heavy metals from aqueous solutions: Role of mineral components in biochars. *Chemosphere* **2013**, *92*, 955–961. [CrossRef]
8. Gao, F.; Xue, Y.; Deng, P.; Cheng, X.; Yang, K. Removal of aqueous ammonium by biochars derived from agricultural residuals at different pyrolysis temperatures. *Chem. Speciat. Bioavailab.* **2015**, *27*, 92–97. [CrossRef]
9. Shang, G.; Li, Q.; Liu, L.; Chen, P.; Huang, X. Adsorption of hydrogen sulfide by biochars derived from pyrolysis of different agricultural/forestry wastes. *J. Air Waste Manag. Assoc.* **2016**, *66*, 8–16. [CrossRef] [PubMed]
10. Glaser, B.; Haumaier, L.; Guggenberger, G.; Zech, W. The 'Terra Preta' phenomenon: A model for sustainable agriculture in the humid tropics. *Naturwissenschaften* **2001**, *88*, 37–41. [CrossRef]
11. Taketani, R.G.; Lima, A.B.; da Conceicao Jesus, E.; Teixeira, W.G.; Tiedje, J.M.; Tsai, S.M. Bacterial community composition of anthropogenic biochar and Amazonian anthrosols assessed by 16s rRNA gene 454 pyrosequencing. *Antonie Van Leeuwenhoek* **2013**, *104*, 233–242. [CrossRef] [PubMed]
12. Biederman, L.A.; Harpole, W.S. Biochar and its effects on plant productivity and nutrient cycling: A meta-analysis. *GCB Bioenergy* **2013**, *5*, 202–214. [CrossRef]
13. Wang, L.; Butterly, C.R.; Wang, Y.; Herath, H.M.S.K.; Xi, Y.G.; Xiao, X.J. Effect of crop residue biochar on soil acidity amelioration in strongly acidic tea garden soils. *Soil Use Manag.* **2014**, *30*, 119–128. [CrossRef]
14. Elbl, J.; Mikajlo, I.; Brtnicky, M.; Kynicky, J. Biochar and organic-waste compost as soil amendments to arable soil: Potential influence on soil reaction, salinity and phytotoxicity. In Proceedings of the 22nd International PhD Students Conference, Brno, Czech Republic, 11–12 November 2015; pp. 212–217.
15. Naveed, M.; Ditta, A.; Ahmad, M.; Mustafa, A.; Ahmad, Z.; Conde-Cid, M.; Tahir, S.; Shah, S.A.A.; Abrar, M.M.; Fahad, S. Processed animal manure improves morpho-physiological and biochemical characteristics of *Brassica napus* L. under nickel and salinity stress. *Environ. Sci. Pollut. Res.* **2021**, *28*, 45629–45645. [CrossRef]
16. Schulz, H.; Dunst, G.; Glaser, B. Positive effects of composted biochar on plant growth and soil fertility. *Agron. Sustain. Dev.* **2013**, *33*, 817–827. [CrossRef]
17. Agegnehu, G.; Bass, A.M.; Nelson, P.N.; Muirhead, B.; Wright, G.; Bird, M.I. Biochar and biochar-compost as soil amendments: Effects on peanut yield, soil properties and greenhouse gas emissions in tropical North Queensland, Australia. *Agric. Ecosyst. Environ.* **2015**, *213*, 72–85. [CrossRef]
18. Dias, B.O.; Silva, C.A.; Higashikawa, F.S.; Roig, A.; Sánchez-Monedero, M.A. Use of biochar as bulking agent for the composting of poultry manure: Effect on organic matter degradation and humification. *Bioresour. Technol.* **2010**, *101*, 1239–1246. [CrossRef]
19. Chen, W.; Liao, X.D.; Wu, Y.B.; Liang, J.B.; Mi, J.D.; Huang, J.J.; Zhang, H.; Wu, Y.; Qiao, Z.F.; Li, X.; et al. Effects of different types of biochar on methane and ammonia mitigation during layer manure composting. *Waste Manag.* **2017**, *61*, 506–515. [CrossRef]
20. Maurer, D.L.; Koziel, J.A.; Kalus, K.; Andersen, D.S.; Opalinski, S. Pilot-Scale Testing of Non-Activated Biochar for Swine Manure Treatment and Mitigation of Ammonia, Hydrogen Sulfide, Odorous Volatile Organic Compounds (VOCs), and Greenhouse Gas Emissions. *Sustainability* **2017**, *9*, 929. [CrossRef]
21. Janczak, D.; Malińska, K.; Czekała, W.; Cáceres, R.; Lewicki, A.; Dach, J. Biochar to reduce ammonia emissions in gaseous and liquid phase during composting of poultry manure with wheat straw. *Waste Manag.* **2017**, *66*, 36–45. [CrossRef]
22. Hagemann, N.; Subdiaga, E.; Orsetti, S.; de la Rosa, J.M.; Knicker, H.; Schmidt, H.-P.; Kappler, A.; Behrens, S. Effect of biochar amendment on compost organic matter composition following aerobic composting of manure. *Sci. Total Environ.* **2018**, *613–614*, 20–29. [CrossRef]
23. Czekała, W.; Malińska, K.; Cáceres, R.; Janczak, D.; Dach, J.; Lewicki, A. Co-composting of poultry manure mixtures amended with biochar—The effect of biochar on temperature and C-CO2 emission. *Bioresour. Technol.* **2016**, *200*, 921–927. [CrossRef] [PubMed]
24. He, X.; Yin, H.; Sun, X.; Han, L.; Huang, G. Effect of different particle-size biochar on methane emissions during pig manure/wheat straw aerobic composting: Insights into pore characterization and microbial mechanisms. *Bioresour. Technol.* **2018**, *268*, 633–637. [CrossRef] [PubMed]

25. Jindo, K.; Sanchez-Monedero, M.A.; Hernandez, T.; Garcia, C.; Furukawa, T.; Matsumoto, K.; Sonoki, T.; Bastida, F. Biochar influences the microbial community structure during manure composting with agricultural wastes. *Sci. Total Environ.* **2012**, *416*, 476–481. [CrossRef]

26. Bird, M.I. Test procedures for biochar analysis in soils. In *Biochar for Environmental Management: Science Technology and Implementation*; Lehmann, J., Joseph, S., Eds.; Routledge: Abingdon, UK, 2015; pp. 677–714.

27. Al-Wabel, M.I.; Hussain, Q.; Usman, A.R.A.; Ahmad, M.; Abduljabbar, A.; Sallam, A.S.; Ok, Y.S. Impact of biochar properties on soil conditions and agricultural sustainability: A review. *Land Degrad. Dev.* **2018**, *29*, 2124–2161. [CrossRef]

28. Agegnehu, G.; Bird, M.I.; Nelson, P.N.; Bass, A.M. The ameliorating effects of biochar and compost on soil quality and plant growth on a Ferralsol. *Soil Res.* **2015**, *53*, 1–12. [CrossRef]

29. Wei, S.; Zhu, M.; Fan, X.; Song, J.; Peng, P.; Li, K.; Jia, W.; Song, H. Influence of pyrolysis temperature and feedstock on carbon fractions of biochar produced from pyrolysis of rice straw, pine wood, pig manure and sewage sludge. *Chemosphere* **2019**, *218*, 624–631. [CrossRef] [PubMed]

30. Malinowski, M.; Wolny-Koładka, K.; Vaverková, M.D. Effect of biochar addition on the OFMSW composting process under real conditions. *Waste Manag.* **2019**, *84*, 364–372. [CrossRef]

31. Li, J.; Bao, H.; Xing, W.; Yang, J.; Liu, R.; Wang, X.; Lv, L.; Tong, X.; Wu, F. Succession of fungal dynamics and their influence on physicochemical parameters during pig manure composting employing with pine leaf biochar. *Bioresour. Technol.* **2020**, *297*, 122377. [CrossRef]

32. Qayyum, M.F.; Liaquat, F.; Rehman, R.A.; Gul, M.; Ul Hye, M.Z.; Rizwan, M.; Rehaman, M.Z.U. Effects of co-composting of farm manure and biochar on plant growth and carbon mineralization in an alkaline soil. *Environ. Sci. Pollut. Res.* **2017**, *24*, 26060–26068. [CrossRef]

33. Khan, N.; Clark, I.; Sanchez-Monedero, M.A.; Shea, S.; Meier, S.; Bolan, N. Maturity indices in co-composting of chicken manure and sawdust with biochar. *Bioresour. Technol.* **2014**, *168*, 245–251. [CrossRef]

34. Agyarko-Mintah, E.; Cowie, A.; Van Zwieten, L.; Singh, B.P.; Smillie, R.; Harden, S.; Fornasier, F. Biochar lowers ammonia emission and improves nitrogen retention in poultry litter composting. *Waste Manag.* **2017**, *61*, 129–137. [CrossRef]

35. Yuan, Y.; Chen, H.; Yuan, W.; Williams, D.; Walker, J.T.; Shi, W. Is biochar-manure co-compost a better solution for soil health improvement and N2O emissions mitigation? *Soil Biol. Biochem.* **2017**, *113*, 14–25. [CrossRef]

36. Wang, Y.; Villamil, M.B.; Davidson, P.C.; Akdeniz, N. A quantitative understanding of the role of co-composted biochar in plant growth using meta-analysis. *Sci. Total Environ.* **2019**, *685*, 741–752. [CrossRef]

37. Brtnicky, M.; Hammerschmiedt, T.; Elbl, J.; Kintl, A.; Skulcova, L.; Radziemska, M.; Latal, O.; Baltazar, T.; Kobzova, E.; Holatko, J. The potential of biochar made from agricultural residues to increase soil fertility and microbial activity: Impacts on soils with varying sand content. *Agronomy* **2021**, *11*, 1174. [CrossRef]

38. *ISO_10390*; Soil Quality—Determination of pH. International Organization for Standardization: Geneva, Switzerland, 2005.

39. *ISO_11465*; Soil Quality—Determination of Dry Matter and Water Content on a Mass Basis—Gravimetric Method. International Organization for Standardization: Geneva, Switzerland, 1993.

40. Egnér, H.; Riehm, H.; Domingo, W.R. Untersuchungen uber die chemische bodenanalyse als grundlage fur die beurteilung des nährstoffzustandes der böden. Ii. Chemische extraktionsmethoden zur phosphor- und kaliumbestimmung. *Kungliga Lantbrukshögskolans Annaler* **1960**, *26*, 199–215.

41. *ISO_14869-3*; Soil Quality—Dissolution for The Determination of Total Element Content—Part 3: Dissolution with Hydrofluoric, Hydrochloric and Nitric Acids Using Pressurised Microwave Technique. International Organization for Standardization: Geneva, Switzerland, 2017.

42. *ISO_10694*; Soil Quality—Determination of Organic and Total Carbon after Dry Combustion (Elementary Analysis). International Organization for Standardization: Geneva, Switzerland, 1995.

43. *ISO_11261*; Soil Quality—Determination of Total Nitrogen—Modified Kjeldahl Method. International Organization for Standardization: Geneva, Switzerland, 1995.

44. *ISO_19822*; Fertilizers and Soil Conditioners—Determination of Humic and Hydrophobic Fulvic Acids Concentrations in Fertilizer Materials. International Organization for Standardization: Geneva, Switzerland, 2018.

45. *ISO_15476*; Fertilizers—Determination of Nitric and Ammoniacal Nitrogen According to Devarda. International Organization for Standardization: Geneva, Switzerland, 2009.

46. Rotthauwe, J.H.; Witzel, K.P.; Liesack, W. The ammonia monooxygenase structural gene amoa as a functional marker: Molecular fine-scale analysis of natural ammonia-oxidizing populations. *Appl. Environ. Microbiol.* **1997**, *63*, 4704–4712. [CrossRef] [PubMed]

47. Kandeler, E.; Deiglmayr, K.; Tscherko, D.; Bru, D.; Philippot, L. Abundance of narG, nirS, nirK, and nosZ Genes of Denitrifying Bacteria during Primary Successions of a Glacier Foreland. *Appl. Environ. Microbiol.* **2006**, *72*, 5957–5962. [CrossRef]

48. *ISO_20130*; Soil Quality—Measurement of Enzyme Activity Patterns in Soil Samples Using Colorimetric Substrates in Micro-Well Plates. International Organization for Standardization: Geneva, Switzerland, 2018.

49. Tabatabai, M.A. Part 2—Methods of soil analysis. In *Chemical and Microbiological Properties*; American Society of Agronomy: Madison, WI, USA; Soil Science Society of America: Madison, WI, USA, 1982. [CrossRef]

50. Campbell, C.D.; Chapman, S.J.; Cameron, C.M.; Davidson, M.S.; Potts, J.M. A Rapid Microtiter Plate Method to Measure Carbon Dioxide Evolved from Carbon Substrate Amendments so as To Determine the Physiological Profiles of Soil Microbial Communities by Using Whole Soil. *Appl. Environ. Microbiol.* **2003**, *69*, 3593–3599. [CrossRef] [PubMed]

51. Hammerschmiedt, T.; Holatko, J.; Pecina, V.; Huska, D.; Latal, O.; Kintl, A.; Radziemska, M.; Muhammad, S.; Gusiatin, Z.M.; Kolackova, M.; et al. Assessing the potential of biochar aged by humic substances to enhance plant growth and soil biological activity. *Chem. Biol. Technol. Agric.* **2021**, *8*, 46. [CrossRef]

52. Lungu, O.I.; Dynoodt, R.F. Acidification from Long-Term Use of Urea and Its Effect on Selected Soil Properties. *Afr. J. Food Agric. Nutr. Dev.* **2008**, *8*, 63–76. [CrossRef]

53. Jien, S.-H.; Wang, C.-C.; Lee, C.-H.; Lee, T.-Y. Stabilization of organic matter by biochar application in compost-amended soils with contrasting ph values and textures. *Sustainability* **2015**, *7*, 13317–13333. [CrossRef]

54. Lentz, R.D.; Ippolito, J.A.; Spokas, K.A. Biochar and Manure Effects on Net Nitrogen Mineralization and Greenhouse Gas Emissions from Calcareous Soil under Corn. *Soil Sci. Soc. Am. J.* **2014**, *78*, 1641–1655. [CrossRef]

55. Kizito, S.; Wu, S.; Kipkemoi Kirui, W.; Lei, M.; Lu, Q.; Bah, H.; Dong, R. Evaluation of slow pyrolyzed wood and rice husks biochar for adsorption of ammonium nitrogen from piggery manure anaerobic digestate slurry. *Sci. Total Environ.* **2015**, *505*, 102–112. [CrossRef] [PubMed]

56. Wu, J.; He, S.; Liang, Y.; Li, G.; Li, S.; Chen, S.; Nadeem, F.; Hu, J. Effect of phosphate additive on the nitrogen transformation during pig manure composting. *Environ. Sci. Pollut. Res.* **2017**, *24*, 17760–17768. [CrossRef] [PubMed]

57. Baninajarian, S.; Shirvani, M. Use of biochar as a possible means of minimizing phosphate fixation and external P requirement of acidic soil. *J. Plant Nutr.* **2020**, *44*, 59–73. [CrossRef]

58. Haghighatjou, M.; Shirvani, M. Sugarcane Bagasse Biochar: Preparation, Characterization, and Its Effects on Soil Properties and Zinc Sorption-desorption. *Commun. Soil Sci. Plant Anal.* **2020**, *51*, 1391–1405. [CrossRef]

59. Brtnicky, M.; Datta, R.; Holatko, J.; Bielska, L.; Gusiatin, Z.M.; Kucerik, J.; Hammerschmiedt, T.; Danish, S.; Radziemska, M.; Mravcova, L.; et al. A critical review of the possible adverse effects of biochar in the soil environment. *Sci. Total Environ.* **2021**, *796*, 148756. [CrossRef]

60. Pignatello, J.J.; Kwon, S.; Lu, Y. Effect of Natural Organic Substances on the Surface and Adsorptive Properties of Environmental Black Carbon (Char): Attenuation of Surface Activity by Humic and Fulvic Acids. *Environ. Sci. Technol.* **2006**, *40*, 7757–7763. [CrossRef] [PubMed]

61. Veeken, A.; Nierop, K.; de Wilde, V.; Hamelers, H. Characterisation of NaOH-extracted humic acids during composting of a biowaste. *Bioresour. Technol.* **2000**, *72*, 33–41. [CrossRef]

62. Raut, N.; Dörsch, P.; Sitaula, B.K.; Bakken, L.R. Soil acidification by intensified crop production in South Asia results in higher $N_2O/(N_2 + N_2O)$ product ratios of denitrification. *Soil Biol. Biochem.* **2012**, *55*, 104–112. [CrossRef]

63. Davidsson, T.E.; Stepanauskas, R.; Leonardson, L. Vertical patterns of nitrogen transformations during infiltration in two wetland soils. *Appl. Environ. Microbiol.* **1997**, *63*, 3648–3656. [CrossRef] [PubMed]

64. Wang, J.; Xiong, Z.; Kuzyakov, Y. Biochar stability in soil: Meta-analysis of decomposition and priming effects. *GCB Bioenergy* **2015**, *8*, 512–523. [CrossRef]

65. Xu, X.; Cao, X.; Zhao, L.; Sun, T. Comparison of sewage sludge- and pig manure-derived biochars for hydrogen sulfide removal. *Chemosphere* **2014**, *111*, 296–303. [CrossRef] [PubMed]

66. Saviozzi, A.; Cardelli, R.; Cipolli, S.; Levi-Minzi, R.; Riffaldi, R. Sulphur mineralization kinetics of cattle manure and green waste compost in soils. *Waste Manag. Res.* **2006**, *24*, 545–551. [CrossRef] [PubMed]

67. Agegnehu, G.; Bass, A.M.; Nelson, P.N.; Bird, M.I. Benefits of biochar, compost and biochar-compost for soil quality, maize yield and greenhouse gas emissions in a tropical agricultural soil. *Sci. Total Environ.* **2016**, *543 Pt A*, 295–306. [CrossRef]

68. Rogovska, N.; Laird, D.; Cruse, R.; Fleming, P.; Parkin, T.; Meek, D. Impact of Biochar on Manure Carbon Stabilization and Greenhouse Gas Emissions. *Soil Sci. Soc. Am. J.* **2011**, *75*, 871–879. [CrossRef]

69. Curci, M.; Pizzigallo, M.D.R.; Crecchio, C.; Mininni, R.; Ruggiero, P. Effects of conventional tillage on biochemical properties of soils. *Biol. Fertil. Soils* **1997**, *25*, 1–6. [CrossRef]

70. Irmak Yilmaz, F. Impact of biochar and animal manure on some biological and chemical properties of soil. *Appl. Ecol. Environ. Res.* **2019**, *17*, 8865–8876. [CrossRef]

71. Whalen, J.K.; Warman, P.R. Arylsulfatase activity in soil and soil extracts using natural and artificial substrates. *Biol. Fertil. Soils* **1996**, *22*, 373–378. [CrossRef]

72. Ekenler, M.; Tabatabai, M. β-glucosaminidase activity of soils: Effect of cropping systems and its relationship to nitrogen mineralization. *Biol. Fertil. Soils* **2002**, *36*, 367–376. [CrossRef]

73. Watson, C.J. Urease Activity and Inhibition—Principles and Practice. In *International Fertiliser Society 2000*; International Fertiliser Society: London, UK, 2000.

74. Lammirato, C.; Miltner, A.; Kaestner, M. Effects of wood char and activated carbon on the hydrolysis of cellobiose by β-glucosidase from aspergillus niger. *Soil Biol. Biochem.* **2011**, *43*, 1936–1942. [CrossRef]

75. Hussain, M.; Farooq, M.; Nawaz, A.; Al-Sadi, A.M.; Solaiman, Z.M.; Alghamdi, S.S.; Ammara, U.O.; Yong, S.S.; Kadambot, H.M. Biochar for crop production: Potential benefits and risks. *J. Soils Sediments* **2016**, *17*, 685–716. [CrossRef]

76. Zhang, Y.L.; Chen, L.J.; Zhang, Y.G.; Wu, Z.J.; Ma, X.Z.; Yang, X.Z. Examining the effects of biochar application on soil phosphorus levels and phosphatase activities with visible and fluorescence spectroscopy. *Guang Pu Xue Yu Guang Pu Fen Xi* **2016**, *36*, 2325–2329. [PubMed]

77. Minhas, W.A.; Hussain, M.; Mehboob, N.; Nawaz, A.; Ul-Allah, S.; Rizwan, M.S.; Hassan, Z. Synergetic use of biochar and synthetic nitrogen and phosphorus fertilizers to improves maize productivity and nutrient retention in loamy soil. *J. Plant Nutr.* **2020**, *43*, 1356–1368. [CrossRef]
78. Jindo, K.; Suto, K.; Matsumoto, K.; Garcia, C.; Sonoki, T.; Sanchez-Monedero, M.A. Chemical and biochemical characterisation of biochar-blended composts prepared from poultry manure. *Bioresour. Technol.* **2012**, *110*, 396–404. [CrossRef] [PubMed]

 *agriculture*

Article

Encapsulated Clove Bud Essential Oil: A New Perspective as an Eco-Friendly Biopesticide

Zoran Milićević [1], Slobodan Krnjajić [2], Milan Stević [3], Jovana Ćirković [2], Aleksandra Jelušić [2], Mira Pucarević [4] and Tatjana Popović [5,*]

[1] Field Test d.o.o., Vinogradska 150b, 11000 Belgrade, Serbia; milicevic1960@gmail.com
[2] Institute for Multidisciplinary Research, University of Belgrade, Kneza Višeslava 1, 11030 Belgrade, Serbia; titanus_serbia@yahoo.com (S.K.); jovana.cirkovic@gmail.com (J.Ć.); jelusic.aleksandra@gmail.com (A.J.)
[3] Faculty of Agriculture, University of Belgrade, Nemanjina 6, 11080 Belgrade, Serbia; stevicm@agrif.bg.ac.rs
[4] Faculty of Environmental Protection, Educons University, Vojvode Putnika 85-87, 21208 Sremska Kamenica, Serbia; mirapucarevic@gmail.com
[5] Department of Plant Diseases, Institute for Plant Protection and Environment, Teodora Drajzera 9, 11040 Belgrade, Serbia
* Correspondence: tanjaizbis@gmail.com

Abstract: In this work by encapsulation technique we have synthetized three new clove bud essential oil (CEO) Emulsifiable Concentrate (EC) formulations depending on the carrier (synthetic zeolite- F-CSZ, nature zeolite- F-CNZ and gelatin- F-CG). The main idea was to develop an eco-friendly biopesticide that can find use in plant protection as an alternative to the use of conventional pesticides. By encapsulation we wanted to enable water solubility and ensure prolonged efficacy of the essential oil. Biological activity of designed CEO formulations was tested on potato tuber moth *Phthorimaea operculella* (fumigant mode of action), gray mold fungal pathogen *Botrytis cinerea* (preserver coatings), and soft rotting bacterial pathogens *Pectobacterium carotovorum* (subsp. *carotovorum* and *brasiliensis*) and *Dickeya dianthicola* (direct competition). CEO formulations evinced a prolonged action on mortality of *P. operculella* during the insects' exposure to the concentration of the emulsions of $40~\mu L~L^{-1}$ air. The mortality gradually decreased from a probability of 100% after the first 24 h to 50% after 5 days for F-CSZ or after 4 days for F-CNZ and F-CG. The most promising formulation is F-CSZ enabling activity during 14 days of exposure, while the effect of the other two formulations lasted 10 days. All three formulations produced a strong fungicidal effect against *B. cinerea* by preventing infection and disease development. The best efficacy was evidenced with F-CSZ (synthetic zeolite as a carrier) showing 100% efficacy when it was used even at the lowest tested concentration of active CEO (1%). The results of in vitro testing against soft rot pathogens determined the MIC value of CEO formulations to be 1% of active CEO. By this research, we present a novel perspective on the use of essential oils as an alternative, environmental biopesticide. CEO formulations can be commercially exploited as a fumigant or preserver coatings to extend the shelf life of stored products or the fresh-fruit market.

Keywords: essential oil; zeolite; gelatin; encapsulation; pest control

Citation: Milićević, Z.; Krnjajić, S.; Stević, M.; Ćirković, J.; Jelušić, A.; Pucarević, M.; Popović, T. Encapsulated Clove Bud Essential Oil: A New Perspective as an Eco-Friendly Biopesticide. *Agriculture* **2022**, *12*, 338. https://doi.org/10.3390/agriculture12030338

Academic Editors: Mubshar Hussain, Sami Ul-Allah, Shahid Farooq and Azucena González Coloma

Received: 13 January 2022
Accepted: 24 February 2022
Published: 27 February 2022

Publisher's Note: MDPI stays neutral with regard to jurisdictional claims in published maps and institutional affiliations.

1. Introduction

Plant protection products known as synthetic pesticides help in increasing the yield of food crops by controlling the loss caused by the pre- and post-harvest pests including insects, pathogens, and weeds. Estimated potential crop yield losses worldwide are 35% in the field and 14% in storage bringing the total loss of up to 50% [1–4].

Chemical-based methods of pest control in stored products are under threat for many reasons including costs, regulatory restrictions, health fears, and environmental dangers [5]. Today's global market has demanded more fresh fruits and vegetables free from pesticide residues. Combating environmental pollution and its impact on the life systems imposed replacement of synthetic chemicals with alternatives. Therefore, the introduction of natural pesticides, both of microbial and plant origin, is aimed to be a promising alternative to conventional pesticides.

Natural pesticides based on plant essential oils are gaining increasing attention in the food and agriculture industry [6–11]. Essential oils (EOs) have widespread use due to their antimicrobial, fungicide, insecticidal, and insect repellent activity [6,8,12]. Clove bud (*Syzygium aromaticum* L., syn. *Eugenia aromaticum, E. caryophyllata*, Myrtaceae family) essential oil (CEO) is known for its insecticide, antimicrobial, and nematicidal activity thanks to eugenol as the main active component [8,13–23].

Although many studies have indicated excellent antimicrobial and insecticidal properties of essential oils, their use is limited due to water insolubility, high volatility, rapid oxidation, and degradation on exposure to air [11,24]. To overcome these problems, different techniques that can provide their physical stability, protection from evaporation, and water solubility are required. Encapsulation of EOs in appropriate carriers is the most promising technique where the bioactive substance is entrapped, surrounded by a carrier matrix, which allows one to control the rate of bioactive release [3,10,24–26]. Polymer coating tends to increase oil stability and efficiency by prolonging the release rate of active components [3,10,24,27–30]. The most common polymers used for food ingredients encapsulation include gelatin, arabic gum, modified starches, and whey protein [10,27,31].

This research is addressed on the synthesis of new, eco-friendly formulations based on CEO encapsulation, which will improve oil solubility in water and biological efficacy by prolonging its release. Zeolites and gelatin were chosen as carriers in formulations. Designed formulations are predicted to be used in agriculture in the field of plant protection, for pests and disease control. Therefore, biological activity will be tested on selected pests and pathogens, which can be found in the open field as well as in storage conditions, (i) potato tuber moth *Phthorimaea operculella* Zeller [32], as one of the most destructive pests of *Solanaceae*, (ii) gray mold fungi pathogen *Botrytis cinerea*, that infects fruits (pome, stone), berries, grapes, and several vegetable species [17,30,33–35], and (iii) soft rot bacterial pathogens *Pectobacterium carotovorum* (subsp. *carotovorum* and *brasiliensis*) and *Dickeya dianthicola* destructive on a wide range of plants [36–40]. All of these harmful organisms with broad-host-range are widely present in Serbia and known to cause severe losses on attacked plant hosts. The harmfulness of *P. operculella* is reflected in the high level of potato tubers damage (up to the complete yield loss), as well as the complexity of its control [32]. The damages are considered to be quite significant at the locations with highest potato yields and production in Serbia [32]. Grey mold is among the most important diseases limiting fruit and vegetable production in Serbia. Total yield loss can exceed 80%, especially during favorable periods for disease development (rainy and wet weather before and during harvest). Development of resistance to fungicides has become a serious problem in the control of *B. cinerea* worldwide; therefore, alternative solutions for pathogen control are required. Soft rot frequently appeared in storage conditions, primarily on potato. On potatoes in the field, these bacteria cause blackleg disease, which can reduce the yield by up to 44.7% in Serbia [39].

The present study aimed to formulate biopesticide based on the encapsulation technique of CEO using different carriers that will improve water solubility and ensure the prolonged efficacy of the oil, and to check their insecticidal (test organism *P. operculella*) and antimicrobial (test organisms *B. cinerea, P. carotovorum, D. dianthicola*) activity.

2. Materials and Methods

2.1. Gas Chromatography–Mass Spectrometry (GC-MS) Analysis

Clove essential oil (CEO) is obtained by steam distillation of clove flowers (*Syzygium aromaticum*) (Probotanic). In order to identify components and to determine their amounts in CEO, GC-MS analysis was performed on a GC Agilent Technologies model 6890N, equipped with an Agilent Technologies capillary HP-5MS column (60 m length; 0.25 mm i.d.; 0.25 μm film thickness), and coupled to a mass selective detector (MSD5975B, ionization voltage 70 eV; all Agilent, Santa Clara, CA, USA). The carrier gas was Helium, used at a 1.2 mL min^{-1} flow rate. The oven temperature program was set as follows: 0 min at 60 °C ramped from 60 to 240 °C at 3 °C/min, and 5 min at 240 °C. The chromatograph was equipped with a split/splitless injector used in the split mode with a split ratio of 50:1. Before the analysis sample was diluted with EtOH in ratio 1:10 (100 μL CEO + 900 μL EtOH). The relative proportion of each component was expressed as the percentage obtained by peak area normalization, while all relative response factors were taken as one. Their Kovats indices were calculated using a homologous series of C8–C25 n-alkanes injected at the same conditions. Identification of components, in the first step, was assigned by matching their mass spectra with NIST05 Mass Spectral Library (National Institute of Standards and Technology) data with NIST MASS Spectral Search program Ver 2.0d (June 2005). Identifications were also made by comparison of their Kovats retention indices with Adams reference library file [41].

2.2. Encapsulation

CEO (Probotanic) formulations were prepared in the form of Emulsifiable Concentrate (EC), using three different active ingredient carriers: synthetic zeolite (FMC Foret, Barcelona, Spain), natural zeolite (ZEO-MEDIC, Belgrade, Serbia), and bovine gelatin (Sigma-Aldrich Inc., Oakville, ON, Canada). The natural zeolite mineral form clinoptilolite is used in this research, with a cation exchange capacity of 180 meq/100 g, and particle size of 20 μm. The synthetic zeolite molecular Sieve type 4A is an alkali alumino silicate; it is the sodium form of the Type A crystal structure. Gelatin is in the form of powder, from bovine skin, Type B (225 g Bloom).

The rapeseed oil (Granum Food, Hajdukovo, Serbia) was used as a solvent, while Tween 20 (Polysorbate 20, Aldrich-Chemie, Steinheim, West Germany) served as an emulsifier (Table 1). CEO EC formulations were prepared by mixing CEO, rapeseed oil, and Tween 20 at room temperature in a certain weight ratio as presented in Table 1. Intensive stirring was applied after the addition of each component. Different carriers at the amount of 0.25% (w/w) were added into solutions, followed by intensive stirring, in order to obtain stable and homogeneous CEO EC formulations.

Table 1. Design of CEO EC formulations.

Formulation Name	Active Ingredient (%)	Carrier (%)	Solvent (%)	Emulsifier (%)
F-CSZ	CEO (20.00)	Synthetic zeolite (0.25)	Rapeseed oil (78.75)	Tween 20 (1.00)
F-CNZ	CEO (20.00)	Natural zeolite (0.25)	Rapeseed oil (78.75)	Tween 20 (1.00)
F-CG	CEO (20.00)	Bovine gelatin (0.25)	Rapeseed oil (78.75)	Tween 20 (1.00)

Physical properties, commonly used for EC formulation, such as color, pH, density, persistent foam, and ignition point were determined on CEO formulations, using the standard procedure given by WHO [42].

2.3. Insecticidal Activity

The insecticidal activity was tested using potato tuber moth (*P. operculella*) adults. Insects were collected from the potato field in the locality Maglić (Vojvodina) and then reproduced and reared in chambers under controlled conditions (temperature at 25 ± 1 °C, relative humidity 65 ± 5%, day-night cycle 16:8 h). Determination of adults was performed using external morphological characteristics. In the insect mortality tests, 0–24 h old moths were used. Per 20 moths were added in each glass volumetric flask (720 mL in volume).

In order to evaluate the effective concentration of pure CEO that kills 95% (LC$_{95}$) or 50% (LC$_{50}$) of *P. operculella* adults, a procedure given by Negahban et al. was performed [43]. Five different concentrations of pure CEO corresponding to the μL L^{-1} air were used: 0.5625, 0.375, 0.25, 0.17and 0.11. Each of these concentrations was dissolved in a certain amount of acetone and placed onto the filter paper strips (5 × 1 cm), fixed on the lid of glass volumetric flasks with insects, and tightly closed. *P. operculella* adults were exposed to CEO vapors for the next 24 h.

To test the prolonged release (Lethal time), adults of *P. operculella* were exposed to the concentration of 8 μL L^{-1} air of pure CEO and 40 μL L^{-1} air of encapsulated formulations for 14 days. The tested concentration was equal to the 10 times higher concentration that was determined to kill 99% of *P. operculella* adults after 24 h of treatment. Each of the three encapsulated formulations and pure CEO were applied on the filter paper strips and fixed on the lid of glass volumetric flasks with *P. operculella* adults, and then were tightly closed. On a daily basis, over 14 days, the treated lids were transferred to the new volumetric flasks with 20 new, 0–24 h old adults. The insect's mortality was determined every 24 h by transferring moths from the glass flasks to opened Petri dishes. During 30 min moths were checked if they were alive, i.e., if they can move or fly. Lethal time LT$_{90}$ and LT$_{50}$ were determined. A negative control was prepared without adding CEO or encapsulated formulations. The experiments were designed in a completely randomized design in four replications and were repeated twice.

2.4. Antimicrobial Activity

2.4.1. Activity on Gray Mold Pathogen

Three encapsulated CEO formulations were diluted in sterilized water in concentrations of 25%, 5%, and 2.5%, corresponding to 5%, 1%, and 0.5% of pure CEO in solution, respectively. The solutions obtained in this way were tested for their efficacy against gray mold pathogen (*B. cinerea*) on raspberry (*Rubus idaeus* L.) fruits. *B. cinerea* strain MSTP-19 from laboratory collection (Faculty of Agriculture, Belgrade, Serbia) originated from raspberry was grown on Potato Dextrose Agar (PDA) at 23 °C for 8–10 days for the experiment. The strain was identified based on the determined morphology and molecularly, using a polymerase chain reaction (PCR) method with a universal primer pair ITS1/ITS4 and sequencing afterwards.

Fresh, healthy fruits with no damage or visible disease symptoms were immersed into solutions of CEO formulations, kept there for 30 s, and then placed on a sterile blotting sheet for 20 min to dry. Upon drying, the fruits were placed on Petri dishes (ø 90 mm) at the rate of ten fruits per dish and inoculated with *B. cinerea* isolate MSTP-19 spore suspension (2×10^5 spore mL^{-1}) using an air brush method. Petri dishes were kept at the temperature of 5 °C. The experiment was performed in three replications and were repeated twice. The effects of treatments were evaluated five days after treatment, and data were rated as the number of diseased fruits per treatment.

2.4.2. Activity on Soft Rot Pathogens

Three soft rot causing bacteria *P. carotovorum* subsp. *carotovorum* (strain Pcc10), *P. carotovorum* subsp. *brasiliensis* (strain Pcb62) and *D. dianthicola* (strain Dd31) were tested in vitro for their susceptibility to CEO encapsulated formulations. All strains belong to the laboratory collection of the Institute for Plant Protection and Environment, and were originated from cabbage in case of *P. carotovorum* subsp. *carotovorum* strain Pcc10 [38], and potato in case of strains of *P. carotovorum* subsp. *brasiliensis* Pcb62 and *D. dianthicola* Dd31 [39]. All strains were molecularly identified [38,39]. For the experiments, strains were grown on Nutrient Agar (NA) for 48 h at 26 °C. The inhibitory effect was evaluated by Agar well diffusion method using holes in the medium [44]. Bacterial suspensions of the tested strains were mixed in Nutrient Agar adjusted to the final concentration of 10^8 cells mL^{-1} and poured in sterile Petri plates (ø 90 mm). After the media solidified, 7 mm diameter holes were punched with a sterile cork borer on the center of each Petri plate. CEO formulations in concentrations of 25% (equal to 5% of pure CEO), 5% (equal to 1% of pure CEO), and 2.5% (equal to 0.5% of pure CEO) were filled with a volume of 100 μL into the well. To compare the antibacterial activity of tested concentrations of CEO formulations, pure CEO was examined at the same and higher concentrations (10%, 25%, 50%, and 100%). Plates treated with sterile distilled water served as negative control treatment. Petri plates were placed for incubation for three days at a temperature of 26 °C. Experiments were performed in two independent assays in three replications using a completely randomized design. The presence/absence and the diameter of inhibition halos were measured (expressed in millimeters) after three days of plates' incubation.

2.5. Statistics

For the pure CEO, LC_{50} (lethal concentration to cause 50% mortality in the population) and LC_{95} (lethal concentration to cause 95% mortality in the population) values were calculated by probit analysis using IBM SPSS version 23 (2015). For the formulated CEO in EC, LT_{50} (lethal time to cause 50% mortality in the population) and LT_{90} (lethal time to cause 90% mortality in the population) values were also calculated by probit analysis using the same statistical software. For modeling the dependence of the probability of insect mortality over time for encapsulated CEO or the concentration of pure CEO, probit regression was used:

$$p_i = \Phi\,(\beta_0 + \beta_1 \log_{10} x_i),$$

where p_i is the probability of realization of the outcome encoded with 1 (mortality) at the value of the predictor of x_i (time or concentration) and Φ is the function of the standard normal cumulative distribution. Sample estimate of this model is:

$$p_i = \Phi\,(b_0 + b_1 \log_{10} x_i).$$

For antimicrobial activity, the data were analyzed using the analysis of variance (One Way ANOVA, Statistica 7 software). The significance of differences was determinate by Duncan's multiple range ($p < 0.05$). The efficacy of tested formulations was calculated using Abbott's formula [45]:

$$EF\,(\%) = (X - Y)/X \times 100$$

where: X = the number of infected fruits in the control; Y = the number of infected fruits in the treatment.

3. Results

3.1. CEO Chemical Composition

GC–MS analyses of CEO (Probotanic) detected ten components accounting for 98.98% among which eugenol (4-allyl-2-ethoxyphenol) was determined as the major component with the content of 79.70%, and is followed by three compounds with abundance higher than 1% (eugenol acetate, caryophyllene, caryophyllene oxide) (Table 2).

Table 2. Chemical compositions of CEO obtained by GC-MS analysis.

Component	Content in CEO (%)	Retention Time (min)	Kovatcs Index
Eugenol	79.70	20.9212	2773.6
Eugenol acetate	11.83	27.3963	3589.1
Caryophyllene (E-)	4.51	22.9065	3023.5
Caryophyllene oxide	1.37	29.2375	3821.0
α-Humulene	0.72	24.2154	3188.5
Vanillin	0.35	22.0565	2916.6
Calamene trans	0.15	26.9725	3535.7
Humilene epoxide II	0.14	30.2112	3943.6
Chavicol	0.11	16.2289	2196.0
Coniferyl alcohol (E)	0.10	40.5097	5240.6
Total	98.98		

3.2. Encapsulated CEO Formulations

The three different, homogeneous and stable CEO formulations were prepared in the form of Emulsifiable Concentrate (EC), using three different active ingredient carriers:

natural zeolite (formulation named as F-CNZ), synthetic zeolite (formulation named as F-CSZ), and bovine gelatin (formulation named as F-CG).

Determined values of physical parameters of designed formulations are given in Table 3. Obtained results fulfill the criteria for EC formulation given by WHO [42].

Table 3. Physical parameters of CEO EC prepared formulations in working solution (0.5%).

Parameter	CEO EC Formulation		
	F-CSZ	**F-CNZ**	**F-CG**
Colour	yellow	yellow	yellow
pH	7.6	7.61	7.65
Density	0.9158 g mL^{-1}	0.9163 g mL^{-1}	0.9169 g mL^{-1}
Persistent foam (0.5%)			
10 s	0 mL	1 mL	2 mL
60 s	0 mL	0 mL	1 mL
180 s	0 mL	0 mL	0 mL
720 s	0 mL	0 mL	0 mL
Ignition point	>100 °C	>100 °C	>100 °C

3.3. Insecticidal Activity

Exposing of *P. operculella* to five concentrations of pure CEO for 24 h resulted in lethal concentrations of 0.225 μL L^{-1} air to cause mortality of 50% (LC$_{50}$) and 0.536 μL L^{-1} air to cause mortality of 95% (LC$_{95}$) of *P. operculella* adults (Table 4). The concentration of 0.768 μL L^{-1} air killed 99% of *P. operculella* adults after 24 h exposure.

Table 4. The lethal concentrations (μL L^{-1}) of pure CEO and the lethal time (day) of encapsulated CEO EC formulations with an appropriate confidence intervals, chi, and *p*-values.

Samples	LC$_{50}$	CI$_{50}$	LC$_{95}$	CI$_{95}$	LT$_{50}$	TI$_{50}$	LT$_{90}$	TI$_{90}$	Chi	*p*
Pure CEO	0.225	0.208–0.243	0.536	0.466–0.645	-	-	-	-	7.435	0.998
F-CSZ	-	-	-	-	5.898	5.539–6.243	0.954	0.210–1.576	30.762	0.995
F-CNZ	-	-	-	-	4.341	4.031–4.635	0.765	0.112–1.294	26.718	0.915
F-CG	-	-	-	-	4.472	4.189–4.744	1.220	0.652–1.683	19.856	0.993

To test the prolonged-release, treatment with pure CEO provided 100% mortality of *P. operculella* adults after 24 h exposure. After 48 h, mortality was not determined, i.e., CEO lost insecticidal activity. When the three encapsulated CEO EC formulations were used, a prolonged action on *P. operculella* adults mortality was recorded. During the insects' exposure to the emulsion concentration of 40 μL L^{-1} air, the mortality gradually decreased from a probability of 100% after the first 24 h to 50% after 5 days for F-CSZ or after 4 days for F-CNZ and F-CG. The most promising formulation is F-CSZ enabling activity during 14 days of exposure, while the effect of the other two formulations (F-CNZ, F-CG) lasted 10 days. For all three CEO formulations, the insect mortality rate was higher than 50% after four days of *P. operculella* adults' exposure.

3.4. Antimicrobial Activity

B. cinerea isolate caused gray mold symptoms on inoculated raspberry fruits in the control treatment. Experimental data revealed completely reduced (100%) infection of raspberry fruits when 5% active ingredient (CEO) was applied for all three formulations (Table 5). When formulations were applied in a concentration equal to 1% of pure CEO, formulations with zeolite as the carrier (F-CSZ and F-CNZ) showed a 100% efficacy, while formulation with gelatin (F-CG) reduced the disease for 93.4%. Use of concentration equal to 0.5% of CEO was 100% effective only for F-CSZ formulation, while other two

CEO formulations showed reduced efficacy of 93.4% and 66.7% for F-CNZ and F-CG, respectively (Table 5). The best efficacy was evidenced with F-CSZ (synthetic zeolite as the carrier), showing 100% efficacy when it was used, even at the lowest tested concentration of active CEO. A statistically significant difference was found in the treatment with the CEO concentration of 0.5% between zeolite-carrier formulations (F-CSZ and F-CNZ) and gelatin-carrier formulation F-CG (Table 5).

Table 5. Efficacy of encapsulated CEO formulations used on raspberry fruits against *B. cinerea*.

Formulation Name	Working Concentration (%)	Equal Concentration of CEO (%)	Replicate *			Average	Efficacy (%)
			I	II	III		
F-CSZ	25	5	0	0	0	0 a	100
	5	1	0	0	0	0 a	100
	2.5	0.5	0	0	0	0 a	100
F-CNZ	25	5	0	0	0	0 a	100
	5	1	0	0	0	0 a	100
	2.5	0.5	0	1	1	0.67 a	93.4
F-CG	25	5	0	0	0	0 a	100
	5	1	1	0	1	0.67 a	93.4
	2.5	0.5	3	3	4	3.3 b	66.7
Control	-	-	10	10	10	10.0 c	0

* number of fruits with gray mold symptom from a total of 10 per replicate; values followed by the same letter within columns are not significantly different ($p < 0.05$); Duncan test > MS = 0.1111; df = 18.

In vitro testing of antibacterial activity of encapsulated formulations and pure CEO determine the concentration of 1% to evince inhibitory effect against soft rot bacteria growth (*P. carotovorum* subsp. *carotovorum*, *P. carotovorum* subsp. *brasiliensis*, *D. dianthicola*) (Figure 1). CEO EC formulations exhibited a similar antibacterial activity in tested concentrations compared with the same concentrations of pure CEO. The concentration of 1% active CEO was determined to be a minimal inhibitory concentration (MIC) for both encapsulated formulations and pure CEO. From the results obtained by the testing of higher concentrations of pure CEO (Figure 1d), it is evident that CEO expressed an almost similar efficacy when used in an undiluted form and reduced concentrations until 5%. These results indicate that the increase of CEO concentration is not in correlation with the efficacy, meaning that a similar trend in efficacy was obtained when CEO was used in undiluted form and in reduced concentrations till 10%.

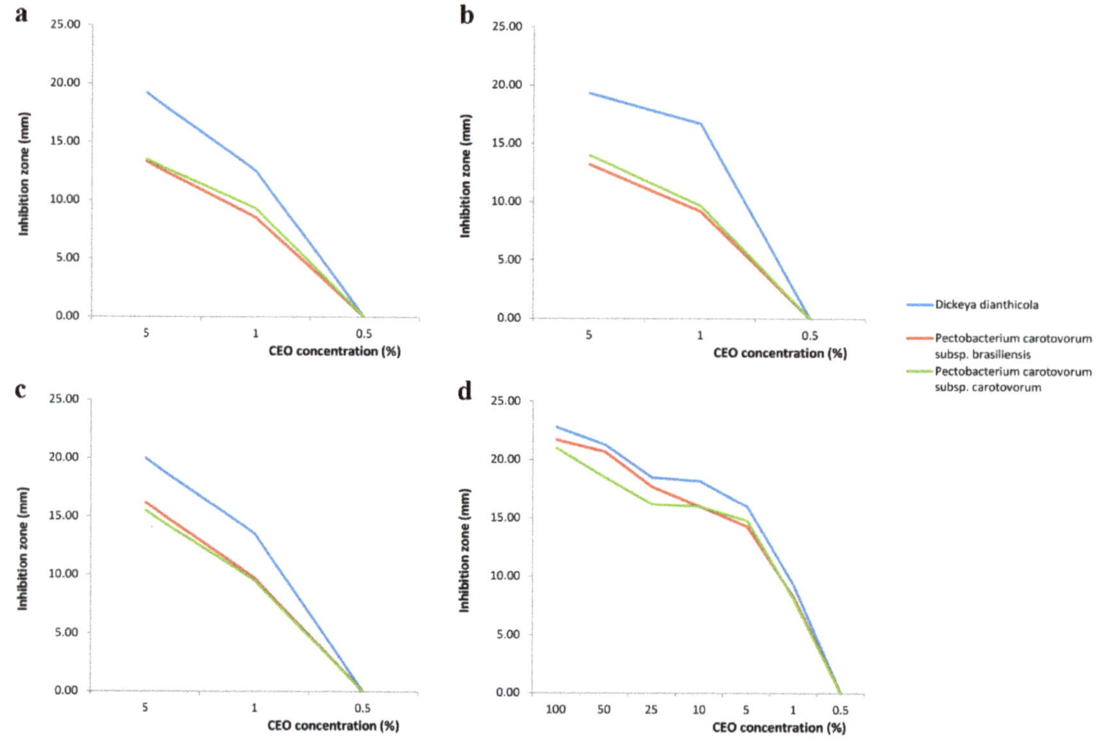

Figure 1. *In vitro* antibacterial activity of encapsulated and pure CEO against soft rot bacteria; (**a**) F-CNZ; (**b**) F-CSZ; (**c**) F-CG; (**d**) pure CEO.

4. Discussion

Instead of using synthetic pesticides, this study is considering a solution for reducing post-harvest insect and pathogen losses with the CEO EC formulations with the aim to have chemical-free, uncontaminated products during storage time. Some of the most harmful storage organisms in Serbia are potato tuber moth *P. operculella*, gray mold pathogen (*B. cinerea*) and soft rot pathogens (*Pectobacterium* spp. and *Dickeya* spp.) Many pesticides are withdrawn from the global market, and their use was prohibited on any plant products such as fruit and vegetables. In contrast, the market of biopesticides is increasing. They are derived from plant, microbial, or mineral origin acting via competition, production of antibiotics, parasitism or induction of plant defense. In this work, we designed three new, eco-friendly CEO formulations with the main idea to improve the bioactivity of oil by enabling its slow release and solubility in water. We choose CEO because it was determined to have a high efficacy against insects and plant pathogens in preliminary testing. Formulations are differing in the used carriers. Bovine Gelatin serves as a polymer matrix where CEO lipid droplets remain embedded. Zeolite adsorbs CEO droplets due to its porosity. Additives, such as Tween 20, are used as surfactants for CEO droplets dispersing in solution. The idea for gelatin arose from our previous work [46] where it was proven as a well exploited career for lemongrass EO. Zeolites were chosen for use because of their ability to control emissions and adsorption capacity for low volatile organic compounds concentrations, thus making them more suitable for working with low end concentrations than activated carbon or polymer [47]. According to De Smedt et al., zeolites can be applied as particle films against pests and diseases and can act as carriers of different active substances, ensuring slow-release applications [48]. Consequently, our assumption was to test zeolites as carriers in CEO formulations and to compare them with formulations where the well-known carrier gelatin (polymer) was used. Some researchers used natural zeolite, clinoptilolite, as a carrier for pesticides, such is the case with a solid form of the synthetic pyrethroid insecticide supercypermetrine [49] or for biology agents such is case of photostabilisation of *Bacillus thuringiensis* [50].

The choice of the material used as a coating or carrier for the encapsulation of EOs is crucial; it must protect the oil, guarantee its controlled release, be water-soluble, biocompatible, and biodegradable [31]. Four categories of compounds are reported to be used to obtain edible polymers: hydrocolloids (proteins and polysaccharides), polypeptides, lipids, and composites [31].

Encapsulation of EOs is, in general, a promising technique for the protection of sensitive bioactive compounds from degradation. Consequently, many researchers who developed several methods such as: films, microencapsulation via spray-drying, nanoprecipitation, electrohydrodynamic processes electrospinning, and electrospraying [51–53]. Encapsulation is most widely used in the food industry [31,51,52], but can find use in the pharmaceutical sector [53], and in some cases in plant protection [46,54]. Using different techniques and/or carriers, many EOs were encapsulated, such are Basil, Cinnamon, Carvacrol, Jasmine, Lime, Thyme, Lemon, Lemongrass, Orange, Oregano, Peppermint, Rosemary, etc. [46,52–55] Jovanović et al. developed a biodegradable, eco-friendly material based on natural polymers and lemongrass (*Cymbopogon citratus* L.), and essential oil (LEO) for application as a green pesticide against the potato tuber moth (*P. operculella*) [46]. Thyme was encapsulated by spray drying and evaluated for efficacy on the mortality and persistence of corn weevil *Sitophilus zeamais* [54]. Hill et al. microencapsulated inclusion complexes of Cinnamon bark extract, trans-cinnamaldehyde, clove bud extract, eugenol, and a *trans*-cinnamaldehyde:eugenol (2:1) mixture by freeze-drying [28]. These complexes were effective at lower active compound concentrations than free oils, likely due to their increased water solubility, which led to increased contact between pathogens and EOs. Kouassi et al. incorporated *Cinnamomum zeylanicum* into commercial citrus waxes (shellac, carnauba, paraffin, and polyethylene) and achieved excellent biological activity with shellac and/or carnauba wax formulations [56]. Encapsulated EOs are promising agents that can be used to increase the antimicrobial and pesticidal activities of EOs in real food systems [10]. The increased bioactivity of essential oils in the encapsulation matrix compared to free oil, even at the same or a lower concentration, was reported [10].

In the present study, we analysed the chemical composition of the CEO (Probotanic). Eugenol was detected as a major compound. Eugenol was also previously reported as a main compound of CEO in the range from 45–90% [18,19,57,58]. Tested CEO (Probotanic) was used for encapsulation, matched to a complex of compounds that are under optimum concentration ranges. Differences in the chemical composition of the CEO can vary depending on the climatic, seasonal, and/or geographic conditions, or the plant part they are extracted from. The significant variations in the content of the main compounds can be correlated to different harvest seasons [59]. Eugenol's antimicrobial activity is based on the ability to permeabilize the cell membrane and interact with proteins [58,60], hence its high content is desirable in CEO. Eugenol demonstrated contact toxicity and affected the food consumption and growth of stored-product insects [61].

Our idea was to check different modes of action of CEO formulations on target organisms; therefore, for insects, we demonstrated a fumigant mode of action. For fungal pathogens, we checked their acting as preserver coatings to extend the shelf life of stored products, and the direct inhibitory effect on the pathogen growth was checked in the case of bacterial pathogens (in vitro assay). In our study, the CEO concentration of 0.768 µL L^{-1} air was found to kill 99% of *P. operculella* adults after 24 h of exposure. When encapsulated CEO formulations were used, 14 days of prolonged action was determined for F-CSZ, and 10 days for the other two, F-CNZ and F-CG. Similarly, in other study, encapsulated lemongrass formulation showed the insecticidal and prolonged effect for up to seven days against the potato tuber moth [46]. The CEO was shown to be promising for applications in insect control. Insecticide potential of CEO for pest control whether by contact, repellence or fumigation is stated by many researchers [57,62]. According to Jairoce et al., CEO caused 100% mortality of bean and maize weevils using concentrations of 17.9 and 35 μL g^{-1}, with LC$_{50}$ ranging from 9.45–10.15 μL g^{-1} [22]. Similarly, Tian et al. showed that CEO exhibited strong contact toxicity against *Cacopsylla chinensis* and the reduced population for 73.01% (with dose 4.80 mg mL^{-1}), 66.18% (with dose 2.40 mg mL^{-1}), and 46.56% (with dose 1.20 mg mL^{-1}) [21].

The determination of gray mold decontamination of raspberry fruits after treatment with encapsulated CEO EC formulations showed that they can reduce the infection and prolong the disease-free period of fruits. According to the obtained results, we can recommend commercial coatings enriched with CEO EC formulations of fruit, berry, grape, or vegetable pre-storage applications, as an effective control treatment against the gray mold pathogen. The efficacy of different essential oils for in vivo stone fruit postharvest treatment in the control of *B. cinerea* was reported [4,33,34,63,64]. Siripornvisal et al. reported antifungal properties of CEO against *B. cinerea* mycelial growth [17]. Daniel et al. found that after direct exposure of apples for *B. cinerea* management, CEO and garlic extracts have a curative effect, either when used individually or in combination, and are proven to be more effective than a protective application [35]. Control of the gray mold pathogen is possible using *Zataria multiflora* encapsulated in chitosan nanoparticles [30] or *Rosmarinus officinalis* (rosemary) and *Mentha piperita* (peppermint) individually, or in combination with hypobaric treatment at 50 kPa [65]. Some commercial formulations containing thymol and carvacrol as active ingredients inhibited the mycelium growth and spore germination of *B. cinerea* [34].

The CEO is known to be effective against soft rot bacteria [36–38], as well as in general against both Gram-positive and Gram-negative bacteria [57]. We found that MIC (or minimum bactericidal concentrations, MBC) of CEO was 1% concentration in controlling soft rot pathogens. CEO with MBC/MIC = 1–2 and eugenol with MBC/MIC = 1–4 were bactericidal for seven Gram-negative and nine Gram-positive fish pathogenic bacteria [66]. Hajian-Maleki et al. conducted research to exploit an innovative approach for potato soft rot disease management by application of three novel EOs extracted from indigenous plants including *Hyssopus officinalis*, *Satureja khuzistanica*, and *Zataria multiflora* under in vitro and in vivo conditions [67]. The highest suppressive effects were displayed by *S. khuzistanica* and *Z. multiflora* with MIC at 0.19 and 0.38 g L^{-1}. The results of the in vivo trial indicated that tuber rot development was more efficiently controlled in preventive than in curative conditions. Disease incidence was reduced by 38.4–70.6% as compared with non-coated samples in preventing assessment.

Current research suggests that CEO and encapsulated EC formulations for their application as natural, eco-friendly biopesticides are an integral part of a new approach to pest and disease control.

5. Conclusions

Within this study, we successfully synthesized three new stable and homogenous CEO EC formulations based on natural biopolymers such as gelatine; zeolite and clove bud essential oil, that fulfill physical properties according to the criteria for the practical usage as a biopesticide. CEO encapsulated in EC formulations provides an environment-friendly alternative to synthetic pesticides in pest and pathogen control management strategies by demonstrating a high biological efficacy against *P. operculella* by providing prolonged oil efficacy of up to 14 days. This is a significant improvement, since the pure CEO lost the bioactivity against the insects after the first day of action. All three CEO formulations showed a fungicidal and bactericidal effect against gray mold and soft rot pathogens, respectively. By enabling better and prolonged oil efficacy without affecting fresh produce quality and storage, CEO formulations can be commercially exploited as fumigant or preserver coatings to extend the shelf life of stored products or the fresh-fruit market. Further work should be focused on experiments in pre-harvest production, to evaluate its possible use, efficiency, and economic justification.

Author Contributions: Conceptualization, T.P., S.K. and M.P.; Methodology, Z.M., S.K. and T.P.; software, S.K., M.S. and T.P.; validation, S.K., M.P. and T.P.; formal analysis, Z.M., S.K., M.S. and T.P.; investigation, Z.M., S.K., M.S., A.J. and T.P.; resources, Z.M. and T.P.; data curation, Z.M., S.K., M.S. and T.P.; writing—original draft preparation, Z.M. and T.P.; writing—review and editing, S.K., J.Ć., M.S., M.P. and A.J.; visualization, J.Ć., A.J. and M.P.; supervision, T.P.; funding acquisition, Z.M., S.K. and T.P. All authors have read and agreed to the published version of the manuscript.

Funding: This research received no external funding.

Institutional Review Board Statement: Not applicable.

Informed Consent Statement: Not applicable.

Data Availability Statement: Not applicable.

Acknowledgments: This work was partially supported by the Ministry of Education, Science and Technological Development of the Republic of Serbia. We greatly appreciate and thank Zora Žagar for helping us in designing formulations.

Conflicts of Interest: The authors declare no conflict of interest.

References

1. Pimentel, D.; McLaughlin, L.; Zepp, A.; Lakitan, B.; Kraus, T.; Kleinman, P.; Selig, G. Environmental and Economic Effects of Reducing Pesticide Use: A substantial reduction in pesticides might increase food costs only slightly. *Bioscience* **1991**, *41*, 402–409. [CrossRef]
2. Oerke, E.C. Crop losses to pests. *J. Agric. Sci.* **2006**, *144*, 31–43. [CrossRef]
3. Devi, N.; Maji, T.K. Neem seed oil: Encapsulation and control. In *Pesticides in the Modern World-Pesticides Use and Management*; Stoytcheva, M., Ed.; In Tech: Rijeka, Croatia, 2011; pp. 191–232.
4. Zamani-Zadeh, M.; Soleimanian-Zad, S.; Sheikh-Zeinoddin, M.; Goli, S.A.H. Integration of *Lactobacillus plantarum* A7 with thyme and cumin essential oils as a potential biocontrol tool for gray mold rot on strawberry fruit. *Postharvest Biol. Technol.* **2014**, *92*, 149–156. [CrossRef]
5. Moore, D.; Lord, J.C.; Smith, S.M. Pathogens. In *Alternatives to Pesticides in Stored-Product IPM*; Subramanyam, B., Hagstrum, D., Eds.; Kluwer Academic: New York, NY, USA, 2000; pp. 193–228.
6. Isman, M.B. Plant essential oils for pest and disease management. *Crop Prot* **2000**, *19*, 603–608. [CrossRef]
7. Isman, M.B. Botanical insecticides, deterrents, and repellents in modern agriculture and an increasingly regulated world. *Annu. Rev. Entomol.* **2006**, *51*, 45–66. [CrossRef]
8. Bakkali, F.; Averbeck, S.; Averbeck, D.; Idaomar, M. Biological effects of essential oils–a review. *Food Chem. Toxicol.* **2008**, *46*, 446–475. [CrossRef] [PubMed]
9. Koul, O.; Walia, S.; Dhaliwal, G.S. Essential oils as green pesticides: Potential and constraints. *Biopestic. Int.* **2008**, *4*, 63–84.
10. Majeed, H.; Bian, Y.Y.; Ali, B.; Jamil, A.; Majeed, U.; Khan, Q.F.; Fang, Z. Essential oil encapsulations: Uses, procedures, and trends. *RSC Adv.* **2015**, *5*, 58449–58463. [CrossRef]
11. Bakry, A.M.; Abbas, S.; Ali, B.; Majeed, H.; Abouelwafa, M.Y.; Mousa, A.; Liang, L. Microencapsulation of oils: A comprehensive review of benefits, techniques, and applications. *Compr. Rev. Food Sci. Food Saf.* **2016**, *15*, 143–182. [CrossRef]
12. Kumar, R.; Srivastava, M.; Dubey, N.K. Evaluation of *Cymbopogon martinii* oil extract for control of postharvest insect deterioration in cereals and legumes. *J. Food Prot.* **2007**, *70*, 172–178. [CrossRef]
13. Taniwiryono, D.; Berg, H.; Riksen, J.A.G.; Rietjens, I.M.C.M.; Djiwantia, S.R.; Kammenga, J.E.; Murk, A.J. Nematicidal activity of plant extracts against the root-knot nematode, *Meloidogyne incognita. Open Nat. Prod. J.* **2009**, *2*, 77–85.
14. Meyer, S.L.; Lakshman, D.K.; Zasada, I.A.; Vinyard, B.T.; Chitwood, D.J. Phytotoxicity of clove oil to vegetable crop seedlings and nematotoxicity to root-knot nematodes. *HortTechnology* **2008**, *18*, 631–638. [CrossRef]
15. Elnabawy, E.S.M.; Hassan, S.; Taha, E.K.A. Repellent and Toxicant Effects of Eight Essential Oils against the Red Flour Beetle, *Tribolium castaneum* Herbst (Coleoptera: Tenebrionidae). *Biology* **2022**, *11*, 3. [CrossRef] [PubMed]
16. Kačániová, M.; Galovičová, L.; Borotová, P.; Valková, V.; Ďúranová, H.; Kowalczewski, P.Ł.; Said-Al Ahl, H.A.H.; Hikal, W.M.; Vukic, M.; Savitskaya, T. Chemical Composition, In Vitro and In Situ Antimicrobial and Antibiofilm Activities of *Syzygium aromaticum* (Clove) Essential Oil. *Plants* **2021**, *10*, 2185. [CrossRef] [PubMed]
17. Siripornvisal, S.; Rungprom, W.; Sawatdikarn, S. Antifungal activity of essential oils derived from some medicinal plants against grey mould (*Botrytis cinerea*). *Asian J. Food Ag-Ind.* **2009**, *2*, S229–S233.
18. Nurdjannah, N.; Bermawie, N. Cloves. In *Handbook of Herbs and Spices*, 2nd ed.; Peter, K.V., Ed.; Woodhead Publishing: Cambridge, UK, 2012; pp. 197–215.
19. Barakat, H. Composition, antioxidant, antibacterial activities and mode of action of clove (*Syzygium aromaticum* L.) buds essential oil. *Brit. J. Appl. Sci. Technol.* **2014**, *4*, 1934–1951. [CrossRef]
20. Hamini-Kadar, N.; Hamdane, F.; Boutoutaou, R.; Kihal, M.; Henni, J.E. Antifungal activity of clove (*Syzygium aromaticum* L.) essential oil against phytopathogenic fungi of tomato (*Solanum lycopersicum* L) in Algeria. *J. Exp. Biol. Agric. Sci.* **2014**, *2*, 447–454.

21. Tian, B.L.; Liu, Q.Z.; Liu, Z.L.; Li, P.; Wang, J.W. Insecticidal potential of clove essential oil and its constituents on *Cacopsylla chinensis* (Hemiptera: Psyllidae) in laboratory and field. *J. Econ. Entomol.* **2015**, *108*, 957–961. [CrossRef]

22. Jairoce, C.F.; Teixeira, C.M.; Nunes, C.F.; Nunes, A.M.; Pereira, C.M.; Garcia, F.R. Insecticide activity of clove essential oil on bean weevil and maize weevil. *Revista Brasileira de Engenharia Agrícola e Ambiental* **2016**, *20*, 72–77. [CrossRef]

23. Mossa, A.T.H. Green pesticides: Essential oils as biopesticides in insect-pest management. *J. Environ. Sci. Tech.* **2016**, *9*, 354. [CrossRef]

24. Vishwakarma, G.S.; Gautam, N.; Babu, J.N.; Mittal, S.; Jaitak, V. Polymeric encapsulates of essential oils and their constituents: A review of preparation techniques, characterization, and sustainable release mechanisms. *Polym. Rev.* **2016**, *56*, 668–701. [CrossRef]

25. Maes, C.; Bouquillon, S.; Fauconnier, M.L. Encapsulation of essential oils for the development of biosourced pesticides with controlled release: A review. *Molecules* **2019**, *24*, 2539. [CrossRef] [PubMed]

26. Liao, W.; Badri, W.; Dumas, E.; Ghnimi, S.; Elaissari, A.; Saurel, R.; Gharsallaoui, A. Nanoencapsulation of Essential Oils as Natural Food Antimicrobial Agents: An Overview. *Appl. Sci.* **2021**, *11*, 5778. [CrossRef]

27. Beirão-da-Costa, S.; Duarte, C.; Bourbon, A.I.; Pinheiro, A.C.; Januário, M.I.N.; Vicente, A.A.; Delgadillo, I. Inulin potential for encapsulation and controlled delivery of Oregano essential oil. *Food Hydrocolloid* **2013**, *33*, 199–206. [CrossRef]

28. Hill, L.E.; Gomes, C.; Taylor, T.M. Characterization of beta-cyclodextrin inclusion complexes containing essential oils (trans-cinnamaldehyde, eugenol, cinnamon bark, and clove bud extracts) for antimicrobial delivery applications. *LWT-Food Sci. Technol.* **2013**, *51*, 86–93. [CrossRef]

29. Sun, P.; Zeng, M.; He, Z.; Qin, F.; Chen, J. Controlled release of fluidized bed-coated menthol powder with a gelatin coating. *Dry. Technol.* **2013**, *31*, 1619–1626. [CrossRef]

30. Mohammadi, A.; Hashemi, M.; Hosseini, S.M. Nanoencapsulation of *Zataria multiflora* essential oil preparation and characterization with enhanced antifungal activity for controlling *Botrytis cinerea*, the causal agent of gray mould disease. *Innov. Food Sci. Emerg. Technol.* **2015**, *28*, 73–80. [CrossRef]

31. Froiio, F.; Ginot, L.; Paolino, D.; Lebaz, N.; Bentaher, A.; Fessi, H.; Elaissari, A. Essential oils-loaded polymer particles: Preparation, characterization and antimicrobial property. *Polymers* **2019**, *11*, 1017. [CrossRef]

32. Miletaković, S.; Stankoviić, S.; Krnjajić, S.; Todorović, M.J.; Tomić, V.; Jovanović, R. Economic justification of biological measures for potato tuber moth control. In Proceedings of the IX International Scientific Agriculture Symposium AGROSYM, Jahorina, Bosnia and Herzegovina, 4–7 October 2018; pp. 1030–1033.

33. Tsao, R.; Zhou, T. Antifungal activity of monoterpenoids against postharvest pathogens *Botrytis cinerea* and *Monilinia fructicola*. *J. Essent. Oil Res.* **2000**, *12*, 113–121. [CrossRef]

34. Adebayo, O.; Dang, T.; Bélanger, A.; Khanizadeh, S. Antifungal studies of selected essential oils and a commercial formulation against *Botrytis cinerea*. *J. Food Res.* **2013**, *2*, 217. [CrossRef]

35. Daniel, C.K.; Lennox, C.L.; Vries, F.A. In vivo application of garlic extracts in combination with clove oil to prevent postharvest decay caused by *Botrytis cinerea*, *Penicillium expansum* and *Neofabraea alba* on apples. *Postharvest Biol. Technol.* **2015**, *99*, 88–92. [CrossRef]

36. Popović, T.; Kostić, I.; Milićević, Z.; Gašić, K.; Kostić, M.; Dervišević, M.; Krnjajić, S. Essential oils as an alternative bactericides against soft-rot bacteria, *Pectobacterium carotovorum* subsp.*carotovorum*.. In Proceedings of the VIII International Scientific Agriculture Symposium Agrosym 2017, Jahorina, Bosnia and Herzegovina, 5–8 October 2017; pp. 1377–1383.

37. Popović, T.; Milićević, Z.; Iličić, R.; Marković, S.; Oro, V.; Jelušić, A.; Krnjajić, S. Antibacterial activities of essential oils of wild oregano, clove bud, rosemary, peppermint, basil and lemongrass against growth of soft rot bacteria. In Proceedings of the 1st International Symposium: Modern Trends in Agricultural Production and Environmental Protection, Tivat, Montenegro, 2–5 July 2019; pp. 230–242.

38. Popović, T.; Jelušić, A.; Marković, S.; Iličić, R. Characterization of *Pectobacterium carotovorum* subsp. *carotovorum* isolates from a recent outbreak on cabbage in Bosnia and Herzegovina. *Pesticidi i Fitomedicina* **2019**, *34*, 211–222. [CrossRef]

39. Marković, S.; Stanković, S.; Jelušić, A.; Iličić, R.; Kosovac, A.; Poštić, D.; Popović, T. Occurrence and Identification of *Pectobacterium carotovorum* subsp. *brasiliensis* and *Dickeya dianthicola* Causing Blackleg in some Potato Fields in Serbia. *Plant Dis.* **2021**, *105*, 1080–1090. [CrossRef] [PubMed]

40. Marković, S.; Milić Komić, S.; Jelušić, A.; Iličić, R.; Bagi, F.; Stanković, S.; Popović, T. First report of *Pectobacterium versatile* causing blackleg of potato in Serbia. *Plant Dis.* **2021**, *106*, 312. [CrossRef]

41. Adams, R.P. *Identification of Essential Oil Components by Gas Chromatography/Mass Spectrometry*; Allured Publishing Corporation: Carol Stream, IL, USA, 2007.

42. WHO. *Manual on Development and Use of FAO and WHO Specifications for Pesticides. November 2010—Second Revision of the First Edition*; FAO: Rome, Italy, 2010; Available online: http://www.fao.org/3/a-y4353e.pdf (accessed on 11 March 2017).

43. Negahban, M.; Moharramipour, S.; Sefidkon, F. Fumigant toxicity of essential oil from *Artemisia sieberi* Besser against three stored-product insects. *J. Stored Prod. Res.* **2007**, *43*, 123–128. [CrossRef]

44. Balouiri, M.; Sadiki, M.; Ibnsouda, S.K. Methods for in vitro evaluating antimicrobial activity: A review. *J. Pharm. Anal.* **2016**, *6*, 71–79. [CrossRef]

45. Abbott, W.S. A Method of Computing the Effectiveness of an Insecticide. *J. Econ. Entomol.* **1925**, *18*, 265–267. [CrossRef]

46. Jovanović, J.; Krnjajić, S.; Ćirković, J.; Radojković, A.; Popović, T.; Branković, G.; Branković, Z. Effect of encapsulated lemongrass (*Cymbopogon citratus* L.) essential oil against potato tuber moth *Phthorimaea operculella*. *Crop Prot.* **2020**, *132*, 105109. [CrossRef]

47. Cox, L. Choosing an Adsorption System for VOC: Carbon, Zeolite or Polymers? Catc Tech. Bull. EPA 456/F-99-004, U.S. Environmental Protec-tion Agency, Research, NC, USA. 1999. Available online: https://www3.epa.gov/ttncatc1/dir1/fadsorb.pdf (accessed on 10 August 2021).
48. De Smedt, C.; Someus, E.; Spanoghe, P. Potential and actual uses of zeolites in crop protection. *Pest Manag. Sci.* **2015**, *71*, 1355–1367. [CrossRef]
49. Sopkova, A.; Janokova, E. An insecticide stabilized by natural zeolite. *J. Therm. Anal. Calorim.* **1998**, *53*, 477–485. [CrossRef]
50. Kvachantiradze, M.; Tvalchrelidze, E.; Kotetishvili, M.; Tsitsishvili, T. Application of clinoptilolite as an additive for the photostabilization of the *Bacillus thuringiensis* formulation. In *Porous Materials in Environmentally Friendly Processes*; Kiricsi, I., Pal-Borbély, G., Nagy, J.B., Karge, H.G., Eds.; Elsevier: Amsterdam, The Netherlands, 1999; pp. 731–735.
51. Froiio, F.; Mosaddik, A.; Morshed, M.T.; Paolino, D.; Fessi, H.; Elaissari, A. Edible polymers for essential oils encapsulation: Application in food preservation. *Ind. Eng. Chem. Res.* **2019**, *58*, 20932–20945. [CrossRef]
52. Munteanu, B.S.; Vasile, C. Encapsulation of Natural Bioactive Compounds by Electrospinning—Applications in Food Storage and Safety. *Polymers* **2021**, *13*, 3771. [CrossRef] [PubMed]
53. Nguyen, T.T.T.; Le, T.V.A.; Dang, N.N.; Nguyen, D.C.; Nguyen, P.T.N.; Tran, T.T.; Nguyen, Q.V.; Bach, L.G.; Thuy Nguyen Pham, T.D. Microencapsulation of Essential Oils by Spray-Drying and Influencing Factors. *J. Food Qual.* **2021**, *2021*, 5525879. [CrossRef]
54. Barros, F.A.; Radünz, M.; Scariot, M.A.; Camargo, T.M.; Nunes, C.F.; de Souza, R.R.; Gilson, I.K.; Hackbart, H.C.; Radünz, L.L.; Oliveira, J.V.; et al. Efficacy of encapsulated and non-encapsulated thyme essential oil (*Thymus vulgaris* L.) in the control of *Sitophilus zeamais* and its effects on the quality of corn grains throughout storage. *Crop Prot.* **2022**, *153*, 105885. [CrossRef]
55. Tangpao, T.; Krutmuang, P.; Kumpoun, W.; Jantrawut, P.; Pusadee, T.; Cheewangkoon, R.; Sommano, S.R.; Chuttong, B. Encapsulation of Basil Essential Oil by Paste Method and Combined Application with Mechanical Trap for Oriental Fruit Fly Control. *Insects* **2021**, *12*, 633. [CrossRef]
56. Kouassi, K.H.S.; Bajji, M.; Jijakli, H. The control of postharvest blue and green molds of citrus in relation with essential oil–wax formulations, adherence and viscosity. *Postharvest Biol. Technol.* **2012**, *73*, 122–128. [CrossRef]
57. Chaieb, K.; Hajlaoui, H.; Zmantar, T.; Kahla-Nakbi, A.B.; Rouabhia, M.; Mahdouani, K.; Bakhrouf, A. The chemical composition and biological activity of clove essential oil, *Eugenia caryophyllata* (*Syzigium aromaticum* L. Myrtaceae): A short review. *Phytother. Res.* **2007**, *21*, 501–506. [CrossRef] [PubMed]
58. Jafri, H.; Ansari, F.A.; Ahmad, I. Prospects of essential oils in controlling pathogenic biofilm. In *New Look to Phytomedicine*; Khan, M.S.A., Ahmad., I., Chattopadhyay, D., Eds.; Academic Press: Cambridge, MA, USA, 2019; pp. 203–236.
59. Santos-Gomes, P.C.; Fernandes-Ferreira, M. Organ-and season-dependent variation in the essential oil composition of *Salvia officinalis* L. cultivated at two different sites. *J. Agric. Food Chem.* **2001**, *49*, 2908–2916. [CrossRef]
60. Hyldgaard, M.; Mygind, T.; Meyer, R.L. Essential oils in food preservation: Mode of action, synergies, and interactions with food matrix components. *Front. Microbiol.* **2012**, *3*, 12. [CrossRef]
61. Huang, Y.; Ho, S.H.; Lee, H.C.; Yap, Y.L. Insecticidal properties of eugenol, isoeugenol and methyleugenol and their effects on nutrition of *Sitophilus zeamais* Motsch. (Coleoptera: Curculionidae) and *Tribolium castaneum* (Herbst)(Coleoptera: Tenebrionidae). *J. Stored Prod. Res.* **2002**, *38*, 403–412. [CrossRef]
62. Kafle, L.; Shih, C.J. Toxicity and repellency of compounds from clove (*Syzygium aromaticum*) to red imported fire ants *Solenopsisinvicta* (Hymenoptera: Formicidae). *J. Econ. Entomol.* **2013**, *106*, 131–135. [CrossRef] [PubMed]
63. Lopez-Reyes, J.G.; Spadaro, D.; Prelle, A.; Garibaldi, A.; Gullino, M.L. Efficacy of plant essential oils on postharvest control of rots caused by fungi on different stone fruits in vivo. *J. Food Prot.* **2013**, *76*, 631–639. [CrossRef]
64. Elshafie, H.S.; Mancini, E.; Camele, I.; De Martino, L.; De Feo, V. In vivo antifungal activity of two essential oils from Mediterranean plants against postharvest brown rot disease of peach fruit. *Ind. Crop Prod.* **2015**, *66*, 11–15. [CrossRef]
65. Servili, A.; Feliziani, E.; Romanazzi, G. Exposure to volatiles of essential oils alone or under hypobaric treatment to control postharvest gray mold of table grapes. *Postharvest Biol. Technol.* **2017**, *133*, 36–40. [CrossRef]
66. Pathirana, H.N.K.S.; Wimalasena, S.H.M.P.; De Silva, B.C.J.; Hossain, S.; Heo, G. Antibacterial activity of clove essential oil and eugenol against fish pathogenic bacteria isolated from cultured olive flounder (*Paralichthys olivaceus*). *Slov. Vet. Res.* **2019**, *56*, 31–38. [CrossRef]
67. Hajian-Maleki, H.; Baghaee-Ravari, S.; Moghaddam, M. Efficiency of essential oils against *Pectobacterium carotovorum* subsp. *carotovorum* causing potato soft rot and their possible application as coatings in storage. *Postharvest Biol. Technol.* **2019**, *156*, 110928. [CrossRef]

Article

Barley-Based Cropping Systems and Weed Control Strategies Influence Weed Infestation, Soil Properties and Barley Productivity

Muhammad Naeem [1], Waqas Ahmed Minhas [1], Shahid Hussain [2], Sami Ul-Allah [3], Muhammad Farooq [4], Shahid Farooq [5] and Mubshar Hussain [1,6,*]

[1] Department of Agronomy, Bahauddin Zakariya University, Multan 60800, Pakistan; naeem_agrarian@yahoo.com (M.N.); waqas.minhas05@gmail.com (W.A.M.)
[2] Department of Soil Science, Bahauddin Zakariya University, Multan 60800, Pakistan; shahid.hussain@bzu.edu.pk
[3] College of Agriculture, Bahauddin Zakariya University, Bahadur Sub-Campus, Layyah 31200, Pakistan; samipbg@bzu.edu.pk
[4] Department of Plant Sciences, College of Agricultural and Marine Sciences, Sultan Qaboos University, Al-Khoud 123, Oman; farooqcp@gmail.com
[5] Department of Plant Protection, Faculty of Agriculture, Harran University, Şanlıurfa 63050, Turkey; csfa2006@gmail.com
[6] School of Veterinary and Life Sciences, Murdoch University, 90 South Street, Murdoch, WA 6150, Australia
* Correspondence: mubashiragr@gmail.com; Tel.: +92-301-7164879

Citation: Naeem, M.; Minhas, W.A.; Hussain, S.; Ul-Allah, S.; Farooq, M.; Farooq, S.; Hussain, M. Barley-Based Cropping Systems and Weed Control Strategies Influence Weed Infestation, Soil Properties and Barley Productivity. *Agriculture* **2022**, *12*, 487. https://doi.org/10.3390/agriculture12040487

Academic Editor: Roberto Mancinelli

Received: 25 December 2021
Accepted: 28 March 2022
Published: 30 March 2022

Publisher's Note: MDPI stays neutral with regard to jurisdictional claims in published maps and institutional affiliations.

Abstract: Barley-based cropping systems (BCS) alter barley production by influencing weed infestation rates and soil nutrient dynamics. This two-year field study evaluated the interactive effects of five BCS and five weed control strategies (WCS) on soil properties and the growth and yield of barley. Barley was planted in five different cropping systems, i.e., fallow-barley (FB), maize-barley (MaB), cotton-barley (CB), mungbean-barley (MuB) and sorghum-barley (SB). Similarly, five different WCS, weed-free (control, WF), weedy-check (control, WC), false seedbeds (FS), chemical control (CC) and use of allelopathic water extracts (AWE), were included in the study. The SB system had the highest soil bulk density (1.48 and 1.47 g cm^{-3} during the period 2017–2018 and 2018–2019, respectively) and lowest total soil porosity (41.40 and 41.07% during the period 2017–2018 and 2018–2019, respectively). However, WCS remained non-significant for bulk density and total soil porosity during both years of the study. Barley with WF had a higher leaf area index (5.28 and 4.75) and specific leaf area (65.5 and 64.9 cm^{-2} g^{-1}) compared with barley grown under WC. The MuB system under WC had the highest values of extractable NH$_4$-N (5.42 and 5.58 mg kg^{-1}), NO$_3$-N (5.79 and 5.93 mg kg^{-1}), P (19.9 and 19.5 mg kg^{-1}), and K (195.6 and 194.3 mg kg^{-1}) with statistically similar NO$_3$-N in the MaB system under WC and extractable K in the MuB system under FS. Grain yield ranged between 2.8–3.2 and 2.9–3.3 t ha^{-1} during the period 2017–2018 and 2018–2019, respectively, among different WCS. Similarly, grain yield ranged between 2.9–3.2 and 3.0–3.2 t ha^{-1} during the period 2017–2018 and 2018–2019, respectively, within different BCS. Among WCS, the highest grain yield (3.29 and 3.32 t ha^{-1}) along with yield-related traits of barley were in WF as compared to WC. Overall, MuB system recorded better yield and yield-related traits, whereas the lowest values of these traits were recorded for FB systems. In conclusion, the MuB system with WF improved soil characteristics and barley yield over other cropping systems. The AWE significantly suppressed weeds and was equally effective as the chemical control. Therefore, MuB and AWE could be used to improve barley productivity and suppress weeds infestation.

Keywords: allelopathy; barley; cropping systems; soil quality; weed biocontrol

---

1. Introduction

The type and sequence of the crops grown over a large area and practices opted for the production of the crops are collectively known as cropping systems [1]. Overall, the cropping system refers to all cropping sequences being practiced over a large area. Successful crop production in a particular cropping system depends on the management of different components of that system. Significant management efforts are needed regarding tillage, crop residue management, cropping sequence, nutrients, irrigation and erosion control to successfully grow a crop in a specific cropping system [1]. Soil and water conservation greatly depend on the management of the cropping system. A balanced and well-designed cropping system increases soil fertility, decreases soil erosion, and improves soil properties. However, a cropping system with poor management results in decreased soil fertility and increased erosion. A narrow cropping system (monocropping) results in reduced crop diversity, poor soil properties, increased use of fertilizers and pesticides, weed and pest invasions, and reduced crop yields [1]. Therefore, a cropping system should be diverse and include the crops which could improve soil properties, decrease weed infestation and have higher economic benefits.

Soil chemical properties are significantly altered by the cropping system through moisture and nutrient uptake and the amount and quality of crop residues [2]. The crops included in a cropping system significantly differ for nutrient removal from the soil; however, the nutrient status of the soil could be improved through returned residues after crop harvest [3,4]. The legumes can fix atmospheric nitrogen (N), and their residues could supply it to the crops sown after them. Mono-cropping of maize (*Zea mays* L.) resulted in lower nutrients in the soil compared to the cropping system containing soybean (*Glycine max* L.) in rotation [3]. Similarly, soil phosphorus contents are increased in diversified cropping systems compared to mono-cropping systems. Hence, diversifying cropping systems with legumes could improve soil properties compared with narrow cropping systems.

Adaptation of a suitable cropping system has a significant role in sustainable agricultural production [5]. The inclusion of various crops in the cropping systems plays a significant role in maintaining soil fertility. Legumes help to lower N_2O emission along with accelerating the decomposition of mineralizable N-containing compounds and hence improve plant nutrition [6,7]. Similarly, the addition of allelopathic crops like millet (*Pennisetum glaucum* (L.) R.Br.), wheat (*Triticum aestivum* L.), sunflower (*Helianthus annuus* L.), maize, sorghum (*Sorghum bicolor* L.), buckwheat (*Fagopyrum esculentum* Moench), rice (*Oryza sativa* L.) and canola (*Brassica napus* L.) in crop rotations can suppress the weed flora and improve the crop yields [8,9].

Barley (*Hordeum vulgare* L.) is one of the oldest domesticated cereals, which currently ranks fourth after maize, wheat, and rice in terms of production globally [10]. It is cultivated for brewing and malting processes, human food, and livestock fodder. Barley was cultivated over 60 thousand ha in Pakistan during the period 2017–2018 which produced 58 thousand tonnes of barley grains [11]. The average barley yield in the country is 0.95 tonnes per ha which is far below the global average of 3 t per ha. The annual barley demand of Pakistan is 100 thousand tonnes. In total, 40 thousand tonnes of barley are imported to fulfill the country's requirements. Weed infestation, low yielding cultivars, cultivation on marginal lands and poor crop management are responsible for the low average yield in the country [11].

Weed infestation significantly reduces the yield and quality of barley [12–15]. Weeds compete with crops for essential resources (moisture, nutrients, and light) resulting in lower crop productivity. Weeds could decrease barley yields by 50% depending upon the nature of weed species and the intensity of infestation [13]. Therefore, the successful management of weeds is necessary to improve the yield of barley. Different weed-controlling methods (chemical, mechanical, allelopathic, cultural, and biological) are used to manage weed infestation in barley. Herbicides are used for weed management in barley; however, concerns are raising on their use due to environmental pollution and negative impacts on human health. The rising herbicide costs and evolution of herbicide resistance require alternative weed management methods in barley.

Allelopathy is a relatively environmentally friendly weed management approach compared to herbicides [16]. Allelopathy is the biological phenomenon in which biochemicals produced by one plant negatively or positively influence the germination, growth, survival, and reproduction of other organisms. Allelopathy has been exploited to suppress weeds in different crops through the use of allelopathic crop water extracts [17–19], inclusion of allelopathic crops in cropping systems [20–22] and mulches [23]. All these methods significantly reduced weed infestation; therefore, they can be combined with other methods for suppressing weed flora in different cropping systems.

False seedbeds are another weed management technique used for suppressing weed infestation in winter and summer crops. The false seedbed technique is the preparation of a seedbed before sowing the crop which results in the emergence of weeds well before a crop is sown. The subsequent tillage operations used to prepare the true seedbed destroy the emerged weeds; thus, providing significant control over weed infestation. False seedbed preparation reduced weed infestation up to 85% as compared to the direct sowing method [12]. However, the efficiency of this method depends on various factors like soil and climatic situations, and the method and time of preparation of seedbeds [24].

Weed management strategies and cropping systems significantly affect weed infestation, soil physico-chemical properties, crop allometry, and crop yield [25–28]. However, the interactive effect of barley-based cropping systems and weed management methods on weed infestations, soil physico-chemical properties, crop allometry, and barley yield has not been investigated. Therefore, this study was conducted to investigate the impact of different barley-based cropping systems (combinations with exhaustive, restorative, and allelopathic crops) and weed management strategies on soil physio-chemical properties, crop allometry, and barley growth and yield.

2. Materials and Methods

2.1. Experimental Site and Treatment

This two-year (2017–2019) field experiment was conducted at Agronomic Research Farm, Bahauddin Zakariya University, Multan (30.26° N, 71.51° E, and 122 m above sea level), Pakistan. The climate of Multan is arid to semi-arid. The weather data during the growth period are given in Figure 1. For pre-sowing soil analysis, soil samples (3 samples from the site designated for each cropping system) were collected (0–15 cm depth), dried, and crushed to pass through a 2 mm sieve. The soil was alkaline in reaction with a pH of 8.20. The other soil properties were: EC (2.78 dS m^{-1}), organic matter content (0.60%), available N (0.03 mg kg^{-1}), available phosphorus (P, 7.25 mg kg^{-1}), and available potassium (K, 240 mg kg^{-1}).

The experiment consisted of five barley-based cropping systems (BCS), i.e., fallow-barley (FB), maize-barley (MaB), cotton-barley (CB), mungbean-barley (MuB), and sorghum-barley (SB). The BCS were factorially combined with five weed control strategies (WCS), i.e., weed-free (WF), weedy-check (WC), false seedbeds (FS), chemical control (CC), and allelopathic weed control (AWE). The experiment was laid out following a randomized complete block design (RCBD) in a split-plot layout (cropping systems in main plots and weed control strategies in sub-plots). In both years, the experiment was replicated three times with a net plot size of 3 m × 5 m. For WF treatment, weeds were manually removed at their emergence during the entire barley growth period. In WC treatment, weeds were allowed to grow with no control measures. In FS, the plots were cultivated to destroy weeds one week before the seedbed preparation. For chemical control, Buctril M by Bayer Crop Science (Bromoxynil+MCPA (60% EC)) was sprayed (at 1.25 L ha^{-1}) after one week of first irrigation to barley. In AWE, a balanced volume-based mixture (1:1:1:1) of water extracts of mulberry, sorghum, eucalyptus, and sunflower was sprayed (at 12 L ha^{-1}; Shahzad et al. [28]) after one week of first irrigation to barley. The leaves and branches of all crops were taken, chopped into small pieces, and dried under the sun for the preparation of AWE. The dried materials were then soaked in distilled water (1:20 ratio), separately for 24 h. The solutions were filtered after 24 h to obtain the extracts. The resulting extracts were then mixed in a 1:1:1:1 ratio, diluted by 10 times and sprayed.

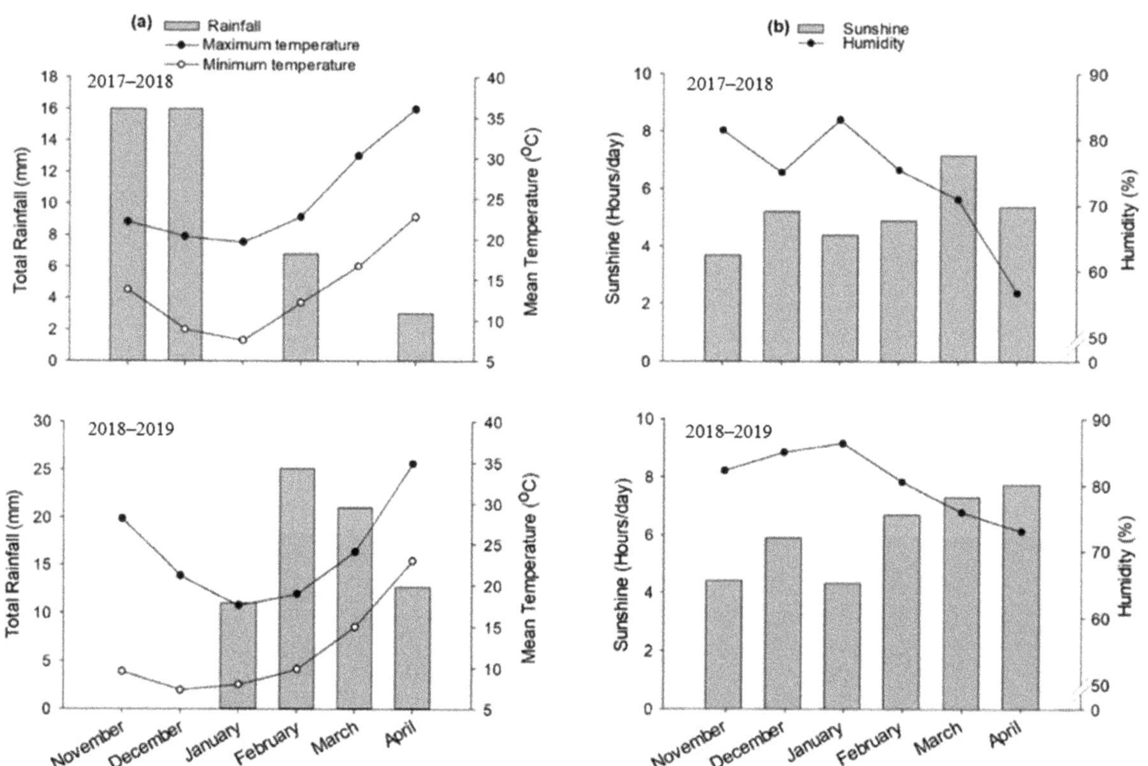

Figure 1. Weather data of the experimental site, i.e., rainfall and temperature (**a**) and sunshinbe hours and relative humidity (**b**) during barley growing seasons of 2017–2018 and 2018–2019. (Source: climate observatory at the Department of Agricultural Engineering 1 Km from the experimental location).

2.2. Crop Husbandry

During both seasons, pre-soaking irrigation of 10 cm was applied to the experimental area. Seedbeds of all crops were prepared once the soil reached field capacity. All crops were cultivated as per recommended production technology for the area and are shown in Table 1. Barley was manually sown in lines with a hand drill. The crop was irrigated according to its requirement (6 cm of water in each irrigation) by following the surface irrigation method. The fertilizers applied were urea and di-ammonium phosphate (DAP). The full dose of P and 1/3rd of N was applied at sowing time. The remaining doses of N were applied at first and second irrigations. Diseases, insects, and pests were controlled in both cropping seasons by following the recommended agronomic and crop protection measures. The barley crop was harvested once all the ear heads turned yellow. After proper cleaning and winnowing, the grain yield of each plot was noted at 12% moisture content. The experiment was conducted in the same field during both years of the study.

Table 1. Crop husbandry of different crops included in barley-based cropping systems of the study (2017–2019).

Crops	Sowing Time	Cultivars	Seed Rate (kg ha^{-1})	Fertilizer NPK (kg ha^{-1})	P–P (cm)	R–R (cm)	Harvesting Time	Harvest Method
Cotton	14 May	IUB-2013	25	250–200–0	20	75	28 October (Last picking)	Manual
Sorghum	13 June	YS-16	10	100–60–0	15	60	29 October	Manual
Mungbean	15 June	NIAB-Mung 2011	20	20–60–0	10	30	27 September	Manual
Maize	26 July	YH-1898	25	200–150–0	22	75	30 October	Manual
Barley	15 and 18 November	Haider-93	80	50–25–0	-	25	7 and 9 April	Manual

P–P = Plant spacing; R–R = Row spacing.

2.3. Post-Harvest Soil Analysis

For post-harvest soil analysis, composite soil samples were collected (0–15 cm depth) from each experimental unit. The samples were oven-dried at 105 °C for 24 h to measure dry weights and then bulk density [29] and total soil porosity [30] were determined. Another batch of composite soil samples was air-dried and sieved through a 2 mm sieve. Plant-available concentrations of NH_4-N, NO_3-N, P, and K in the soil were estimated by ammonium bicarbonate-DTPA method (AB-DTPA) as detailed by Soltanpour and Workman [31].

2.4. Weed Dry Biomass

Data for weed biomass were collected at the booting stage of the barley crop (Zadok stage 4.5). An area of 1 m^2 was selected randomly from each experimental unit, and all weeds present in the area were collected. Collected weeds were sun-dried followed by oven drying until a constant weight was reached. The weight of dried weeds was recorded on an electronic balance.

2.5. Allometric Traits of Barley

Allometric traits of barley were estimated 105 days after sowing (DAS). Three places (0.5 m row of barley) were harvested to ground level, leaves were separated from harvested plants and immediately weighed to record fresh leaf weight. Leaf area was then assessed by using a leaf area meter (DT Area Meter, model MK2). The measured leaf area was converted to the total leaf area of the harvested samples. After that, the leaf area index (LAI) was calculated as a ratio of total leaf area to ground area [32]. The leaf samples were then dried in an oven and weighed. Specific leaf area (SLA) was calculated as total leaf area to leaf dry weight [33].

2.6. Agronomic and Yield-Related Traits of Barley

The number of productive tillers was counted from three random positions (1 m^2) within each experimental unit and averaged. The number of grains per spike was noted from each experimental unit at crop maturity by choosing twenty spikes at random. The 1000-grain weight was averaged from five samples of 1000 grains from each experimental unit. Crops from each experimental unit were harvested manually, dried under the sun, and weighed to record biological yield. The harvested samples were threshed to record grain yield. Harvest index (%) was determined as the proportion of seed yield to biological yield.

2.7. Statistical Analysis

The collected data on allometric and yield-related traits of barley, weed dry biomass and soil properties were analyzed in four different steps. First, the differences among years were tested, which were significant. Therefore, data from both years were analyzed and interpreted separately. Normality in the dataset was tested in the second step through a Shapiro–Wilk normality test [34] and variables with non-normal distribution were normalized by the Arcsine transformation technique. Two-way analysis of variance (ANOVA) was used in the third step to infer the significance in the data. The least significant difference (LSD) post hoc test at a 5% level of probability [35] was used to rank the means of different treatments where ANOVA indicated significant differences. The interactive effect of BCS and WCS was non-significant for most of the recorded traits, except soil NH4-N, soil NO3-N, number of productive tillers per plant, number of grains per spike and weed dry biomass. Therefore, individual effects of BCS and WCS were presented and interpreted for the traits having non-significant BCS by WCS interaction. For the traits having significant BCS by WCS interaction, each BCS was analyzed for its effects on all WCS. Similarly, all WCS were individually analyzed for their impact on different BCS. All statistical computations were done on SPSS statistical software version 21 [36].

3. Results

3.1. Soil Properties

Different WCS had non-significant impact on soil bulk density during both years of the study; however, BCS significantly altered it. The SB cropping system recorded the highest soil bulk density during both years of study, whereas FB recorded the lowest values for bulk density during both years (Figure 2).

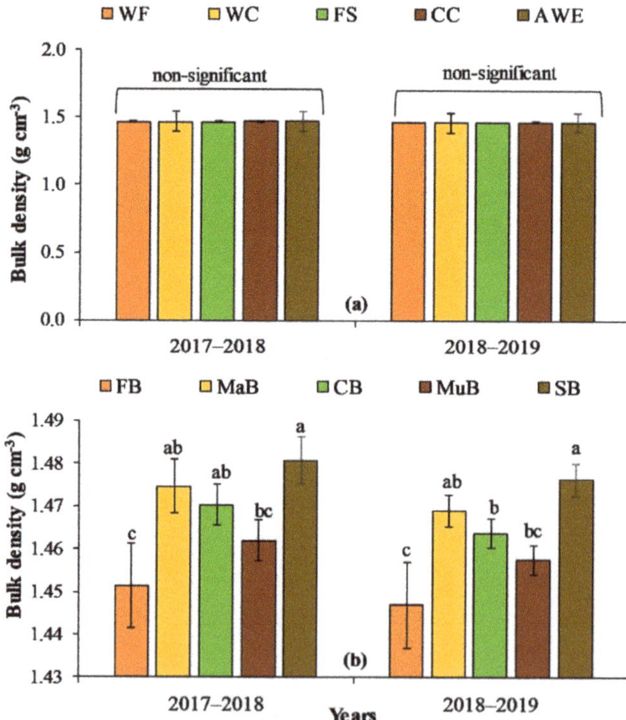

Figure 2. Influence of various weed control strategies (**a**) and barley-based cropping systems (**b**) on soil bulk density. WF = weed-free, WC = weedy-check, FS = false-seedbed, CC = chemical-control, AWE = allelopathic-water-extract, FB = fallow-barley, MaB = maize-barley, CB = cotton-barley, MuB = mungbean-barley and SB = sorghum-barley, the bar indicates the means ± standard errors of the means. Any two means sharing different letters are statistically different ($p \leq 0.05$) from each other.

Different WCS had a non-significant impact on soil porosity during both years of the study; however, BCS significantly altered it. The FB, CB and MuB cropping systems recorded the highest values for soil porosity during both years of study, whereas SB recorded the lowest values during both years (Figure 3).

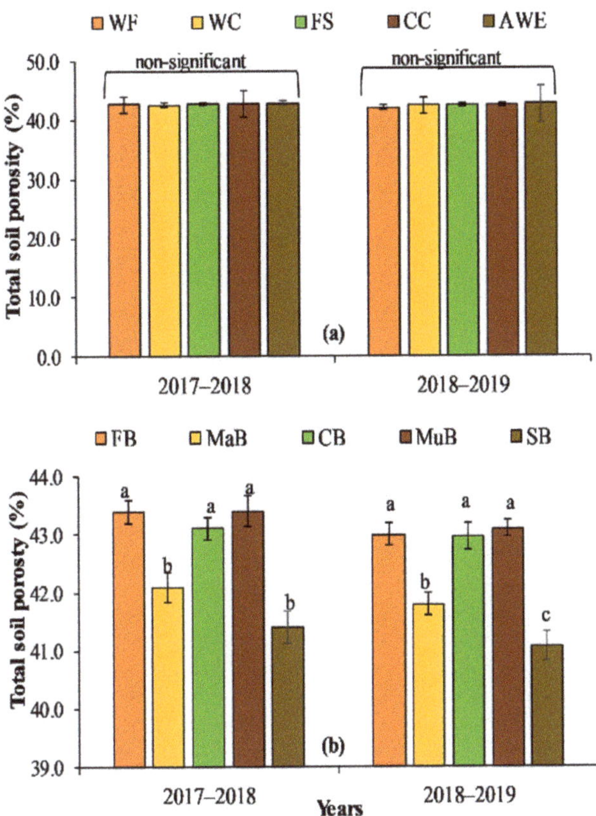

Figure 3. Influence of various weed control strategies (**a**) and barley-based cropping systems (**b**) on soil porosity. WF = weed-free, WC = weedy-check, FS = false-seedbed, CC = chemical-control, AWE = allelopathic-water extract, FB = fallow-barley, MaB = maize-barley, CB = cotton-barley, MuB = mungbean-barley and SB = sorghum-barley, the bar indicates the means ± standard errors of the means. Any two means sharing different letters are statistically different ($p \leq 0.05$) from each other.

Soil available K was significantly altered by WCS and BCS during both years of the study. It ranged from 178 to 199 mg/kg during the period 2017–2018 and 179 to 197 mg/kg during the period 2018–2019 among different WCS included in the study. The WC and FS resulted in the highest values of soil available K during both years, whereas WF and CC treatments resulted in the lowest values of soil available K during both years (Figure 4). Similarly, soil available K ranged from 183 to 195 and 182 to 194 mg/kg during the periods 2017–2018 and 2018–2019, respectively. The highest and the lowest values for soil available K were recorded for MuB and FB cropping systems, respectively, during both years of the study (Figure 4).

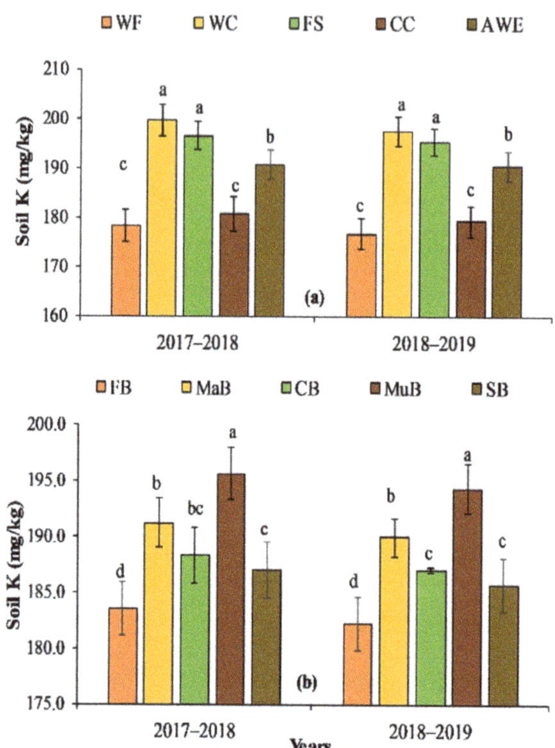

Figure 4. Influence of various weed control strategies (**a**) and barley-based cropping systems (**b**) on soil available potassium. WF = weed-free, WC = weedy-check, FS = false-seedbed, CC = chemical control, AWE = allelopathic-water-extract, FB = fallow-barley, MaB = maize-barley, CB = cotton-barley, MuB = mungbean-barley and SB = sorghum-barley, the bar indicates the means $\pm$ standard errors of the means. Any two means sharing different letters are statistically different ($p \leq 0.05$) from each other.

Soil available P was significantly altered by WCS and BCS during both years of the study. It ranged from 14.9 to 20.6 and 14.4 to 19.4 mg/kg during the periods 2017–2018 and 2018–2019, respectively, for different WCS. The highest and the lowest soil available P was noted for WC and WF treatments, respectively, during both years (Figure 5). Similarly, soil available P ranged from 15.0 to 19.9 and 14.6 to 19.5 mg/kg during the periods 2017–2018 and 2018–2019, respectively, among different BCS. The highest and the lowest values for soil available P were recorded for MuB and FB cropping systems, respectively, during both years of the study (Figure 5).

The interactive effect of BCS and WCS significantly altered NH_4-N during both years of the study. The NH_4-N ranged from 3.74 to 5.42 mg kg^{-1} during the period 2017–2018, and 3.88 to 5.58 mg kg^{-1} during the period 2018–2019 (Table 2). All WCS recorded the highest (4.20–5.42 mg kg^{-1}) NH_4-N in the MuB cropping system during the period 2017–2018, whereas the lowest (3.74–4.98 mg kg^{-1}) NH_4-N was noted for the FB cropping system with all WCS. Similarly, all WCS recorded the highest (4.34–5.58 mg kg^{-1}) NH_4-N in the MuB cropping system against the lowest (3.88–4.98 mg kg^{-1}) in the FB cropping system during the period 2018–2019 (Table 2). All BCS recorded the highest NH_4-N in WC treatment, whereas the lowest values were recorded for WF treatment during both years of the study (Table 2).

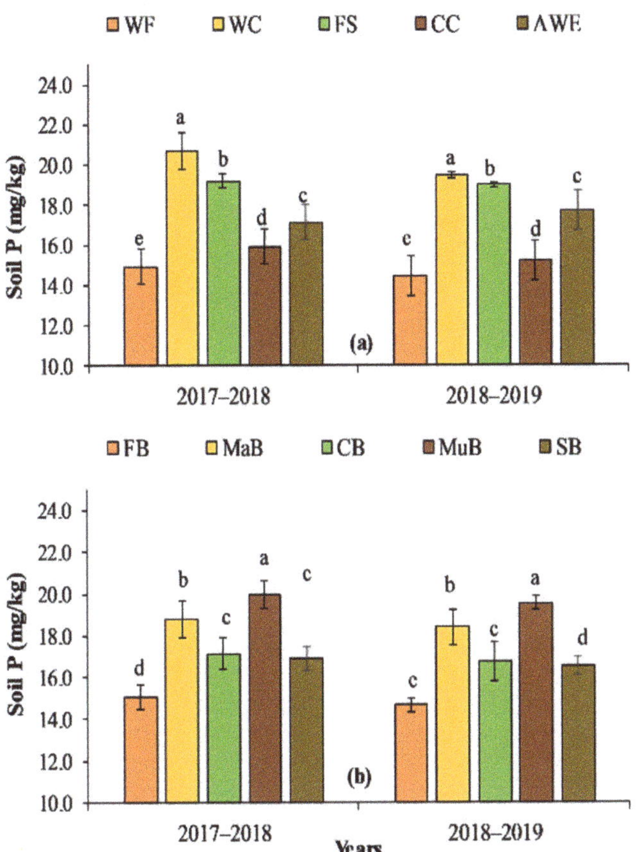

Figure 5. Influence of various weed control strategies (**a**) and barley-based cropping systems (**b**) on soil available phosphorus. WF = weed-free, WC = weedy-check, FS = false-seedbed, CC = chemical control, AWE = allelopathic-water-extract, FB = fallow-barley, MaB = maize-barley, CB = cotton-barley, MuB = mungbean-barley and SB = sorghum-barley, the bar indicates the means ± standard errors of the means. Any two means sharing different letters are statistically different ($p \leq 0.05$) from each other.

Table 2. Influence of different barley-based cropping systems and weed control strategies on soil NH_4-N (mg kg^{-1}) during the periods 2017–2018 and 2018–2019.

	FB	MaB	CB	MuB	SB	LSD WCS
			2017–2018			
WF	3.74 ± 0.02 c E *	4.07 ± 0.03 b E	3.96 ± 0.05 b E	4.20 ± 0.02 a E	4.07 ± 0.04 b E	0.11
WC	4.98 ± 0.02 d A	5.30 ± 0.03 ab A	5.23 ± 0.03 bc A	5.42 ± 0.03 a A	5.14 ± 0.07 c A	0.12
FS	4.73 ± 0.03 c B	4.93 ± 0.05 b B	4.84 ± 0.02 bc B	5.09 ± 0.04 a B	4.74 ± 0.03 c B	0.12
CC	4.06 ± 0.05 c D	4.23 ±0.02 b D	4.16 ± 0.04 bc D	4.35 ± 0.02 a D	4.21 ± 0.03 b D	0.10
AWE	4.38 ± 0.03 d C	4.73 ± 0.03 ab C	4.65 ± 0.05 bc C	4.81 ± 0.03 a C	4.56 ± 0.02 c C	0.11
LSD CS	0.09	0.11	0.13	0.10	0.12	

Table 2. *Cont.*

	FB	MaB	CB	MuB	SB	LSD WCS
2018–2019						
WF	3.88 ± 0.03 d E	4.18 ± 0.02 bc E	4.11 ± 0.02 c E	4.34 ± 0.05 a E	4.20 ± 0.01 b E	0.08
WC	5.14 ± 0.02 d A	5.48 ± 0.03 ab A	5.44 ± 0.03 b A	5.58 ± 0.04 a A	5.30 ± 0.05 c A	0.10
FS	4.98 ± 0.02 c B	5.09 ± 0.02 b B	5.18 ± 0.05 b B	5.34 ± 0.03 a B	4.98 ± 0.04 c B	0.10
CC	4.20 ± 0.07 b D	4.29 ± 0.05 b D	4.33 ± 0.06 b D	4.49 ± 003 a D	4.34 ± 0.04 ab D	0.15
AWE	4.57 ± 0.03 d C	4.84 ± 0.03 bc C	4.92 ± 0.04 ab C	5.00 ± 0.02 a C	4.75 ± 0.04 c C	0.09
LSD WCS	0.11	0.09	0.12	0.10	0.11	

* The lowercase letters denote how respective weed control strategy varied among barley-based cropping systems, whereas uppercase letters indicate how a cropping system differed among various weed control strategies included in the study. Means followed by different lower or uppercase letters significantly ($p \leq 0.05$) differ from each within a row and column, respectively. WF = weed-free, WC = weedy-check, FS = false-seedbed, CC = chemical-control, AWE = allelopathic-water-extract, FB = fallow-barley, MaB = maize-barley, CB = cotton-barley, MuB = mungbean-barley and SB = sorghum-barley, CS = cropping-systems, WCS = weed-control strategies.

The NO_3-N was significantly affected by the interactive effect of BCS and WCS during both years of the study. The NO_3-N ranged from 4.08 to 5.79 mg kg^{-1} during the period 2017–2018, and 4.22 to 5.93 mg kg^{-1} during the period 2018–2019 (Table 3). All WCS recorded the highest (4.54–5.79 mg kg^{-1}) NO_3-N in the MuB cropping system during the period 2017–2018, whereas the lowest (4.40–5.35 mg kg^{-1}) NO_3-N was noted for the FB cropping system with all WCS. Similarly, all WCS recorded the highest (4.68–5.93 mg kg^{-1}) NH_4-N in the MuB cropping system against the lowest (4.22–5.48 mg kg^{-1}) in the FB cropping system during the period 2018–2019 (Table 3). All BCS recorded the highest NO_3-N in WC treatment, whereas the lowest values were recorded for WF treatment during both years of the study (Table 3).

Table 3. Influence of different barley-based cropping systems and weed control strategies on soil NO_3-N (mg kg^{-1}) during the periods 2017–2018 and 2018–2019.

	FB	MaB	CB	MuB	SB	LSD WCS
2017–2018						
WF	4.08 ± 0.02 d E	4.38 ± 0.03 bc E	4.31 ± 0.05 c E	4.54 ± 0.02 a E	4.41 ± 0.04 b E	0.07
WC	5.35 ± 0.02 d A	5.73 ± 0.03 ab A	5.64 ± 0.03 b A	5.79 ± 0.03 a A	5.51 ± 0.07 c A	0.12
FS	5.07 ± 0.03 c B	5.18 ± 0.05 bc B	5.27 ± 0.02 b B	5.43 ± 0.04 a B	5.08 ± 0.03 c B	0.11
CC	4.40 ± 0.05 c D	4.50 ± 0.02 b D	4.53 ± 0.04 b D	4.69 ± 0.02 a D	4.55 ± 0.03 b D	0.06
AWE	4.77 ± 0.03 e C	5.04 ± 0.03 c C	5.12 ± 0.05 b C	5.20 ± 0.03 a C	4.95 ± 0.02 d C	0.07
LSD CS	0.08	0.09	0.10	0.10	0.08	
2018–2019						
WF	4.22 ± 0.04 d E	4.55 ± 0.03 b E	4.44 ± 0.04 c E	4.68 ± 0.03 a E	4.55 ± 0.03 bc E	0.10
WC	5.48 ± 0.03 d A	5.80 ± 0.03 ab A	5.70 ± 0.05 bc A	5.93 ± 0.06 a A	5.65 ± 0.03 c A	0.12
FS	5.32 ± 0.04 c B	5.53 ± 0.04 b B	5.45 ± 0.04 b B	5.68 ± 0.05 a B	5.32 ± 0.02 c B	0.12
CC	4.54 ± 0.02 c D	4.68 ± 0.04 b D	4.62 ± 0.03 bc D	4.83 ± 0.03 a D	4.68 ± 0.02 b D	0.09
AWE	4.91 ± 0.03 d C	5.28 ± 0.03 ab C	5.22 ± 0.04 b C	5.34 ± 0.03 a C	5.09 ± 0.02 c C	0.09
LSD WCS	0.10	0.11	0.12	0.12	0.07	

The lowercase letters denote how respective weed control strategy varied among barley-based cropping systems, whereas uppercase letters indicate how a cropping system differed among various weed control strategies included in the study. Means followed by different lower or uppercase letters significantly ($p \leq 0.05$) differ from each within a row and column, respectively. WF = weed-free, WC = weedy-check, FS = false-seedbed, CC = chemical-control, AWE = allelopathic-water-extract, FB = fallow-barley, MaB = maize-barley, CB = cotton-barley, MuB = mungbean-barley and SB = sorghum-barley, CS = cropping-systems, WCS = weed-control-strategies.

3.2. Weed Dry Biomass

The interactive effect of BCS and WCS significantly altered the dry biomass of observed weed species during both years of the study. The dry biomass ranged from 0.31 to 18.10 g/m^{-2} during the period 2017–2018, and 1.22 to 41.41 g/m^{-2} during the period 2018–2019 (Table 4). All BCS recorded the lowest dry biomass of weeds with CC, which was followed by AWE during both years of the study (Table 4). All BCS recorded the highest dry biomass of weeds with WC treatment, whereas the lowest values were recorded for the SB cropping system in both years of the study (Table 4). Overall, FB and CB were highly infested BCS, whereas SB was the least infested cropping system during both years of the study (Table 4).

Table 4. Influence of different barley-based cropping systems and weed control strategies on total dry biomass (g m^{-2}) of weed species recorded in barley crop during the periods 2017–2018 and 2018–2019.

	FB	MaB	CB	MuB	SB	LSD CS
			2017–2018			
WC	18.10 ± 0.29 a A	7.85 ± 0.26 a C	17.87 ± 0.12 a A	15.87 ± 0.48 a B	6.85 ± 0.13 a D	0.85
FS	10.15 ± 0.15 b A	3.52 ± 0.17 b D	8.56 ± 0.25 c B	6.03 ± 0.48 b C	1.99 ± 0.25 b E	0.89
CC	0.97 ± 0.16 d A	0.33 ± 0.01 c B	1.17 ± 0.14 d A	0.86 ± 0.13 c A	0.31 ± 0.02 c B	0.35
AWE	9.15 ± 0.47 c A	3.67 ± 0.20 b C	9.46 ± 0.33 b A	6.71 ± 0.11 b B	2.16 ± 0.47 b D	1.09
LSD WCS	0.96	0.51	0.74	1.13	0.88	
			2018–2019			
WC	41.41 ± 0.94 a B	22.13 ± 0.92 a D	38.85 ± 0.02 a C	47.14 ± 0.40 a A	19.45 ± 0.53 a E	2.08
FS	23.75 ± 0.84 b B	13.47 ± 0.38 b C	22.33 ± 0.69 b B	27.55 ± 1.12 b A	10.59 ± 1.72 b C	3.31
CC	4.59 ± 0.17 c AB	2.96 ± 0.40 c BC	4.83 ± 1.01 c AB	6.04 ± 0.84 c A	1.22 ± 0.06 c C	1.94
AWE	23.62 ± 0.43 b B	13.69 ± 0.59 b C	22.62 ± 0.23 b B	26.74 ± 0.73 b A	11.12 ± 0.43 b D	1.60
LSD WCS	2.19	2.00	2.02	2.64	3.01	

The lowercase letters denote how respective cropping system differed among various weed control strategies, whereas uppercase letters indicate how a weed control strategy varied among barley-based cropping systems included in the study. Means followed by different lower or uppercase letters significantly ($p \leq 0.05$) differ from each within a column and row, respectively. WF = weed-free, WC = weedy-check, FS = false-seedbed, CC = chemical-control, AWE = allelopathic-water extract, FB = fallow-barley, MaB = maize-barley, CB = cotton-barley, MuB = mungbean-barley and SB = sorghum-barley, CS = cropping-systems, WCS = weed-control-strategies.

3.3. Crop Allometry

The LAI was significantly altered by WCS and BCS during both years of the study. It ranged from 4.74 to 5.28 and 4.28 to 4.76 during the periods 2017–2018 and 2018–2019, respectively, among different WCS. The highest and the lowest LAI was noted for WF and WC treatments, respectively, during both years (Figure 6). Similarly, LAI ranged from 4.87 to 5.23 and 4.36 to 4.72 during the periods 2017–2018 and 2018–2019, respectively, within different BCS. The highest and the lowest values for LAI were recorded for SB and FB cropping systems, respectively, during both years of the study (Figure 6).

Different WCS and BCS significantly altered the SLA of barley crops during both years of the study. It ranged from 63.2 to 65.4 and 61.2 to 65.1 cm^{-2} g^{-1} during the periods 2017–2018 and 2018–2019, respectively, among different WCS. The highest and the lowest SLA was noted for WF and CC, and WC treatments, respectively, during both years (Figure 7). Similarly, SLA ranged between 62.7–65.0 and 62.2–64.8 cm^{-2} g^{-1} during the periods 2017–2018 and 2018–2019, respectively, within different BCS. The highest and the lowest values for SLA were recorded for MaB and MuB, and SB cropping systems, respectively, during both years of the study (Figure 7).

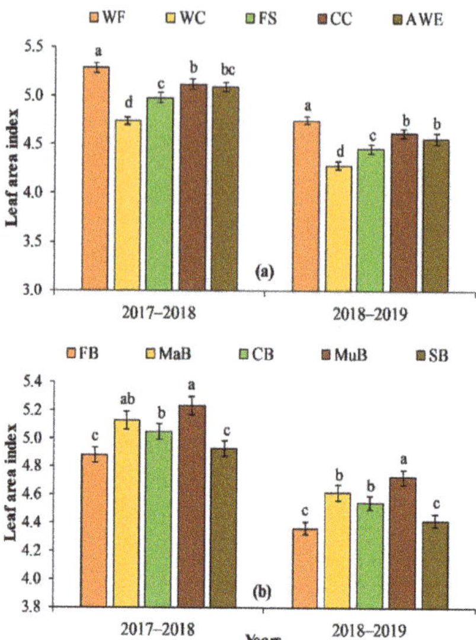

Figure 6. Influence of various weed control strategies (**a**) and barley-based cropping systems (**b**) on leaf area index of barley. WF = weed-free, WC = weedy-check, FS = false-seedbed, CC = chemical-control, AWE = allelopathic-water-extract, FB = fallow-barley, MaB = maize-barley, CB = cotton-barley, MuB = mungbean-barley and SB = sorghum-barley, the bar indicates the means ± standard errors of the means. Any two means sharing different letters are statistically different ($p \leq 0.05$) from each other.

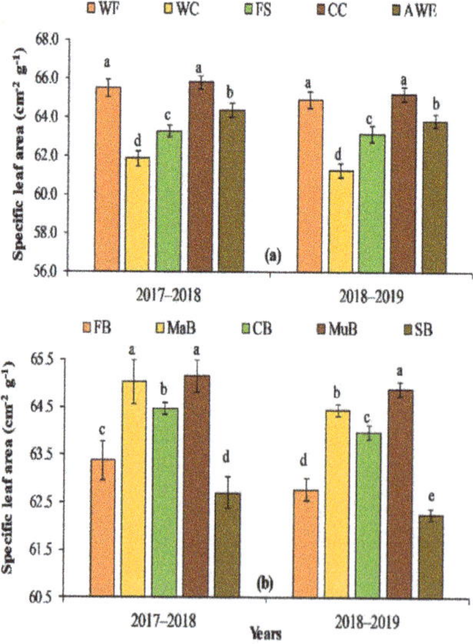

Figure 7. Influence of various weed control strategies (**a**) and barley-based cropping systems (**b**) on specific leaf area of barley. WF = weed-free, WC = weedy-check, FS = false-seedbed, CC = chemical-control, AWE = allelopathic-water-extract, FB = fallow-barley, MaB = maize-barley, CB = cotton-barley, MuB = mungbean-barley and SB = sorghum-barley, the bar indicates the means ± standard errors of the means. Any two means sharing different letters are statistically different ($p \leq 0.05$) from each other.

3.4. Yield and Associated Traits

The number of grains per spike was significantly affected by the interactive effect of BCS and WCS during both years of the study. The number of grains per spike ranged from 171 to 244 mg/kg^{-1} during the period 2017–2018, and 241 to 251 mg/kg^{-1} during the period 2018–2019 (Table 5). All WCS recorded a similar number of grains per spike under WF treatment. Similarly, SB and MuB WCS recorded the highest number of grains per spike with the rest of the WCS during both years of the study, whereas the FB cropping system recorded the lowest number of grains per spike (Table 5). All BCS (except FB) recorded a similar number of grains per spike with all WCS except WC, which recorded the lowest values (Table 5). The FB cropping system had a similar number of grains per spike under WF and CC treatments, whereas the lowest number of grains per spike was recorded for WC treatment (Table 5).

Table 5. Influence of different barley-based cropping systems and weed control strategies on number of grains per spike of barley during the periods 2017–2018 and 2018–2019.

	FB	MaB	CB	MuB	SB	LSD WCS
			2017–2018			
WF	238.51 ± 1.91 a A	242.23 ± 3.58 a A	238.52 ± 3.01 a A	244.67 ± 2.59 a A	239.92 ± 2.07 a A	8.50
WC	171.35 ± 2.45 c D	206.23 ± 2.23 ab C	199.84 ± 3.69 b C	206.18 ± 3.55 ab B	209.61 ± 1.32 a C	8.79
FS	206.89 ± 2.99 b C	231.51 ± 2.81 a B	225.01 ± 2.76 a B	233.84 ± 4.53 a A	230.12 ± 1.66 a B	9.73
CC	234.03 ± 3.52 a A	234.24 ± 3.13 a AB	233.19 ± 3.93 a AB	242.39 ± 3.43 a A	242.12 ± 4.60 a A	11.83
AWE	216.73 ± 1.56 b B	236.44 ± 2.17 a AB	233.71 ± 3.20 a AB	239.87 ± 4.04 a A	236.19 ± 1.75 a AB	8.54
LSD CS	8.13	8.93	10.54	11.61	8.09	
			2018–2019			
WF	241.90 ± 3.47 a A	245.24 ± 2.90 a A	247.28 ± 2.39 a A	251.06 ± 4.07 a A	244.98 ± 3.33 a A	10.34
WC	174.74 ± 1.11 b C	205.53 ± 3.10 a B	209.62 ± 2.45 a B	210.29 ± 1.71 a B	211.67 ± 4.03 a B	8.45
FS	218.44 ± 2.14 b B	238.56 ± 3.86 a A	241.73 ± 2.91 a A	245.39 ± 3.77 a A	241.67 ± 2.20 a A	9.66
CC	240.75 ± 2.59 a A	242.58 ± 4.36 a A	240.96 ± 2.54 a A	247.11 ± 2.06 a A	247.18 ± 3.42 a A	9.77
AWE	224.23 ± 2.43 b B	243.55 ± 3.90 a A	242.60 ± 1.80 a A	245.04 ± 3.39 a A	243.69 ± 3.56 a A	9.82
LSD WCS	7.77	11.55	7.70	9.91	10.66	

The lowercase letters denote how respective weed control strategy varied among barley-based cropping systems, whereas uppercase letters indicate how a cropping system differed among various weed control strategies included in the study. Means followed by different lower or uppercase letters significantly ($p \leq 0.05$) differ from each within a row and column, respectively. WF = weed-free, WC = weedy-check, FS = false-seedbed, CC = chemical-control, AWE = allelopathic-water-extract, FB = fallow-barley, MaB = maize-barley, CB = cotton-barley, MuB = mungbean-barley and SB = sorghum-barley, CS = cropping-systems, WCS = weed-control-strategies.

The number of productive tillers was significantly affected by the interactive effect of BCS and WCS during both years of the study. All WCS recorded a similar number of tillers in the MuB cropping system. Similarly, all BCS recorded the highest number of productive tillers in WF treatment, which were followed by CC and AWE during both years of study (Table 6).

Different WCS and BCS significantly altered the 1000-grain weight of the barley crop during both years of the study. It ranged between 37.0–39.9 and 38.2–40.1 g during the periods 2017–2018 and 2018–2019, respectively, among different WCS. The highest and lowest 1000-grain weight was noted for WF and WC treatments, respectively, during both years (Figure 8). Similarly, the 1000-grain weight ranged between 37.3–39.1 and 37.6–39.3 g during the periods 2017–2018 and 2018–2019, respectively, within different BCS. The highest and the lowest values for the 1000-grain weight were recorded for MuB and FB cropping systems, respectively, during both years of the study (Figure 8).

Table 6. Influence of different barley-based cropping systems and weed control strategies on number of productive tillers (m^{-2}) of barley during the periods 2017–2018 and 2018–2019.

	FB	MaB	CB	MuB	SB	LSD WCS
			2017–2018			
WF	49.54 ± 0.91 c A	51.83 ± 0.65 b AB	52.65 ± 0.41 ab A	54.23 ± 0.62 a A	53.60 ± 0.77 ab A	2.18
WC	46.43 ± 0.88 b BC	48.66 ± 0.68 ab C	47.93 ± 0.67 ab B	47.76 ± 0.62 b C	50.06 ± 0.70 a B	2.25
FS	45.53 ± 0.70 b C	51.68 ± 0.78 a AB	50.92 ± 1.03 a A	50.68 ± 0.88 a B	47.41 ± 0.33 b C	2.45
CC	48.92 ± 1.15 c AB	52.38 ± 0.76 ab A	50.61 ± 0.98 abc A	52.81 ± 0.68 a AB	49.70 ± 0.80 bc B	2.81
AWE	48.32 ± 1.13 b ABC	49.71 ± 0.97 ab BC	50.82 ± 0.15 ab A	52.20 ± 0.73 a AB	48.97 ± 0.86 b BC	2.64
LSD CS	3.05	2.44	2.30	2.24	2.26	
			2018–2019			
WF	50.58 ± 0.78 c A	53.69 ± 0.50 ab A	52.55 ± 0.63 bc A	55.27 ± 0.96 a A	54.57 ± 1.10 ab A	2.59
WC	48.14 ± 0.50 a AB	48.55 ± 0.83 a B	47.97 ± 0.38 a B	47.27 ± 0.40 a C	47.69 ± 0.67 a C	1.82
FS	47.90 ± 0.83 b B	51.95 ± 0.87 a A	52.40 ± 0.52 a A	51.71 ± 0.87 a B	50.89 ± 0.49 a B	2.32
CC	50.41 ± 0.53 b A	52.10 ± 0.82 ab A	51.78 ± 0.82 ab A	53.87 ± 0.69 a AB	51.12 ± 0.64 b B	2.22
AWE	49.69 ± 1.11 b AB	52.13 ± 0.85 ab A	50.83 ± 0.76 b A	53.57 ± 0.41 a AB	50.27 ± 0.82 b B	2.59
LSD WCS	2.46	2.47	2.01	2.22	2.42	

The lowercase letters denote how respective weed control strategy varied among barley-based cropping systems, whereas uppercase letters indicate how a cropping system differed among various weed control strategies included in the study. Means followed by different lower or uppercase letters significantly ($p \leq 0.05$) differ from each within a row and column, respectively. WF = weed-free, WC = weedy-check, FS = false-seedbed, CC = chemical-control, AWE = allelopathic-water-extract, FB = fallow-barley, MaB = maize-barley, CB = cotton-barley, MuB = mungbean-barley and SB = sorghum-barley, CS = cropping-systems, WCS = weed-control-strategies.

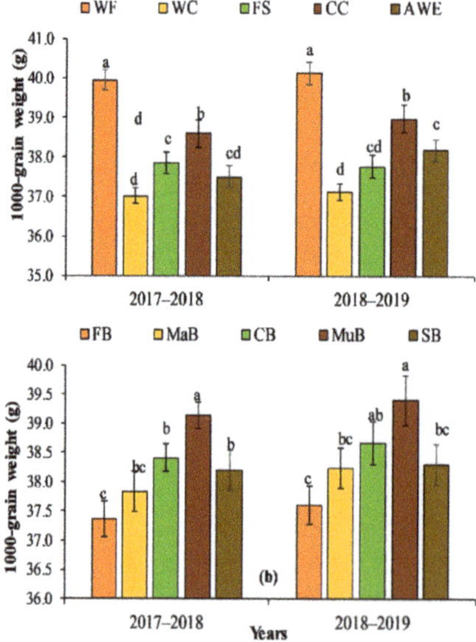

Figure 8. Influence of various weed control strategies (**a**) and barley-based cropping systems (**b**) on 1000-grain weight of barley. WF = weed-free, WC = weedy-check, FS = false-seedbed, CC = chemical-control, AWE = allelopathic-water-extract, FB = fallow-barley, MaB = maize-barley, CB = cotton-barley, MuB = mungbean-barley and SB = sorghum-barley, the bar indicates the means ± standard errors of the means. Any two means sharing different letters are statistically different ($p \leq 0.05$) from each other.

Different WCS and BCS significantly altered the grain yield of barley crops during both years of the study. Grain yield ranged between 2.8–3.2 and 2.9–3.3 t/ha during the periods 2017–2018 and 2018–2019, respectively, among different WCS. The highest and the lowest grain yield was noted for WF and CC treatments, and WC and FS treatments, respectively, during both years (Figure 9). Similarly, grain yield ranged between 2.9–3.2 and 3.0–3.2 t/ha during the periods 2017–2018 and 2018–2019, respectively, within different BCS. The highest and the lowest values for grain yield were recorded for MuB and FB cropping systems, respectively, during both years of the study (Figure 9).

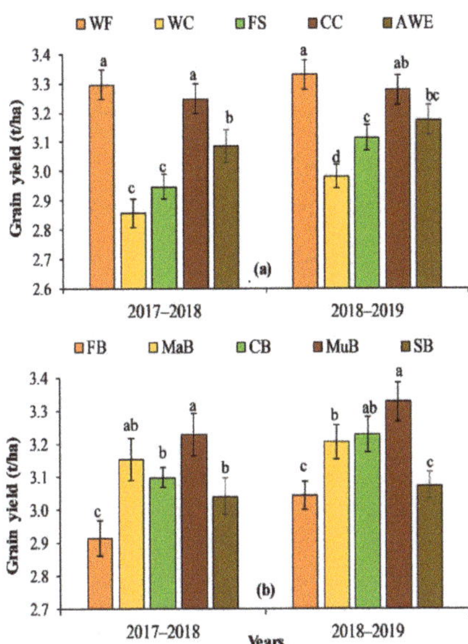

Figure 9. Influence of various weed control strategies (**a**) and barley-based cropping systems (**b**) on grain yield of barley. WF = weed-free, WC = weedy-check, FS = false-seedbed, CC = chemical-control, AWE = allelopathic-water-extract, FB = fallow-barley, MaB = maize-barley, CB = cotton-barley, MuB = mungbean-barley and SB = sorghum-barley, the bar indicates the means ± standard errors of the means. Any two means sharing different letters are statistically different ($p \leq 0.05$) from each other.

Different WCS and BCS significantly altered the biological yield of barley crops during both years of the study. Biological yield ranged between 9.7–10.4 and 9.8–10.5 t/ha during the periods 2017–2018 and 2018–2019, respectively, among different WCS. The highest and the lowest biological yield was noted for WF and CC treatments, and WC treatments, respectively, during both years (Figure 10). Similarly, biological yield ranged between 9.7–10.3 and 9.9–10.4 t/ha during the periods 2017–2018 and 2018–2019, respectively, within different BCS. The highest and the lowest values for biological yield were recorded for MuB and FB cropping systems, respectively, during both years of the study (Figure 10). The MaB and CB also produced comparable biological yields to MuB cropping system during both years of the study.

Different WCS and BCS significantly altered the harvest of barley crops during both years of the study. Harvest index ranged between 29.2–31.5 and 30.2–31.5% during the periods 2017–2018 and 2018–2019, respectively, among different WCS. The highest and the lowest harvest index was noted for WF and CC treatments, and WC treatment, respectively, during the period 2017–2018. However, all WCS (except WC) had a similar harvest index during the period 2018–2019 (Figure 11). Similarly, the harvest index ranged between 29.7–31.2 and 30.5–31.7% during the periods 2017–2018 and 2018–2019, respectively, within different BCS. The highest and the lowest values for harvest index were recorded for MuB and FB cropping systems, respectively, during both years of the study (Figure 11).

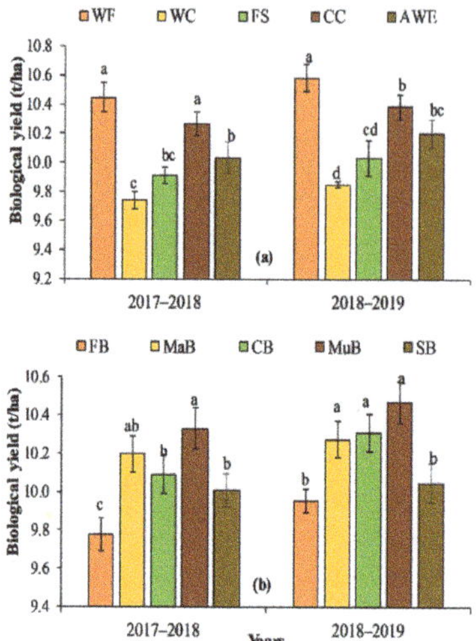

Figure 10. Influence of various weed control strategies (**a**) and barley-based cropping systems (**b**) on biological yield of barley. WF = weed-free, WC = weedy-check, FS = false-seedbed, CC = chemical-control, AWE = allelopathic-water-extract, FB = fallow-barley, MaB = maize-barley, CB = cotton-barley, MuB = mungbean-barley and SB = sorghum-barley, the bar indicates the means ± standard errors of the means. Any two means sharing different letters are statistically different ($p \leq 0.05$) from each other.

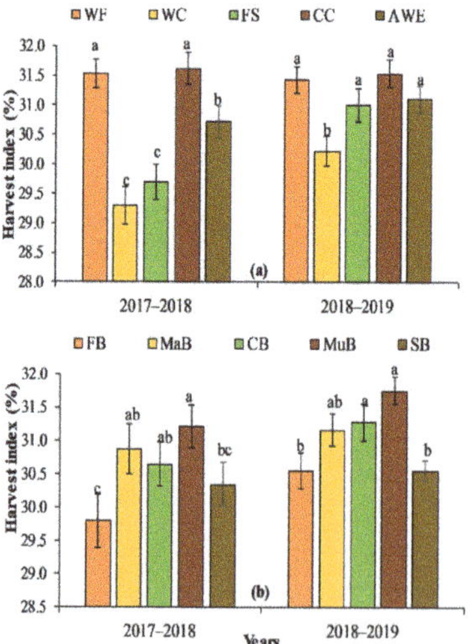

Figure 11. Influence of various weed control strategies (**a**) and barley-based cropping systems (**b**) on harvest index of barley. WF = weed-free, WC = weedy-check, FS = false-seedbed, CC = chemical-control, AWE = allelopathic-water-extract, FB = fallow-barley, MaB = maize-barley, CB = cotton-barley, MuB = mungbean-barley and SB = sorghum-barley, the bar indicates the means ± standard errors of the means. Any two means sharing different letters are statistically different ($p \leq 0.05$) from each other.

4. Discussion

Soil is a key component for plant growth and development as it provides physical support, water, and nutrients. Different crops have different root systems that significantly impact soil properties. The present study revealed that soil bulk density and porosity were significantly influenced by BCS (Figures 2 and 3). The change in soil bulk density and soil porosity can be attributed to the differences in the root system of the crops [37]. Burr-Hersey et al. [38] reported that crops with a high root surface area (like fibrous root systems of barley and mungbean) proliferate more in the soil; thus, increasing soil porosity and reducing soil bulk density. Changes in soil bulk density and porosity due to different cropping systems also suggest the role of crop rotation in sustainable soil management [39,40].

Different BCS significantly influenced the nutrient status of the soil (Figures 4 and 5). The macronutrients (P, K) were higher in the MuB cropping system followed by CB (Figures 4 and 5). The high availability of soil macronutrients in the MuB system can be attributed to the restorative nature of mungbean crop [41–43] as it transorms atmospheric N_2 into plant-available N. Moreover, the efficiency of soil microbes, which enhance P and K availability, also increases in legumes [41,43]. It may also be attributed to the influence of crop residue retention and management practices on soil nutrient dynamics. In WCS, the highest values for available NH_4-N, NO_3-N, P, and K contents were noted in WC, whereas the WF and CC had the lowest values (Tables 2 and 3). It may be due to more weed–crop competition for essential resources and ultimately crops utilized more nutrients in WF treatment [25].

All cropping systems with WF and CC resulted in better allometric traits of barley as compared to WC (Figures 6 and 7). The better allometric traits of barley crops may be due to less weed–crop competition for essential resources (i.e., water, light, and nutrients) in WF or CC; hence, improving crop performance. Moreover, herbicides improved crop performance by reducing weed dry biomass [44,45]. Babiker et al. [46] found that weeds in maize crops were reduced by 97% by using a herbicidal mixture (Gesaprim @1.6 kg ha^{-1} + Stomp @1.5 L ha^{-1}). As compared to direct sowing, the preparation of false seedbeds is an efficient technique to control annual weeds and improve barley yield [12]. Similarly, allelopathic crops, e.g., sorghum, significantly reduced the dry biomass of weeds in the current study. Some crops like sunflower, barley, rice, sorghum and wheat have allelopathic potential to suppress weeds [47]. This potential is due to the presence of hydrophobic compounds (e.g., *sorgoleone*), phenolic acids and hydrophilic substances [48]. Weeds like *Avena fatua* L., *Chenopodium album* L., *Phalaris minor* Retz., etc. were significantly controlled by the use of aqueous extracts of allelopathic crops (*Moringa oleifera* Lam., 1785, *Cannabis sativa* L., and *Parthenium hysterophorus* L.), which increased crop yields [49]. Better weed control was achieved in this study due to the foliar application of allelopathic water extracts (eucalyptus, sunflower, mulberry, and sorghum).

Among BCS, the MuB system performed better, whereas the SB and FB systems resulted in poor allometric traits of barley (Table 5). The better allometric (Figures 2–4) traits of barley in the MuB system are attributed to an increase in soil physical properties and fertility (Tables 2 and 3), which ultimately enhanced the LAI, SLA and CGR [25,27]. Therefore, the MuB system improved the soil's physical and fertility status, hence enhancing barley allometry (Figures 6 and 7). Nonetheless, the SB system negatively influenced barley allometry. It may be due to the allelopathic potential of sorghum crops which reduced the weed's population (as discussed above) and forthcoming barley performance. It was reported by Shirgapure and Ghosh [50] that the allelochemicals released from any crop can influence the growth of weeds and upcoming crops, and this effect was also observed in the current study.

The yield and related traits of barley were significantly affected by the different WCS (Tables 5 and 6 and Figures 8–11). Barley sown in WF and CC under MuB systems significantly improved the yield and related traits which may be attributed to improvement in soil physio-chemical properties under the MuB system. It may be due to better soil conditions which improved the crop performance as was described by Shahzad et al. [51] The sowing of cereals after legumes can get N through biological N fixation [52] because the atmospheric N can be fixed by the bacteria existing in root nodules of legumes. However, barley yield was negatively influenced by sorghum crop in the SB system due to its allelopathic ability [53]. As discussed above, better allometric traits of barley in WF control significantly improved the yield-related traits of the crop. The interception of more light in WF control improved the LAI [54] also assimilate more carbohydrates which turn into productivity. Therefore, crop yield was better in WF control as compared to other weed management methods. Kandhro et al. [55] and Khaliq et al. [56] reported that AWE, FS and CC significantly improved the crop yield by decreasing the weed's dry weight. The lowest crop yield in WC was due to competition of weeds with crops for necessary amounts of light, nutrients, and space [57].

5. Conclusions

Barley-based cropping systems and weed control systems interacted to improve soil physiochemical properties, weed control, growth and yield of barley crop. The MuB system under WF proved better for soil quality and crop yield. Moreover, AWE with MuB cropping system improved soil physiochemical properties, weed control, growth and yield of barley nearly equal to CC; hence, it is an eco-friendly approach which can be opted for sustainable barley production.

Author Contributions: Conceptualization, S.H., S.U.-A., M.F., S.F. and M.H.; Data curation, M.N. and W.A.M.; Formal analysis, M.N. and W.A.M.; Funding acquisition, M.H.; Methodology, S.U.-A.; Project administration, M.H.; Supervision, M.F. and M.H.; Validation, M.N. and S.H.; Visualization, M.N. and W.A.M.; Writing—original draft, M.N.; Writing—review and editing, W.A.M., S.H., S.F., S.U.-A., M.F. and M.H. All authors have read and agreed to the published version of the manuscript.

Funding: The first author acknowledges the financial grant from Higher Education Commission, Islamabad, Pakistan to complete the doctoral studies through grant number 17-6/HEC/HRD/IS-II-5/2019.

Institutional Review Board Statement: Not applicable.

Informed Consent Statement: Not applicable.

Data Availability Statement: All data are within the manuscript.

Acknowledgments: This paper forms part of the Ph.D. thesis of Muhammad Naeem (2021).

Conflicts of Interest: The authors declare no conflict of interest.

References

1. Blanco-Canqui, H.; Lal, R. Cropping Systems. In *Principles of Soil Conservation and Management*; Springer: Dordrecht, The Netherlands, 2010; pp. 165–193.
2. Sainju, U.M.; Alasinrin, S.Y. Changes in soil chemical properties and crop yields with long-term cropping system and nitrogen fertilization. *Agrosyst. Geosci. Environ.* **2020**, *3*, e20019. [CrossRef]
3. Ashworth, A.J.; Allen, F.L.; DeBruyn, J.M.; Owens, P.R.; Sams, C. Crop Rotations and Poultry Litter Affect Dynamic Soil Chemical Properties and Soil Biota Long Term. *J. Environ. Qual.* **2018**, *47*, 1327–1338. [CrossRef] [PubMed]
4. Sainju, U.M.; Allen, B.L.; Caesar-TonThat, T.; Lenssen, A.W. Dryland soil chemical properties and crop yields affected by long-term tillage and cropping sequence. *Springerplus* **2015**, *4*, 320. [CrossRef]
5. Metherell, A.K.; Cambardella, C.A.; Parton, W.J.; Peterson, G.A.; Harding, L.A.; Cole, C. V Simulation of soil organic matter dynamics in dryland wheat-fallow cropping systems. In *Soil Management and Greenhouse Effect*; CRC Press: Boca Raton, FL, USA, 2018; pp. 259–270. ISBN 0203739310.
6. Duchene, O.; Vian, J.-F.; Celette, F. Intercropping with legume for agroecological cropping systems: Complementarity and facilitation processes and the importance of soil microorganisms. A review. *Agric. Ecosyst. Environ.* **2017**, *240*, 148–161. [CrossRef]
7. Scalise, A.; Pappa, V.A.; Gelsomino, A.; Rees, R.M. Pea cultivar and wheat residues affect carbon/nitrogen dynamics in pea-triticale intercropping: A microcosms approach. *Sci. Total Environ.* **2017**, *592*, 436–450. [CrossRef] [PubMed]
8. Jabran, K. Allelopathy: Introduction and concepts. In *Manipulation of Allelopathic Crops for Weed Control*; Springer: Berlin/Heidelberg, Germany, 2017; pp. 1–12.
9. Mennan, H.; Jabran, K.; Zandstra, B.H.; Pala, F. Non-Chemical Weed Management in Vegetables by Using Cover Crops: A Review. *Agronomy* **2020**, *10*, 257. [CrossRef]

10. FAO. Available online: www.faostat.fao.org (accessed on 10 November 2021).
11. GOP. *Economic Survey of Pakistan*; Govt. of Pakistan, Economic Advisory Wing: Islamabad, Pakistan, 2020.
12. Kanatas, P.J.J.; Travlos, I.S.S.; Gazoulis, J.; Antonopoulos, N.; Tsekoura, A.; Tataridas, A.; Zannopoulos, S. The combined effects of false seedbed technique, post-emergence chemical control and cultivar on weed management and yield of barley in Greece. *Phytoparasitica* **2020**, *48*, 131–143. [CrossRef]
13. Woźniak, A. Effect of various systems of tillage on winter barley yield, weed infestation and soil properties. *Appl. Ecol. Environ. Res.* **2020**, *18*, 3483–3496. [CrossRef]
14. Watson, P.R.; Derksen, D.A.; Van Acker, R.C. The ability of 29 barley cultivars to compete and withstand competition. *Weed Sci.* **2006**, *54*, 783–792. [CrossRef]
15. Mahajan, G.; Hickey, L.; Chauhan, B.S. Response of barley genotypes to weed interference in Australia. *Agronomy* **2020**, *10*, 99. [CrossRef]
16. Mushtaq, W.; Siddiqui, M.B.; Hakeem, K.R. Allelopathic control of native weeds. In *Allelopathy*; Springer: Berlin/Heidelberg, Germany, 2020; pp. 53–59.
17. Jabran, K.; Cheema, Z.A.; Farooq, M.; Hussain, M. lower doses of pendimethalin mixed with allelopathic crop water extracts for weed management in canola (*Brassica napus*). *Int. J. Agric. Biol.* **2010**, *12*, 335–340.
18. Khan, M.B.; Ahmad, M.; Hussain, M.; Jabran, K.; Farooq, S.; Waqas-Ul-Haq, M. Allelopathic plant water extracts tank mixed with reduced doses of atrazine efficiently control *Trianthema portulacastrum* L. in *Zea mays* L. *J. Anim. Plant Sci.* **2012**, *22*, 339–346.
19. Tanveer, A.; Jabbar, M.K.; Kahliq, A.; Matloob, A.; Abbas, R.N.; Javaid, M.M. Allelopathic effects of aqueous and organic fractions of Euphorbia dracunculoides Lam. on germination and seedling growth of chickpea and wheat. *Chil. J. Agric. Res.* **2012**, *72*, 495–501. [CrossRef]
20. Farooq, M.; Khan, I.; Nawaz, A.; Cheema, M.A.; Siddique, K.H.M. Using sorghum to suppress weeds in autumn planted maize. *Crop Prot.* **2020**, *133*, 105162. [CrossRef]
21. Mahmood, A.; Cheema, Z.A.; Mushtaq, M.N.; Farooq, M. Maize–sorghum intercropping systems for purple nutsedge management. *Arch. Agron. Soil Sci.* **2013**, *59*, 1279–1288. [CrossRef]
22. Hussain, M.I.; Danish, S.; Sánchez-Moreiras, A.M.; Vicente, Ó.; Jabran, K.; Chaudhry, U.K.; Branca, F.; Reigosa, M.J. Unraveling Sorghum Allelopathy in Agriculture: Concepts and Implications. *Plants* **2021**, *10*, 1795. [CrossRef] [PubMed]
23. Riaz Marral, M.W.; Khan, M.B.; Ahmad, F.; Farooq, S.; Hussain, M. The influence of transgenic (Bt) and non-transgenic (non-Bt) cotton mulches on weed dynamics, soil properties and productivity of different winter crops. *PLoS ONE* **2020**, *15*, e0238716. [CrossRef]
24. Singh, R. Weed management in major kharif and rabi crops. In Proceedings of the National Training on Advances in Weed Management, Jabalpur, India, 13–18 December 2021; pp. 31–40.
25. Naeem, M.; Farooq, M.; Farooq, S.; Ul-Allah, S.; Alfarraj, S.; Hussain, M. The impact of different crop sequences on weed infestation and productivity of barley (*Hordeum vulgare* L.) under different tillage systems. *Crop Prot.* **2021**, *149*, 105759. [CrossRef]
26. Naeem, M.; Hussain, M.; Farooq, M.; Farooq, S. Weed flora composition of different barley-based cropping systems under conventional and conservation tillage practices. *Phytoparasitica* **2021**, *49*, 751–769. [CrossRef]
27. Naeem, M.; Mehboob, N.; Farooq, M.; Farooq, S.; Hussain, S.; Ali, H.M.; Hussain, M. Impact of Different Barley-Based Cropping Systems on Soil Physicochemical Properties and Barley Growth under Conventional and Conservation Tillage Systems. *Agronomy* **2021**, *11*, 8. [CrossRef]
28. Shahzad, M.; Jabran, K.; Hussain, M.; Raza, M.A.S.; Wijaya, L.; El-Sheikh, M.A.; Alyemeni, M.N. The impact of different weed management strategies on weed flora of wheat-based cropping systems. *PLoS ONE* **2021**, *16*, e0247137. [CrossRef]
29. Blake, G.R.; Hartge, K.H. Bulk density. In *Methods of Soil Analysis. Part 1 Physical and Mineralogical Methods, 5.1*, 2nd ed.; American Society of Agronomy: Madison, WI, USA, 1986; pp. 363–375.
30. Danielson, R.E.; Sutherland, P.L. *Porosity. Methods Soil Anal. Part 1 Physical and Mineralogical Methods, 5.1*, 2nd ed.; American Society of Agronomy: Madison, WI, USA, 1986; pp. 443–461.
31. Soltanpour, P.N.; Workman, S. Modification of the NH4HCO3-DTPA Soil Test to Omit Carbon Black1. *Commun. Soil Sci. Plant Anal.* **1979**, *10*, 1411–1420. [CrossRef]
32. Watson, D.J. Comparative physiological studies on the growth of field crops: I. Variation in net assimilation rate and leaf area between species and varieties, and within and between years. *Ann. Bot.* **1947**, *11*, 41–76. [CrossRef]
33. Garnier, E.; Shipley, B.; Roumet, C.; Laurent, G. A standardized protocol for the determination of specific leaf area and leaf dry matter content. *Funct. Ecol.* **2001**, *15*, 688–695. [CrossRef]
34. Shapiro, S.S.; Wilk, M.B. An analysis of variance test for normality (complete samples). *Biometrika* **1965**, *52*, 591–611. [CrossRef]
35. Steel, R.; Torrei, J.; Dickey, D. *Principles and Procedures of Statistics A Biometrical Approach*; McGraw-Hill: New York, NY, USA, 1997.
36. Ibm, C. *SPSS Statistics for Windows*; IBM Corp. Released, Version 20; IBM Corporation: Armonk, NY, USA, 2012; pp. 1–8.
37. Çerçioğlu, M.; Anderson, S.H.; Udawatta, R.P.; Alagele, S. Effect of cover crop management on soil hydraulic properties. *Geoderma* **2019**, *343*, 247–253. [CrossRef]
38. Burr-Hersey, J.E.; Mooney, S.J.; Bengough, A.G.; Mairhofer, S.; Ritz, K. Developmental morphology of cover crop species exhibit contrasting behaviour to changes in soil bulk density, revealed by X-ray computed tomography. *PLoS ONE* **2017**, *12*, e0181872. [CrossRef]

39. Haruna, S.I.; Nkongolo, N.V. Tillage, cover crop and crop rotation effects on selected soil chemical properties. *Sustainability* **2019**, *11*, 2770. [CrossRef]

40. de Moura, M.S.; Silva, B.M.; Mota, P.K.; Borghi, E.; de Resende, A.V.; Acuña-Guzman, S.F.; Araújo, G.S.S.; da Silva, L.D.C.M.; de Oliveira, G.C.; Curi, N. Soil management and diverse crop rotation can mitigate early-stage no-till compaction and improve least limiting water range in a Ferralsol. *Agric. Water Manag.* **2021**, *243*, 106523. [CrossRef]

41. Naz, S.; Fatima, Z.; Iqbal, P.; Khan, A.; Zakir, I.; Noreen, S.; Younis, H.; Abbas, G.; Ahmad, S. Agronomic crops: Types and uses. In *Agronomic Crops*; Springer: Berlin/Heidelberg, Germany, 2019; pp. 1–18.

42. Ali, H.Q.Z.; Choudhary, F.A.; Hayat, S.; Iqbal, R.; Khaliq, T.; Ahmad, A. Viable alternatives to cotton-wheat crop rotation for semi-arid climatic conditions. *J. Dev. Agric. Econ.* **2019**, *11*, 57–62.

43. Ahmad, M.; Adil, Z.; Hussain, A.; Mumtaz, M.Z.; Nafees, M.; Ahmad, I.; Jamil, M. Potential of phosphate solubilizing Bacillus strains for improving growth and nutrient uptake in mungbean and maize crops. *Pak. J. Agric. Sci.* **2019**, *56*, 283–289.

44. Singh, T.; Satapathy, B.S.; Gautam, P.; Lal, B.; Kumar, U.; Saikia, K.; Pun, K.B. Comparative efficacy of herbicides in weed control and enhancement of productivity and profitability of rice. *Exp. Agric.* **2018**, *54*, 363–381. [CrossRef]

45. Alsaadawi, I.S.; Khaliq, A.; Farooq, M. Integration of allelopathy and less herbicides effect on weed management in field crops and soil biota: A Review. *Plant Arch.* **2020**, *20*, 225–237.

46. Babiker, M.M.; Salah, A.E.; Mukhtaret, M.U. Impact of herbicides Pendimethalin, Gesaprim and their combination on weed control under maize (*Zea mays* L.). *JAIS* **2013**, *1*, 17–22.

47. Farooq, N.; Abbas, T.; Tanveer, A.; Jabran, K. Allelopathy for weed management. In *Co-Evolution of Secondary Metabolites*; Springer: Cham, Switzerland, 2020; pp. 505–519.

48. Czarnota, M.A.; Paul, R.N.; Weston, L.A.; Duke, S.O. Anatomy of Sorgoleone-Secreting Root Hairs of Sorghum Species. *Int. J. Plant Sci.* **2003**, *164*, 861–866. [CrossRef]

49. Gurmani, A.R.; Khan, S.U.; Mehmood, T.; Ahmed, W.; Rafique, M. Exploring the allelopathic potential of plant extracts for weed suppression and productivity in wheat (*Triticum aestivum* L.). *Gesunde Pflanz.* **2021**, *73*, 29–37. [CrossRef]

50. Shirgapure, K.H.; Ghosh, P. Allelopathy a tool for sustainable weed management. *Arch. Curr. Res. Int.* **2020**, *20*, 17–25. [CrossRef]

51. Shahzad, M.; Farooq, M.; Hussain, M. Weed spectrum in different wheat-based cropping systems under conservation and conventional tillage practices in Punjab, Pakistan. *Soil Tillage Res.* **2016**, *163*, 71–79. [CrossRef]

52. Alarcón, R.; Hernández-Plaza, E.; Navarrete, L.; Sánchez, M.J.; Escudero, A.; Hernanz, J.L.; Sánchez-Giron, V.; Sánchez, A.M. Effects of no-tillage and non-inversion tillage on weed community diversity and crop yield over nine years in a Mediterranean cereal-legume cropland. *Soil Tillage Res.* **2018**, *179*, 54–62. [CrossRef]

53. Bachheti, A.; Sharma, A.; Bachheti, R.K.; Husen, A.; Pandey, D.P. *Plant Allelochemicals and Their Various Applications. Co-Evolution of Secondary Metabolites*; Springer: Cham, Switzerland, 2020; pp. 441–465.

54. Drews, S.; Neuhoff, D.; Köpke, U. Weed suppression ability of three winter wheat varieties at different row spacing under organic farming conditions. *Weed Res.* **2009**, *49*, 526–533. [CrossRef]

55. Kandhro, M.N.; Tunio, S.; Rajpar, I.; Chachar, Q. Allelopathic impact of sorghum and sunflower intercropping on weed management and yield enhancement in cotton. *Sarhad J. Agric.* **2014**, *30*, 311–318.

56. Khaliq, A.; Matloob, A.; Ihsan, M.Z.; Abbas, R.N.; Aslam, Z.; Rasool, F. Supplementing herbicides with manual weeding improves weed control efficiency, growth and yield of direct seeded rice. *Int. J. Agric. Biol.* **2013**, *15*, 191–199.

57. Ibrahim, M.; Ahmad, N.; Shinwari, Z.K.; Bano, A.; Ullah, F. Allelopathic assessment of genetically modified and non modified maize (*Zea mays* L.) on physiology of wheat (*Triticum aestivum* L.). *Pak. J. Bot* **2013**, *45*, 235–240.

Article

Regression Modeling Strategies to Predict and Manage Potato Leaf Roll Virus Disease Incidence and Its Vector

Yasir Ali [1], Ahmed Raza [2,*], Hafiz Muhammad Aatif [1], Muhammad Ijaz [1], Sami Ul-Allah [1], Shafeeq ur Rehman [3], Sabry Y. M. Mahmoud [4,5], Eman Saleh Hassan Farrag [6,7], Mahmoud A. Amer [8,9] and Mahmoud Moustafa [10,11]

[1] College of Agriculture, Bahauddin Zakariya University, Multan, Bahadur Sub-Campus Layyah, Layyah 31200, Pakistan; yasirklasra.uca@gmail.com (Y.A.); aatif.pak@gmail.com (H.M.A.); muhammad.ijaz@bzu.edu.pk (M.I.); samipbg@bzu.edu.pk (S.U.-A.)

[2] Department of Plant Pathology, University of Agriculture, Faisalabad, Depalpur Campus, Okara 56300, Pakistan

[3] School of Environmental and Civil Engineering, Dongguan University of Technology, Dongguan 523820, China; malikshafeeq1559@gmail.com

[4] Biology Department, College of Sciences, University of Hafr Al-Batin, P.O. Box 1803, Hafr Al-Batin 31991, Saudi Arabia; symohamed@uhb.edu.sa

[5] Agricultural Microbiology Department, Faculty of Agriculture, Sohag University, El-Kawameel 82524, Egypt

[6] Department of Clinical Laboratory Sciences, College of Applied Medical Science, University of Hafr Al-Batin, P.O. Box 1803, Hafr Al-Batin 31991, Saudi Arabia; esfarrag@uhb.edu.sa

[7] Botany Department, Faculty of Agriculture, South Valley University, Qena 83523, Egypt

[8] Plant Protection Department, College of Food and Agriculture Sciences, King Saud University, Riyadh 11362, Saudi Arabia; ruamerm@ksu.edu.sa

[9] Viruses and Phytoplasma Research Department, Plant Pathology Research Institute, Agricultural Research Center, Giza 12619, Egypt

[10] Department of Biology, Faculty of Science, King Khalid University, Abha 61421, Saudi Arabia; mfmostfa@kku.edu.sa

[11] Department of Botany and Microbiology, Faculty of Science, South Valley University, Qena 83523, Egypt

* Correspondence: ahmed.ipfp.11740@pern.edu.pk; Tel.: +92-333-6992446

Citation: Ali, Y.; Raza, A.; Aatif, H.M.; Ijaz, M.; Ul-Allah, S.; Rehman, S.u.; Mahmoud, S.Y.M.; Farrag, E.S.H.; Amer, M.A.; Moustafa, M. Regression Modeling Strategies to Predict and Manage Potato Leaf Roll Virus Disease Incidence and Its Vector. *Agriculture* **2022**, *12*, 550. https://doi.org/10.3390/agriculture12040550

Academic Editor: Francesco Vinale

Received: 6 February 2022
Accepted: 23 March 2022
Published: 12 April 2022

Publisher's Note: MDPI stays neutral with regard to jurisdictional claims in published maps and institutional affiliations.

Abstract: The potato leaf roll virus (PLRV) disease is a serious threat to successful potato production and is mainly controlled by integrated disease management; however, the use of chemicals is excessive and non-judicious, and it could be rationalized using a predictive model based on meteorological variables. The goal of the present investigation was to develop a disease predictive model based on environmental responses viz. minimum and maximum temperature, rainfall and relative humidity. The relationship between epidemiological variables and PLRV disease incidence was determined by correlation analysis, and a stepwise multiple regression was used to develop a model. For this purpose, five years (2010–2015) of data regarding disease incidence and epidemiological variables collected from the Plant Virology Section Ayub Agriculture Research Institute (AARI) Faisalabad were used. The model exhibited 94% variability in disease development. The predictions of the model were evaluated based on two statistical indices, residual (%) and root mean square error (RMSE), which were $\leq \pm 20$, indicating that the model was able to predict disease development. The model was validated by a two-year (2015–2017) data set of epidemiological variables and disease incidence collected in Faisalabad, Pakistan. The homogeneity of the regression equations of the two models, five years ($Y = -47.61 - 0.572x_1 + 0.218x_2 + 3.78x_3 + 1.073x_4$) and two years ($Y = -28.93 - 0.148x_1 + 0.510x_2 + 0.83x_3 + 0.569x_4$), demonstrated that they validated each other. Scatter plots indicated that minimum temperature (5–18.5 °C), maximum temperature (19.1–34.4 °C), rainfall (3–5 mm) and relative humidity (35–85%) contributed significantly to disease development. The foliar application of salicylic acid alone and in combination with other treatments significantly reduced the PLRV disease incidence and its vector population over control. The salicylic acid together with acetamiprid proved the most effective treatment against PLRV disease incidence and its vector *M. persicae*.

Keywords: regression model; epidemiological variable; *M. persicae*; PLRV; management

1. Introduction

Potato (*Solanum tuberosum* L.) is one of the most important foods and vegetable crops in the world [1]. It is cultivated on 19.1 million hectares all over the world, with 381.7 million tons of tuber production, whereas, in Pakistan, 2.9 million tons of potatoes are produced from 0.15 million hectares of harvested area [2]. Its production is highly influenced by the attack of two viroids and 40 viruses [3]. One of the most severe viral diseases is caused by a potato leaf roll virus (PLRV), which is widely distributed in the potato growing regions of the world [4]. The virus is the type species of genus *Polerovirus*; it belongs to family *Solemoviridae* and was first identified by Somera et al. in 2021 [5]. It is efficiently transmitted by aphid species, particularly the green peach aphid *M. persicae*, in a circulative non-propagative manner and is restricted to the phloem tissues of infected plants [6]. The pathogen is responsible for 50% yield reduction in individual plants and over 20 million tons yield losses all over the world [7]. The primary symptoms of PLRV infections include rolling and yellowing of leaves, which may later roll inward. The secondary symptoms in the plant grown from infected tubers are the stunted growth of shoots and leaves rolling upward, starting from the oldest leaves [8]. The PLRV also causes net necrosis in the tubers and reduces crop quality. In Pakistan, 90% yield losses have been reported due to the PLRV disease incidence [9].

Efforts have been made by the plant pathologists and breeders to control PLRV disease incidence by adopting various techniques to ensure the production of virus-free seed potato stocks. These methods include specific growth strategies for seed production and storage, tissue culture and thermotherapy. The control of the virus vector by biopesticides, mineral oils and insecticides has been implemented successfully [10]. None of the varieties/advance lines have shown durable resistance against PLRV disease incidence in the country [11]. This is mainly due to the recurrent occurrence of the vector, continuous introductions of the viruses through imported seeds and the presence of diverse virus strains [12]. As a result, the use of insecticides to control the vector population has become an indispensable element for farmers all over the world, particularly in developing countries. A comprehensive study of the epidemiology of PLRV and its vector population is essential for justifying the application of insecticides. As an analytical tool, a predictive model provides an advanced prediction for vector populations and consequently helps in decisions making as to whether there is a need for insecticide application or not.

Epidemiology deals with the pathogen population on host plants under the impact of the environment at a particular time. Therefore, it is essential to investigate the influence of all the epidemiological variables that are involved in the development of a disease epidemic. For this purpose, detailed information regarding the pathogen, the host and the epidemiological variables, which may lead to the build-up of an epidemic, is of fundamental importance. Understanding the epidemiology of PLRV disease enables accurate prediction of its epidemic and determining the precise timing of application of chemicals in the light of most conducive environmental conditions. This would ultimately decrease pesticide use and thus promote environmentally friendly disease management. Hence, the main goal of the present study was to develop the epidemiological models based on environmental conditions of Faisalabad to predict PLRV disease incidence and to test the plant extracts/biopesticides/chemicals against PLRV disease incidence and *Myzus persicae*.

2. Materials and Methods

2.1. Development of Disease Predictive Model Based on Five-Year Data Set (2010–2015)

For the development of a disease predictive model, five years of data of PLRV disease incidence on three potato varieties, namely Desiree, Cardinal and Diamont, continuously cultivated for five years, and epidemiological variables data comprising minimum and maximum temperature, rainfall and relative humidity (RH) over six months, from November 2010 to April 2015, were collected from Plant Virology Section, Ayub Agriculture Research Institute (AARI), Faisalabad.

2.2. Model Evaluation

The model was evaluated based on a method described by Chatterjee and Hadi [13]. The following three steps were used during model evaluation: (i) comparison of physical theory with dependent variables and regression coefficients; (ii) comparison between observed and predicted values; and (iii) collection of new data to check predictions. The assessment of predictions was conducted through the root mean square error (*RMSE*) and error percentage as described by Chatterjee and Hadi [13]:

$$RMSE = \left[\frac{\sum_{i=1}^{n}(p_i - o_i)^2}{n} \right]^{0.5} \tag{1}$$

$$Error\ Percentage = (p_i - o_i)\,100$$

P_i and O_i are the predicted and observed values for the studied variables, respectively, whereas n is the total number of observations.

2.3. Collection of New Data Set

For the collection of a new data set, an experiment was conducted in the research field of the Department of Plant Pathology, University of Agriculture Faisalabad (UAF), during the autumn and spring crop seasons of 2015–2017 following the same procedures utilized in the preceding five years, as the soil type and environmental conditions of both places are almost identical. Three susceptible potato varieties—Cardinal, Diamant and Desiree—were sown during the winter planting in mid-October and spring planting in mid-January periods on the 25 × 25 experimental plots under randomized complete block design, with row-to-row and plant-to-plant distance of 75 and 20 cm, respectively. The crop was maintained in good conditions by following the recommended agronomic practices.

The disease incidence in the PLRV-infected plants was determined through visual inspection at every line in each plot after 15-day intervals during the 2015–2017 study period [14]. In each row, 10 plants demonstrating PLRV disease symptoms were selected and tagged, and the disease incidence was calculated using the expression given below.

$$Disease\ Incidence\ (\%) = \frac{Number\ of\ infected\ plants}{Total\ number\ of\ plants} \times 100 \tag{2}$$

2.4. Model Validation

For model validation, the PLRV disease incidence pertaining to the three potato varieties sown at the UAF experimental site noted during the 2015–2017 study period was used to develop a two-year model. This model was used to validate the five-year model by comparing the regression coefficients (R^2) yielded by the F-test [15]. The data related to the epidemiological variables, namely minimum and maximum temperature, rainfall and relative humidity for the period covering November to April 2015–2017, were collected from UAF's meteorological station (9610-B-1 Orion LX Weather Station).

2.5. Statistical Analysis

All obtained data were analyzed using Minitab V.17 (Minitab Inc., State College, PA, USA) and SPSS V.17 commercial software tools. The PLRV disease incidence and epidemiological variables were subjected to pairwise correlation and analysis of variance [16]. The least significant difference (LSD) test was adopted for the means separation (at $p \leq 0.05$). A predictive model for PLRV disease incidence was developed on the basis of the epidemiological variables by performing stepwise multiple regression analysis [17]. Using the expressions below, coefficient of determination (R^2) was calculated along with Adj. R^2 to determine the strength of the relationship between individual epidemiological variables and the PLRV disease incidence and to test the model's prediction accuracy [16]:

$$R^2 = \frac{Regression\ sum\ of\ square}{Total\ sum\ of\ square} = 1 - \frac{Error\ sum\ of\ square}{Total\ sum\ of\ square}$$

$$R^2_{adj.} = 1 - \frac{\left(1-R^2\right)(n-1)}{(n-k-1)} \tag{3}$$

where n denotes the sample size, and k is the number of independent variables. Mean square error and Mallows' C_p were also calculated to evaluate the influence of the independent variables included in the model using the following expressions [16]:

$$C_p = (n-p)\left[\frac{MSE\ (Reduced)}{MSE(Full)} - s\right] + p$$

$$MSE = \frac{1}{n}\sum_{i=1}^{n}(y_1 - \overline{y}_i) \tag{4}$$

where p and n in the C_p equation are the number of beta coefficients and sample size in the model, respectively, whereas n and y_i in mean square error (MSE) equation show the number of data values, observed values and predicted values, respectively. Average monthly values of all epidemiological variables and the PLRV disease incidence values were graphically plotted, and critical ranges conducive for disease development were determined.

2.6. Management Strategies of PLRV Disease Incidence and Its Vector

For effective management of PLRV disease and its vector *Myzus persicae*, biopesticides, mineral oils and insecticides namely SA (Salicylic acid) @ 200 mM or 27.4 g/L (T$_1$), SA + Chemical (Acetameprid) @ 15 mL/20 L (T$_2$), SA + Biocontrol (Tracer) @ 8 mL/20 L (T$_3$), SA + Plant Extract (concentrated extract of Neem, a product from China) @ 5 mL/L (T$_4$), SA + Mineral Oil (Dicer) @ 125 mL/20 L (T$_5$) were sprayed on all three susceptible potato varieties—Cardinal, Diamant and Desiree—cultivated in the Research Area of Department of Plant Pathology, University of Agriculture Faisalabad (UAF), during the autumn and spring crop seasons of 2015–2017 with the help of hand knapsack sprayer on the PLRV-infested plants in the experimental area of UAF. The application of only distilled water served as control treatment (T$_6$). The data on aphid population (apterae and alate aphids) and PLRV disease incidence were recorded before and after the 7-day application of treatments until the end of season by using the method described by Khan et al. [10]. The data were subjected to ANOVA, and the treatment means were compared with LSD test at $p \leq 0.05$ [16].

3. Results

3.1. Development of PLRV Disease Predictive Model Based on Five-Year Data Set (2010–2015)

The data from five growing seasons (2010–2015) showed that all the epidemiological variables significantly contributed to PLRV disease incidence (Table 1).

Table 1. Estimated Pearson's correlation coefficients of the relationships between PLRV disease incidence in potatoes in Pakistan and environmental variables during the 2010–2015 field seasons.

Epidemiological Variables	2010–2011	2011–2012	2012–2013	2013–2014	2014–2015
Minimum Temperature (°C)	−0.937 **	−0.791 **	−0.686 *	−0.666 *	−0.625 *
	0.001	0.002	0.014	0.018	0.030
Maximum Temperature (°C)	0.965 **	0.975 **	0.961 **	0.974 **	0.964 **
	0.001	0.001	0.001	0.001	0.001
Rainfall (mm)	0.946 **	0.972 **	0.961 **	0.976 **	0.962 **
	0.001	0.001	0.001	0.001	0.001
Relative Humidity (%)	0.977 **	0.894 **	0.864 **	0.794 **	0.802 **
	0.001	0.001	0.001	0.001	0.001

Upper values show Pearson's correlation coefficients, and lower values indicate the level of probability at $p = 0.05$; * = Significant ($p < 0.05$); ** = Highly significant ($p < 0.01$).

A stepwise multiple regression model ($Y = -47.61 - 0.572x_1 + 0.218x_2 + 3.78x_3 + 1.073x_4$) based on a five-year data set exhibited 94% variability in the PLRV disease development (Table 2). This model could be used for PLRV disease prediction.

Table 2. Summary of stepwise multiple regression model to predict PLRV disease incidence during 2010–2015.

Parameter	No. in Model	Model R^2 (%)	Mallows' C_p	Mean Square Error	F Value	Prb. > F
Minimum Temperature (°C)	1	0.94	15.17	5.11230	5.75	0.018 *
Maximum Temperature (°C)	2	0.94	9.57	5.02378	2.19	0.041 *
Rainfall (mm)	3	0.94	5.19	4.94970	7.76	0.006 *
Relative Humidity (%)	4	0.94	5.00	4.93306	68.48	0.001 *

* = Significant at $p < 0.05$.

3.2. Model Evaluation: Comparison of Physical Theory with Dependent Variables and Regression Coefficients

The model exhibited higher R^2 (94.57%) and Adj. R^2 (94.44%) values with lower standard error value ≤ 20 (Table 3).

Table 3. Regression statistics of PLRV disease incidence during 2010–2015.

Regression Statistics	
R^2	94.57%
Adj. R^2	94.44%
Pred. R^2	94.23%
Std. Error	4.93
Total Observations	179

The F-distribution of the disease predictive model indicated significant regression statistics (Table 4).

Table 4. ANOVA of PLRV disease predictive model based on five-year data set (2010–2015).

Source	[a] DF	[b] SS	[c] MS	F-Value	*p*-Value
Regression	4	74,114.8	18,528.7	761.40	0.001
Residual Error	175	4258.6	24.3		
Total	179	78,373.5			

[a] Degree of freedom; [b] Sum of square; [c] Mean sum of square

The minimum and maximum temperature, rainfall and relative humidity showed significant association with the PLRV disease model at $p \leq 0.05$ (Table 5). The higher coefficient of regression (R^2) value, lower standard error value and the significance of regression statistics exhibited that the model was able to predict PLRV disease incidence (Tables 3–5).

Table 5. Coefficients of estimates, their standard error, t Stat and significance of multiple regression model during 2010–2015.

Parameters	Coefficients	Std. Error	T-Value	*p*-Value
Constant	−47.61	3.37	−14.13	0.001
Minimum Temperature (°C)	−0.572	0.239	−2.40	0.018
Maximum Temperature (°C)	0.218	0.148	1.48	0.031
Rainfall (mm)	3.78	1.36	2.79	0.006
Relative Humidity (%)	1.073	0.130	8.28	0.001

3.3. Model Evaluation

For the evaluation of the model, predictions were obtained using a regression model and evaluated based on two criteria: error (%) and root means square error (RMSE). The normal probability plot of the five-year model showed that most of the data points were around the reference line, while only a few data points, both at the higher and lower sides, deviated from the reference line, affecting the normal distribution of data. Overall, 15% residual was recorded, indicating a fair degree of matching between the observed and predicted data points (Figure 1).

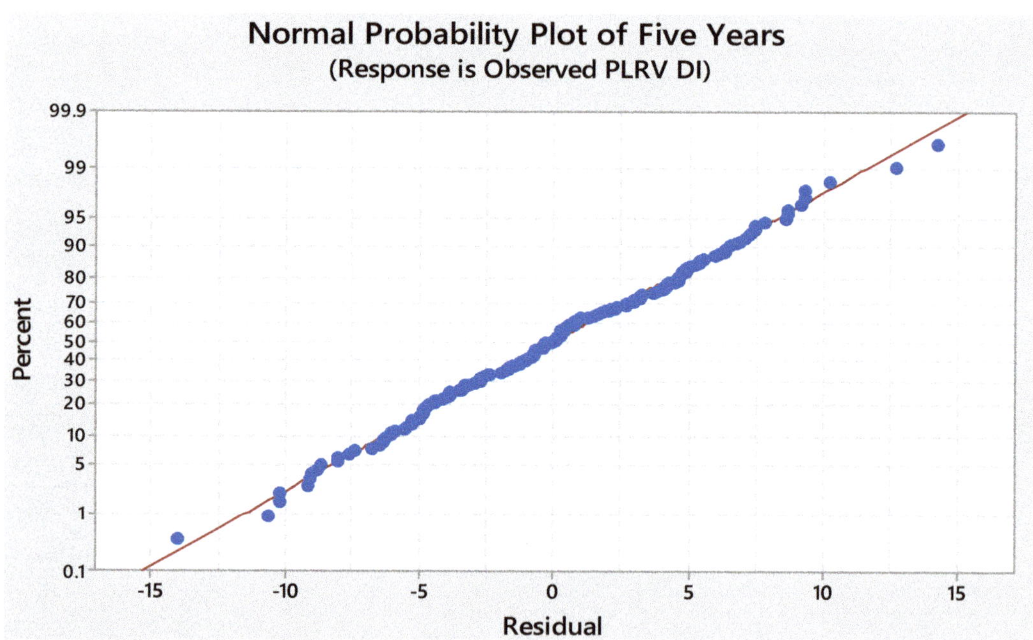

Figure 1. Normal probability plot for five-year (2010–2015) model of potato leaf roll virus (PLRV) disease incidence.

The higher R^2 values > 90% and smaller RMSE values $\leq$ 20 of all three potato genotypes showed the close conformation between observed and predicted data points, indicating that the model was good at predicting PLRV disease incidence (Figure 2).

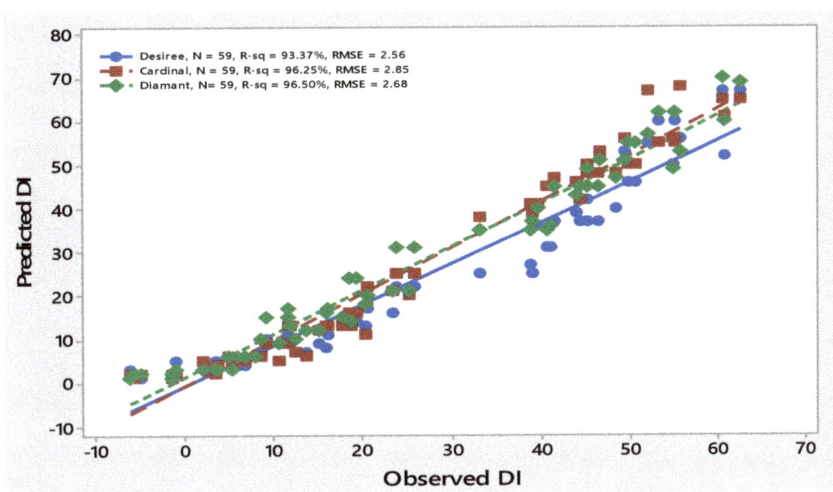

Figure 2. Comparison of observed and predicted data points of PLRV disease incidence of three potato varieties—Desiree, Cardinal and Diamant—during years 2010–2015.

3.4. Model Validation

The stepwise multiple regression model based on a five-year data set was validated on the two-year data set collected from the UAF. The coefficients of determination (R^2) of both models I and II indicated that environmental factors had significantly 94 and 89% impact on PLRV disease incidence, respectively. The regression equations of the two models demonstrated good proximity (Table 6).

Table 6. Comparison of two multiple regression models for validation of PLRV disease incidence.

No. of Model	Regression Equations	R^2 (%)	Adj. R^2 (%)	Prob. > F
I	$Y = -47.61 - 0.572x_1 + 0.218x_2 + 3.78x_3 + 1.073x_4$	94.57	94.44	<0.0001 *
	vs.			
II	$Y = -28.93 - 0.148x_1 + 0.510x_2 + 0.83x_3 + 0.569x_4$	89.31	88.67	<0.0001 *

Model (I) = five-year model; Model (II) = two-year model; Y = PLRV disease incidence; x_1 = minimum temperature; x_2 = maximum temperature; x_3 = rainfall; x_4 = relative humidity; * Sig. at $p < 0.05$.

3.5. Characterization of Environmental Conditions Conducive for PLRV Disease during 2015–2017

Three potato varieties, namely Desiree, Cardinal and Diamant, were employed for regression analysis to characterize critical ranges of epidemiological variables conducive for PLRV disease development. A significant relationship was observed between disease incidence and all environmental variables during both rating seasons. The maximum temperature contributed significantly to the development of PLRV disease on all potato varieties during 2015–2017. It was observed that with an increase in maximum temperature from 19.1–32.1 °C in 2015–2016 and 20.2–34.4 °C during 2016–2017, disease incidence also increased. This relationship was best explained by the linear regression model, as indicated by their correlation coefficient (r) values (Figure 3).

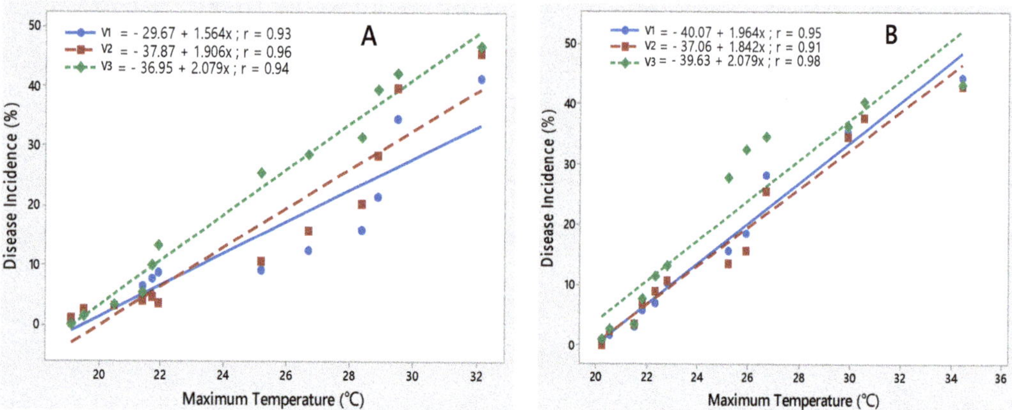

Figure 3. Relationship between maximum temperature and PLRV disease incidence recorded on potato varieties V1 (Desiree), V2 (Cardinal) and V3 (Diamant) during 2015–2016 (**A**) and 2016–2017 (**B**).

A negative linear relationship was observed between minimum temperature and PLRV disease incidence on all three potato varieties during both rating seasons of 2015–2017, indicating that with an increase in minimum temperature from 5 to 18.5 °C, disease incidence decreased (Figure 4).

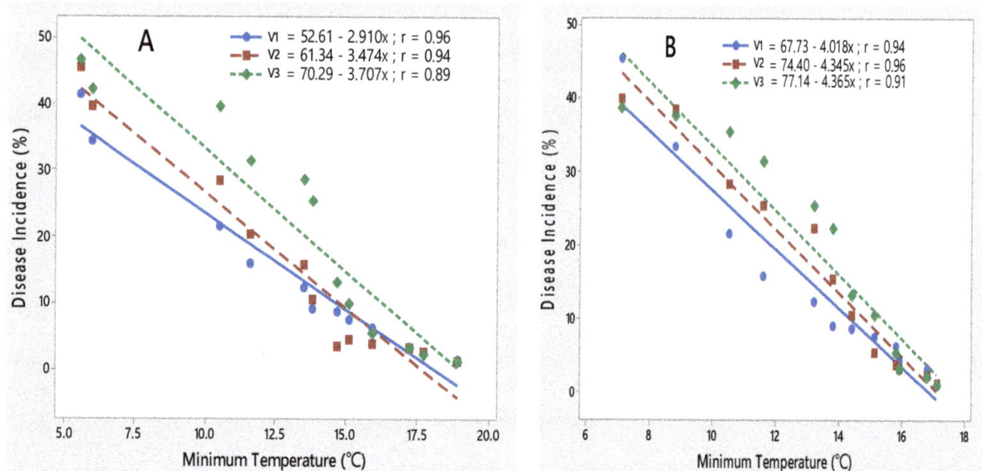

Figure 4. Relationship between minimum temperature and PLRV disease incidence recorded on potato varieties V1 (Desiree), V2 (Cardinal) and V3 (Diamant) during 2015–2016 (**A**) and 2016–2017 (**B**).

The impact of rainfall was recorded as significant with the PLRV disease development. The maximum disease was noted at 3–5 (mm) during both crop seasons; it demonstrated that disease incidence increased with an increase in rainfall, as demonstrated by their r values 0.74, 0.70 and 0.83 during 2015–2016 and 0.50, 0.48 and 0.69 during 2016–2017, respectively (Figure 5).

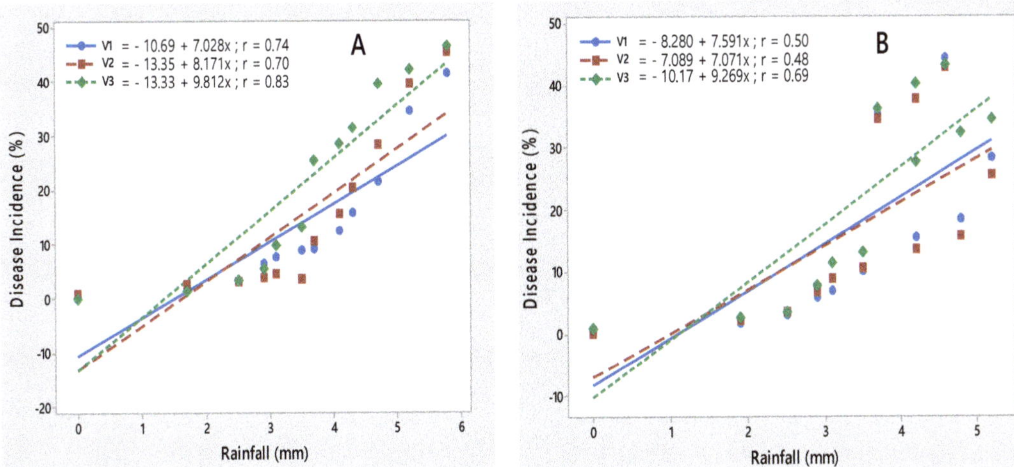

Figure 5. Relationship between rainfall and PLRV disease incidence recorded on potato varieties V1 (Desiree), V2 (Cardinal) and V3 (Diamant) during 2015–2016 (**A**) and 2016–2017 (**B**).

Relative humidity was positively correlated with disease incidence. During both rating seasons, disease incidence increased with an increase in relative humidity from 35 to 85% (Figure 6).

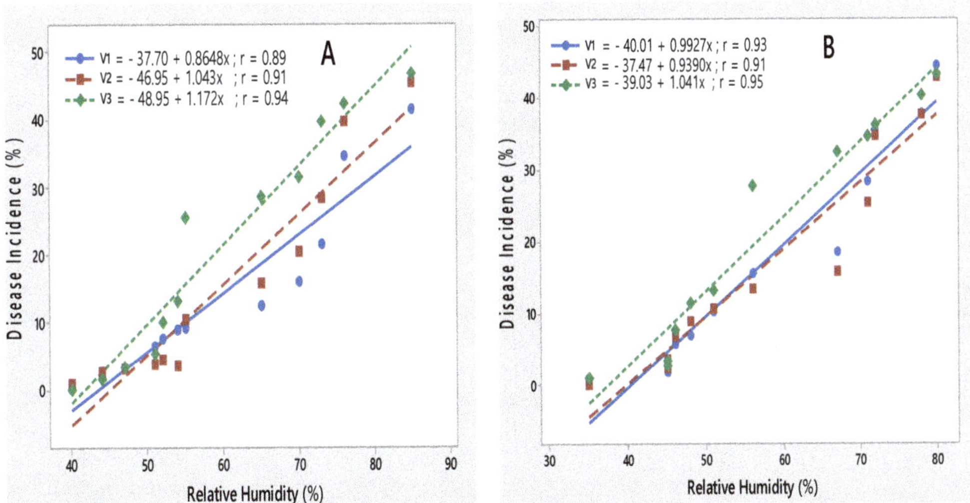

Figure 6. Relationship between relative humidity and PLRV disease incidence recorded on potato varieties V1 (Desiree), V2 (Cardinal) and V3 (Diamant) during 2015–2016 (**A**) and 2016–2017 (**B**).

3.6. Management Strategies for PLRV and Its Vector

The exogenous application of salicylic acid alone and its combination with other treatments indicated a significant effect in controlling PLRV disease and its vector *M. persicae* as compared to the control. The salicylic acid in combination with acetamiprid proved the most effective in controlling PLRV disease incidence and its vector populations during both crop seasons of 2015–2017, followed by SA in combination with biocontrol (tracer), plant extract (neem), mineral oil, as compared to the salicylic acid alone and control (Tables 7 and 8).

Table 7. Comparison of treatments to control aphid populations during 2015–2017.

Sr. No.	Treatments	1st Year 2015–2016	2nd Year 2016–2017
1	Control (T_6)	37.39 [A]	46.55 [A]
2	S.A. (Salicylic acid) (T_1)	31.04 [B]	43.17 [B]
3	S.A + Mineral oil (T_5)	14.45 [C]	15.30 [C]
4	S.A + Plant extract neem (T_4)	9.79 [D]	8.95 [D]
5	S.A + Tracer (T_3)	7.88 [D]	7.35 [D]
6	S.A + Acetamiprid (T_2)	4.42 [E]	3.44 [E]

Means that do not share a letter are significantly different.

Table 8. Comparison of treatments to control PLRV disease incidence during 2015–2017.

Sr. No.	Treatments	1st Year 2015–2016	2nd Year 2016–2017
1	Control (T_6)	55.07 [A]	55.94 [A]
2	S.A (Salicylic acid (T_1)	23.38 [B]	29.17 [B]
3	S.A + Mineral oil (T_5)	17.28 [C]	10.83 [C]
4	S.A + Plant extract neem (T_4)	13.10 [D]	10.06 [C]
5	S.A + Tracer (T_3)	8.33 [E]	8.44 [C]
6	S.A + Acetamiprid (T_2)	6.63 [E]	5.52 [D]

Means that do not share a letter are significantly different.

4. Discussion

Environmental conditions played a significant role in the development of pathogens on any crop; therefore, quantifying the relationship between PLRV disease incidence and epidemiological variables is important in early warning of its onset [10]. PLRV is significantly influenced by the epidemiological variables; however, the degree of correlation changes greatly by varieties and years. A significant correlation between epidemiological variables and PLRV disease incidence was observed in this investigation, in line with the findings of Khan and Abbas [18], who demonstrated a significant correlation of temperature (minimum and maximum), rainfall and relative humidity with PLRV disease incidence.

The significant correlation of temperature with PLRV disease incidence can be explained by the fact that it has a critical role in different aspects of disease development. The expression of viral disease symptoms was delayed at low temperatures in several plant species [19–22]. Szittya et al. (2003) described that temperature effect of plant–pathogen interactions and high temperature can either increase or decrease the disease resistance [21]. This reflects the effects of the same temperature variation on various plant–pathogen systems [23]. Virus resistance was compressed in plants at a higher temperature. For example, Capsicum chinense plants carrying the Tsw gene and tobacco plants carrying the N gene developed systemic infections of tomato spotted wilt virus (TSWV) and tobacco mosaic virus (TMV) at above 28 and 32 °C, respectively [24,25]. The increasing temperature alters the host plant physiology, phenology, morphology, nutritional status and metabolic pathways [26,27]. The rising heat stress and mean temperature reduced the effectiveness of temperature-sensitive single-gene resistance and increased general plant vulnerability to virus infection. Increased temperature also changes the virus multiplication, seed transmission and systemic movement of individual viruses present in mixed infection [27]. Jones (2014) showed that potato yellow vein virus (PYVV) and PLRV best adapted to hot regions; conversely, potato mop-top virus (PMTV) and Andean potato latent virus are projected for regions too cold for growth and development [28]. The significant relationship of relative humidity and rainfall with PLRV disease incidence was due, in part, to its key role in the survival, population growth, behavior and movement of virus vector [29]. Virus dispersal in crops is favored by the soft tender leaves and lush plant growth that develop under conditions of high relative humidity. Such plants are more vulnerable to viral infection as compared to the hard-leaved plants of low-humidity conditions. This is because wounds develop more readily when growth is soft, and viruses have to penetrate a plant's protective cuticle through wounds before they can invade damaged cells [30].

The maximum temperature (19.1–34.4 °C), minimum temperature (5–18.5 °C), (rainfall (3–5 mm) and relative humidity (35–85%) appeared to be the main contributing epidemiological variables in the disease development, as these variables were retained after stepwise regression. The present multiple regression model explained 94% variability in PLRV disease development, whereas only 6% variability remained unexplained. The models that explain >80% variability are considered reliable and provide relatively accurate predictions [31]. The reason behind not explaining 100% variability might be due to the fact that regression models are empirical models. Khan and Abbas (2008) developed the multiple regression models and reported 60% unexplained variability in PLRV disease development when only environmental variables were used [18]. However, by including the primary source of virus inoculum and other biological factors as independent variables, the unexplained variability may be reduced [31]. Further, the present study was laid out under natural environmental conditions where the amounts of inoculum and infection efficiency were uncontrolled; an explanation of 100% variability was not possible. However, the current investigation remained successful in predicting PLRV disease because the model, with a large data set of five years, validated with a two-year data set, generated approximately precise predictions. The high coefficient of determination (R^2) value 0.94 of the model indicated that it can be used in future for accurate prediction of PLRV disease.

Considering the management strategies of PLRV disease and its vector aphid, salicylic acid (SA) alone and its combination with other treatments, such as biopesticides, chemicals, mineral oils and neem extracts, significantly decreased the PLRV disease incidence and aphid population over control. It means that the application of salicylic acid is effective in controlling PLRV disease incidence by inducing systemic resistance in plants. Koo et al. [32] showed that exogenous application of SA provides tolerance to plants against several plant pathogens [33]. In tobacco, the foliar application of SA induced resistance against tobacco mosaic virus (TMV) [34]. The pathogenicity-related proteins are activated by the foliar application of SA against many plant viruses. After the application of SA, potato plants develop systemic acquired resistance (SAR), which results in the activation of plant defense mechanism [35]. The SA in combination with pesticide acetamiprid proved the most effective in controlling PLRV disease incidence and *M. persicae* populations. Acetamiprid has the ability to decrease the infection and dispersal rate of plant viruses during the pre-mortality phase [36]. Acetamiprid is very selective and provides an effective control against sucking pests, such as whiteflies and aphids, without negative impact on non-target insects [37]. The tracer in combination with SA and azadirachtin extracted from the seeds of the neem tree (*Azadirechta indica*) disturbs the feeding behavior of aphid and fecundity through repellent and antifeedant activity [38]. Mineral oil, which was the least effective in controlling PLRV disease incidence, does not kill aphids *Myzus persicae* but reduces the transmission by altering its behavior. Yang et al. [38] described that after 30 min of oil application, *M. persicae* was unable to transmit PVY in plants but could do so after 24 h, although with diminished ability.

5. Conclusions

It was concluded that a five-year model validated with a two-year data set exhibited 94% variability in the PLRV disease development. All environmental variables indicated a significant relationship with PLRV disease incidence. Regression analysis proved that there was a significant effect of average seasonal minimum temperature (5–18.5 °C), maximum temperature (19.1–34.4 °C), rainfall (3–5 mm) and relative humidity (35–85%) on PLRV disease development. The study concluded that PLRV disease can be managed when its vector is controlled. As the environmental conditions play crucial role in the development of the disease, the disease predictive models would be helpful for farmers in the proper management of the disease. The disease forecast model helps them decide whether to spray a crop right away or to wait for more days. Bio-pesticides, insecticides, oils, plant extracts and other chemicals often provide only short-term virus disease control; these materials can be more effectively utilized when the epidemiological components are understood. Thus, understanding the epidemiology of PLRV disease will enable us to predict its development, which will ultimately help farmers to improve plant protection measures more accurately.

Author Contributions: Conceptualization, Y.A., A.R. and M.I.; methodology, Y.A., A.R., H.M.A. and S.U.-A.; software, S.u.R.; validation, Y.A., A.R. and S.u.R.; formal analysis, Y.A. and H.M.A.; investigation, Y.A. and M.I.; resources, A.R., M.A.A. and M.M.; data curation, Y.A. and S.U.-A.; writing—original draft preparation, Y.A., A.R. and M.I.; writing—review and editing, S.Y.M.M., E.S.H.F., M.A.A. and M.M.; supervision, A.R. and M.A.A.; project administration, S.Y.M.M. and M.M.; funding acquisition, A.R., M.A.A. and M.M. All authors have read and agreed to the published version of the manuscript.

Funding: This research was funded by the Deanship of Scientific Research, King Khalid University, Saudi Arabia under grant no. (RGP.-2/135/42).

Institutional Review Board Statement: Not applicable.

Informed Consent Statement: Not applicable.

Data Availability Statement: Not applicable.

Acknowledgments: The authors extend their appreciation to the Deanship of Scientific Research at King Khalid University, Saudi Arabia, for funding this work through research grant no. (RGP.-2/135/42).

Conflicts of Interest: The authors declare that they have no known competing financial interests or personal relationships that could have appeared to influence the work reported in this paper.

Abbreviations

AARI	Ayub Agriculture Research Institute
MSE	Mean Square Error
PLRV	Potato Leaf Roll Virus
RH	Relative Humidity
RMSE	Root Mean Square Error
SA	Salicylic Acid
UAF	University of Agriculture Faisalabad

References

1. He, Z.; Larkin, R.; Honeycutt, W. *Sustainable Potato Production: Global Case Studies*; Springer Science & Business Media: Dordrecht, The Netherlands, 2012.
2. FAOSTAT. Food and Agricultural Organization Statistical Database, Crop Production. 2017. Available online: http://faostat3.fao.org/download/Q/QC/E (accessed on 22 July 2017).
3. Jeffries, C.; Barker, H.; Khurana, S.M. Viruses and viroids. In *Handbook of Potato Production, Improvement, and Postharvest Management*; Food Products Press: New York, NY, USA, 2006; pp. 387–448.
4. Gillen, A.M.; Novy, R.G. Molecular characterization of the progeny of Solanum etuberosum identifies a genomic region associated with resistance to potato leafroll virus. *Euphytica* **2007**, *155*, 403–415. [CrossRef]
5. Sõmera, M.; Fargette, D.; Hébrard, E.; Sarmiento, C.; ICTV Report Consortium. ICTV virus taxonomy profile: *Solemoviridae*. *J. Gen. Virol.* **2021**, *102*, 001707.
6. Lee, L.; Kaplan, I.B.; Ripoll, D.R.; Liang, D.; Palukaitis, P.; Gray, S.M. A surface loop of the potato leafroll virus coat protein is involved in virion assembly, systemic movement, and aphid transmission. *J. Virol.* **2005**, *79*, 1207–1214. [CrossRef] [PubMed]
7. Wales, S.; Platt, H.W.; Cattlin, N. *Diseases, Pests and Disorders of Potatoes*; Manson Publ. Ltd.: London, UK, 2008; pp. 75–76.
8. Abbas, A.; Sohail, M.A.; Mubeen, M.; Alami, M.M.; Umer, M.; Khan, S.U. Plant Viruses in Gilgit-Baltistan (GB) Pakistan: Potential Future Research Direction. *Plant Pathol. Microbiol.* **2020**, *10*, 486. [CrossRef]
9. Bhutta, A.R.; Bhatti, M.F. *Seed Potato Certification in Pakistan*; Federal Seed Certification and Registration Department Ministry of Food Agriculture and Livestock: Islamabad, Pakistan, 2002; pp. 60–66.
10. Khan, M.A.; Hannan, A.; Naqvi, S.A.; Khan, A.A.; Zulfiqar, M.A. Development and validation of potato leaf roll virus disease prediction model based on environmental factors for Faisalabad, Pakistan. *Pak. J. Agric. Res.* **2016**, *29*, 1–14.

11. Qamar, N.; Khan, M.A.; Rashid, A. Screening of potato germplasm against potato virus X (PVX) and potato virus Y (PVY). *Pak. J. Phytopathol.* **2003**, *2*, 189–190.

12. Kreuze, J.F.; Souza-Dias, J.A.; Jeevalatha, A.; Figueira, A.R.; Valkonen, J.P.; Jones, R.A. Viral Diseases in Potato. In *The Potato Crop*; Springer: Cham, Switzerland, 2020; pp. 389–430.

13. Chatterjee, S.; Hadi, A.S. *Regression Analysis by Example*; John Wiley & Sons: Hoboken, NJ, USA, 2015.

14. Gabriel, W. The influence of temperature on the spread of aphid-borne potato virus diseases. *Ann. Appl. Biol.* **1965**, *56*, 461–475. [CrossRef]

15. Harrell, F.E., Jr. *Regression Modeling Strategies: With Applications to Linear Models, Logistic and Ordinal Regression, and Survival Analysis*; Springer: New York, NY, USA, 2015.

16. Steel, R.G.O.; Torrie, J.H.; Dickey, D. *Principles and Procedures of Statistics: A Biometrical Approach*; McGraw-Hill: New York, NY, USA, 1987.

17. Meyer, M.; Woodroofe, M. On the degrees of freedom in shape-restricted regression. *Ann. Stat.* **2000**, *28*, 1083–1104. [CrossRef]

18. Khan, M.A.; Abbas, W. Evaluation of multiple regression models based on epidemiological factors to predict *M. persicae* population and PLRV disease incidence. In *International Conference of Plant Scientists*; The Pakistan Botanical Society: Faisalabad, Pakistan, 2008; pp. 155–168.

19. Saied, M.S.; Medany, M.A.; Abo El-Abbas, F.; El-Hammaday, M. Forecasting incidence potato leaf roll virus disease in Egypt. *Egypt. J. Virol.* **2005**, *2*, 201–213.

20. Velázquez, K.; Renovell, A.; Comellas, M.; Serra, P.; García, M.L.; Pina, J.A.; Navarro, L.; Moreno, P.; Guerri, J. Effect of temperature on RNA silencing of a negative-stranded RNA plant virus: *Citrus psorosis* virus. *Plant Pathol.* **2010**, *59*, 982–990. [CrossRef]

21. Szittya, G.; Silhavy, D.; Molnár, A.; Havelda, Z.; Lovas, Á.; Lakatos, L.; Bánfalvi, Z.; Burgyán, J. Low temperature inhibits RNA silencing-mediated defence by the control of siRNA generation. *EMBO J.* **2003**, *22*, 633–640. [CrossRef] [PubMed]

22. Romon, M.; Soustre-Gacougnolle, I.; Schmitt, C.; Perrin, M.; Burdloff, Y.; Chevalier, E.; Mutterer, J.; Himber, C.; Zervudacki, J.; Montavon, T.; et al. RNA silencing is resistant to low-temperature in grapevine. *PLoS ONE* **2013**, *8*, e82652. [CrossRef] [PubMed]

23. Wang, Y.; Bao, Z.; Zhu, Y.; Hua, J. Analysis of temperature modulation of plant defense against biotrophic microbes. *Mol. Plant Microbe Interact.* **2009**, *22*, 498–506. [CrossRef] [PubMed]

24. Erickson, F.L.; Dinesh-Kumar, S.P.; Holzberg, S.; Ustach, C.V.; Dutton, M.; Handley, V.; Corr, C.; Baker, B.J. Interactions between tobacco mosaic virus and the tobacco N gene. Philosophical Transactions of the Royal Society of London. *Ser. B Biol. Sci.* **1999**, *354*, 653–658. [CrossRef]

25. Roggero, P.; Pennazio, S.; Masenga, V.; Tavella, L. Resistance to tospoviruses in pepper. In *Thrips and Tospoviruses: Proceedings of the 7th International Symposium on Thysanoptera (CD-ROM)*; ANIC: Canberra, Australia, 2002; pp. 105–110.

26. Canto, T.; Aranda, M.A.; Fereres, A. Climate change effects on physiology and population processes of hosts and vectors that influence the spread of hemipteran-borne plant viruses. *Glob. Chang. Biol.* **2009**, *15*, 1884–1894. [CrossRef]

27. Jones, R.A.; Barbetti, M.J. Influence of climate change on, plant disease infections and epidemics caused by viruses and bacteria. *Rev. Plant Sci.* **2012**, *22*, 1–31. [CrossRef]

28. Jone, R.A. Virus disease problems facing potato industries worldwide: Viruses found, climate change implications, rationalizing virus strain nomenclature, and addressing the Potato virus Y issue. In *The Potato: Botany, Production and Uses*; CABI: Wallingford, UK, 2014; pp. 202–224.

29. Were, H.K.; Kabira, J.N.; Kinyua, Z.M.; Olubayo, F.M.; Karinga, J.K.; Aura, J.; Lees, A.K.; Cowan, G.H.; Torrance, L. Occurrence and distribution of potato pests and diseases in Kenya. *Potato Res.* **2013**, *56*, 325–342. [CrossRef]

30. Jones, R.A. Future scenarios for plant virus pathogens as climate change progresses. *Adv. Virus Res.* **2016**, *95*, 87–147.

31. Kumar, P.V. Development of weather-based prediction models for leaf rust in wheat in the Indo-Gangetic plains of India. *Eur. J. Plant Pathol.* **2014**, *140*, 429–440. [CrossRef]

32. Koo, Y.M.; Heo, A.Y.; Choi, H.W. Salicylic acid as a safe plant protector and growth regulator. *Plant Pathol. J.* **2020**, *36*, 1–10. [CrossRef]

33. Murphy, A.M.; Carr, J.P. Salicylic acid has cell-specific effects on tobacco mosaic virus replication and cell-to-cell movement. *Plant Physiol.* **2002**, *128*, 552–563. [CrossRef] [PubMed]

34. Ali, S.; Mir, Z.A.; Tyagi, A.; Bhat, J.A.; Chandrashekar, N.; Papolu, P.K.; Rawat, S.; Grover, A. Identification and comparative analysis of Brassica juncea pathogenesis-related genes in response to hormonal, biotic and abiotic stresses. *Acta Physiol. Plant.* **2017**, *39*, 268. [CrossRef]

35. Bethke, J.A.; Blua, M.J.; Redak, R.A. Effect of selected insecticides on Homalodisca coagulata (Homoptera: Cicadellidae) and transmission of oleander leaf scorch in a greenhouse study. *J. Econ. Entomol.* **2001**, *94*, 1031–1036. [CrossRef] [PubMed]

36. Bambhaniya, V.S.; Khanpara, A.V.; Patel, H.N. Bio-Efficacy of insecticides against sucking pests; whitefly and aphid infesting tomato. *J. Pharmacogn. Phytochem.* **2018**, *7*, 2051–2059.

37. Nisbet, A.J.; Woodford, J.A.; Strang, R.H. The effects of azadirachtin on feeding by Myzus persicae. In *Proceedings of the 8th International Symposium on Insect-Plant Relationships*; Springer: Dordrecht, The Netherlands, 1992; pp. 179–180.

38. Yang, Q.; Arthurs, S.; Lu, Z.; Liang, Z.; Mao, R. Use of horticultural mineral oils to control potato virus Y (PVY) and other non-persistent aphid-vectored viruses. *Crop Prot.* **2019**, *118*, 97–103. [CrossRef]

Article

Phosphorus and Potassium Application Improves Fodder Yield and Quality of Sorghum in Aridisol under Diverse Climatic Conditions

Atique-ur-Rehman [1], Rafi Qamar [2], Muhammad Mohsin Altaf [3,4], Mona S. Alwahibi [5], Rashid Al-Yahyai [6,7] and Mubshar Hussain [1,8,*]

1   Department of Agronomy, Bahauddin Zakariya University, Multan 60800, Pakistan; dr.atique@bzu.edu.pk
2   Department of Agronomy, College of Agriculture, University of Sargodha, Sargodha 40100, Pakistan; rafi.qamar@uos.edu.pk
3   College of Ecology and Environment, Hainan University, Haikou 570228, China; mohsinaltaf641@gmail.com
4   State Key Laboratory of Marine Resource Utilization in South China Sea, College of Ecology and Environment Hainan University, Haikou 570228, China
5   Department of Botany and Microbiology, College of Science, King Saud University, Riyadh 11451, Saudi Arabia; malwhibi@ksu.edu.sa
6   Department of Plant Sciences, College of Agricultural and Marine Sciences, Sultan Qaboos University, Muscat 123, Oman; alyahyai@squ.edu.om
7   Department of Crop Science, University of Reading, Whiteknights, P.O. Box 217, Reading, Berkshire RG6 6AH, UK
8   School of Veterinary and Life Sciences, Murdoch University, 90 South Street, Murdoch, WA 6150, Australia
*   Correspondence: mubashir.hussain@bzu.edu.pk; Tel.: +92-301-7164879

Citation: Atique-ur-Rehman; Qamar, R.; Altaf, M.M.; Alwahibi, M.S.; Al-Yahyai, R.; Hussain, M. Phosphorus and Potassium Application Improves Fodder Yield and Quality of Sorghum in Aridisol under Diverse Climatic Conditions. *Agriculture* **2022**, *12*, 593. https://doi.org/10.3390/agriculture12050593

Academic Editor: Markku Yli-Halla

Received: 20 March 2022
Accepted: 20 April 2022
Published: 23 April 2022

Publisher's Note: MDPI stays neutral with regard to jurisdictional claims in published maps and institutional affiliations.

Abstract: Fodder yield and quality must be improved for sustainable livestock production. A lack of or low application of phosphorus (P) and potassium (P) are among the leading constraints of lower fodder yield and quality of sorghum [most cultivated fodder crop during kharif season (crop cultivation in summer and harvesting during winter] in Aridisol of Pakistan. Therefore, this two-year field study evaluated the role of different P and K levels on fodder yield and quality of sorghum cultivar '*Ijar-2002*' planted in Multan and Okara districts, Punjab, Pakistan. Seven P-K (kg ha^{-1}) levels, i.e., T_1 (40–0), T_2 (80–0), T_3 (0–40), T_4 (0–60), T_5 (40–40), T_6 (80–40), T_7 (60–80) and an untreated T_0 (control) were included in the study. Results indicated that individual effects of years, locations and P-K levels had a significant effect on fodder yield and quality. All treatments received an equal amount of nitrogen (i.e., 120 kg ha^{-1}). Application of P-K in Aridisols at both locations significantly improved fodder yield, dry matter yield, and ether contents during both years. The T_6 (80–40 kg ha^{-1}) significantly improved yield and quality traits of sorghum fodder except for crude fiber (CF) and acid and neutral detergent fiber (ADF and NDF) at both locations during both years of study. Moreover, fodder harvested from Multan observed significantly higher CF, ADF, NDF, cellulose and hemicellulose contents than Okara. However, sorghum grown in Okara harvested more fodder yield due to more plant height and ether contents. In conclusion, planting sorghum in Aridisols, fertilized with 80–40 kg ha^{-1} P-K seemed a viable option to harvest more fodder yield of better quality.

Keywords: sorghum; locations; P-K levels; fodder quality; yield traits; tropical conditions

1. Introduction

Sorghum (*Sorghum bicolor* L. Moench) is widely cultivated for forage purposes during summer season and has a significant role in livestock production [1]. Livestock share in agriculture and gross domestic product of Pakistan is 61% and 12%, respectively [2]. In Punjab, Pakistan, total production of summer fodder is 11,939 thousand tons from an area of 902 thousand hectares with an average fodder production of 55 t ha^{-1} [2]. Recently, sorghum fodder production has declined by 20% to 120 thousand tons [2]. About 60% of livestock depends on the feeds of wheat straw, maize stalks, sorghum, and millet [3]. In Pakistan, sorghum ranks fourth among cereals with respect to area under cultivation and produce fodder and grains, especially under harsh environmental conditions of the country. Availability of high yielding cultivars, imbalance and lower nutrients' application, low plant population, and poor weed management practices are among the leading constraints of low sorghum productivity [4].

Climate variability has significant impact on temperature, whereas the optimum temperature range for vegetative and reproductive growth of sorghum is 26–34 °C and 25–28 °C, respectively [5]. When temperature rises >36 °C it causes abortion of the entire inflorescence [6]. Predictable changes in climatic conditions, especially rising air mean maximum and minimum temperatures and fluctuations in rainfall pattern, had adversely affected fodder yield and quality traits in most parts of the world [7]. Climate change exerted multidimensional stresses on crop plants and changed rainfall patterns in the last few decades. The growth and yield are reduced when crops face these environmental fluctuations [8]. The ultimate solution to cope with these climate changes is timely adaptation of techniques that can mitigate these changes and boost plant production [9]. Failure in the adaptation of these novel techniques will create hurdles in the production of nutritious food and fodder for livestock in sufficient quantity [10]. Tropical and subtropical areas of the world are facing lower average fodder yield of sorghum than the potential of 50–100 tons per hectare. The main causes of severe fodder shortage during winter and summer seasons are low rainfall and high temperature during sowing.

Nutritional composition of fodder is highly sensitive to variation in balanced fertilizer application, drought/limited irrigation, genotypic characteristics, and higher population per unit area etc. [11,12]. Sorghum fodder contains higher digestible nutrients, consisting of 8% protein, 3% fat and 45% nitrogen-free extract [13]. Generally, fodder quality is dependent on the percentage of two main quality characteristics, i.e., crude protein (CP) and crude fiber (CF). The higher the percentage of CF, the lower the value of CP that leads towards lower fodder quality. Acid detergent fiber (ADF), neutral detergent fiber (NDF) and acid detergent lignin (ADL) are mostly applied for standard testing of fiber in any fodder. Fodder digestibility is assessed by determining ADF, while eating prospective is checked through NDF [14].

An appropriate nutrients application is a key constituent for enriching fodder production and quality [15,16]. Balanced application of fertilizers has a vital role in enhancing fodder production. However, lower application of phosphorus (P) than nitrogen (N) in calcareous soils is the main constraint. Calcareous Aridisol is deficient in many essential nutrients and tend to adsorb most of the applied P at exchange sites and limited portion goes to soil solution which is taken by plants [17]. Balanced placement/incorporation of fertilizers into the soil during sowing can play a key role in improving root growth that would lead towards healthy plant growth [18]. Unfortunately, soils remain deficient to P and K due to the negligence of growers for the application of these nutrients, which reduces fodder production in calcareous soils [19].

Fodder growers in Pakistan also apply lower P and K without focusing soil fertility status and mainly focus on N application. Moreover, P contents in Pakistan soils are in the range of 0.02 to 0.5% at surface layer due to alkalinity. There are two main reasons of low P application for fodder production. The first is higher price of P fertilizers [20] and the second is the lack of awareness/unavailability of these fertilizers during the peak demand period [21]. Unfortunately, no or low application of P is a major yield-limiting factor since it is an essential constituent of nucleic acid and plays an important role in cellular respiration and plant metabolism [22]. Involvement of P in enzymatic reactions, atmospheric CO_2 fixation, sugar digestion, and energy storage and transfer have direct contribution in fodder quantity and quality [23]. The second neglected nutrient in Pakistan is K, which reduces crop yield [24,25]. Potassium plays several significant roles in osmoregulation, plant water relations, cell expansion, stomatal conductance, membrane stability, cation–anion balancing, solute transport, protein synthesis, initiating and stimulation enzymes, modifying protein and starch and production of adenosine triphosphate (ATP) [26].

Owing to their roles in plant growth and development, fodder production and quality of sorghum can be improved by P and K application [27]. The information regarding combined application of P and K on fodder production and quality of sorghum is rarely studied or reported [28]. Therefore, this 2-years field study was conducted at two locations with the hypothesis that fodder yield and quality of sorghum can be improved by combined application of P and K on Hyperthermic, Sodic Haplocambids, Haplic Aridisols in Pakistan. The major objective of the study was to determine the P-K level which would improve yield and quality of fodder sorghum on Aridisol.

2. Materials and Methods

2.1. Description of Experimental Sites

This two-year field study was conducted at two distinct locations, i.e., [Agronomy Research Farm, Bahauddin Zakariya University, Multan (30.10° N, 71.25° E and 128.3 m altitude above sea level) and Farmer's field, Okara (30.81° N, 73.45° E and 105 m altitude above sea level)], Punjab, Pakistan during sorghum-fodder seasons (i.e., August-December) in 2015 and 2016. Weather data of both experimental locations are given in Table 1. August was the hottest month during both years with mean daily temperature ranging from 33.2 °C (Multan) to 31.3 (Okara) during 2015 and 36.7 °C (Multan) to 34.1 °C (Okara) during 2016, whereas December was the coolest month with mean daily temperature ranging from 17.7 °C (Multan) to 15.8 °C (Okara) during 2015 and 28.5 °C (Multan) to 23.9 °C (Okara) (Table 1). The growing season of 2015 received 55 mm (Multan) and 45 mm (Okara) rainfall, while 13- and 42-mm rainfall was received during 2016 at Multan and Okara, respectively (Table 1). Before sowing, physico-chemical analysis of soil was conducted to judge initial soil fertility status at both locations. Soil texture of Multan and Okara was determined using Hydrometer method. Multan soil was silty-clay-loam and belonged to Sindhalianwali textural class, was Hyperthermic, Sodic Haplocambids/Haplic Aridisol, whereas Okara soil was silt-loam and belonged to Kasur soil series and was Typic Camborthids according to USDA and FAO classifications, respectively. Soils of Multan and Okara had pH 8.3 and 7.8 and EC 12 and 11 dS m^{-1}, respectively, that were determined through pH meter (Beckman 45 Modal, Gurnee, IL, USA) and EC meter (VWR Conductivity Meter DIG2052, Radnor, PA, USA). Moreover, Multan and Okara soils had 0.78% and 0.85% organic matter, 0.04% and 0.09% total N, 7.6 and 5.9 mg kg^{-1} available-P (NaHCO$_3$-DTPA), and 165 and 285 mg kg^{-1} extractable-K, respectively.

Table 1. Weather data of both study years (2015 and 2016) at Multan and Okara locations.

Months	Multan				Okara			
	2015		2016		2015		2016	
	Rainfall	Temperature	Rainfall	Temperature	Rainfall	Temperature	Rainfall	Temperature
	mm	°C	mm	°C	mm	°C	mm	°C
August	32	34.1	11.4	36.7	33.0	31.3	22.0	33.2
September	10	35.2	1.2	37.9	12.0	29.1	18.0	32.7
October	3	33.6	0	37.2	0	24.8	0	26.3
November	4	26.6	0	32.5	0	21.2	2.0	23.8
December	6	23.9	0	28.5	0	15.8	0	17.7

Sources: Agricultural Meteorology Cell, Department of Agronomy, Bahauddin Zakariya University Multan and Meteorological Department, Railway Road, Okara, Punjab, Pakistan.

2.2. Experimental Details

Seeds of sorghum cultivar *'Ijar-2002'* (widely cultivated for fodder and grain purpose in the irrigated areas of Punjab, Pakistan) were obtained from the Fodder Research Institute, Sargodha, Pakistan. Crop was sown on 5th August 2015 and 2016 at Multan and Okara.

Different P-K levels (kg ha^{-1}) included in the study were T_0 (control); T_1 (40–0); T_2 (80–0); T_3 (0–40); T_4 (0–60); T_5 (40–40); T_6 (80–40) and T_7 (60–80). The fertilizer treatments were applied during sowing. Triple super phosphate (46% P_2O_5) and sulphate of potash (50% K_2SO_4) were used as sources of P and K, respectively. Experiment was laid out following randomized complete block design in a split-split plot arrangement (year as main plots, locations as subplots, and P-K levels as sub-subplots). The experiment was replicated four times with a net sub-plot size of 1.5 m $\times$ 4 m. Before this study, experimental fields were under cotton-wheat and maize-potato cropping systems at Multan and Okara, respectively.

2.3. Crop Husbandry

Before seedbed preparation, pre-soaking irrigation of ~10 cm was applied using tube well water. After attaining workable moisture level, the fields were tilled twice followed by planking with the help of tractor-drawn cultivator to achieve a fine seedbed at both locations. Sowing was carried out by using seed rate of 60 kg ha^{-1} in 30 cm spaced rows on 5th August during 2015 and 2016. Recommended dose (120 kg ha^{-1}) of N was applied using urea (46% N) as source. Thus, all fertilizer treatments received equal amount of N and P-K levels varied across the treatments. The full dose of P-K according to treatments and $^1/_3$ of the N was applied at the time of sowing as basal application, while remaining N was side-dressed at the time of first and second irrigations in equal splits. Three irrigations (each ~7.5 cm) were applied to the crops at both locations.

2.4. Data Collection
2.4.1. Fodder-Related Traits and Yield

Plant heights (cm) of ten randomly selected plants were measured at harvesting by using meter rod and averaged. The crop was harvested 65 days after sowing when the plants attained maximum biomass. Harvested crop bundles were weighed by using bench scales (Model Number: TCS-602) and converted into Mg (mega gram) ha^{-1}. Samples from these bundles were used to calculate the dry matter yield and converted into Mg per hectare.

2.4.2. Quality Attributes of Sorghum Fodder

The whole plant, leaf, and stem fractions of sorghum fodder were run through a fodder cutter and cut into 2 to 3 cm pieces. After cutting, the fodder was mixed and representative samples were drawn. The dry matter contents of fodder were recorded by following the method of Helrich [29]. Oven-dried samples were grinded in a laboratory mill and passed through 4 mm screen [30,31]. Dried fodder samples were used for determination of different quality parameters. The procedures of Association of Official Agricultural Chemists (AOAC) were adopted to determine different fodder quality parameters, i.e., crude protein percentage, crude fiber percentage, ash percentage, ether percentage, acid detergent fiber (ADF) and neutral detergent fiber (NDF) percentage. The cellulose and hemicellulose contents were determined through adopting procedure of Van Soest [32].

2.5. Statistical Procedure

The collected data using standard procedure were verified for normality by Shapiro-Wilk normality test, which directed a normal distribution. The analysis was completed on original data. Data were analyzed following three-way analysis of variance (ANOVA) on SAS software (Version 9.1; SAS Institute, Cary, NC, USA) [33], and means were compared by applying Tukey's honestly significant difference test at 95% probability level where ANOVA showed significant differences [34].

3. Results

Plant height, dry matter and fodder yield of sorghum were significantly affected by P and K application. Plant height significantly differed among years (F = 1431.80, p = 0.000), locations (F = 13356, p = 0.000), P-K levels (F = 2538.43, p = 0.000) and their interactions (Table 2). Plant height was 9% and 25% taller Okara compared to Multan during 2015 and 2016, respectively. The T_6 (80–40 kg ha^{-1}) produced 40.8% taller plants than T_0 (Table 3). Likewise, dry matter yield was significantly influenced by P-K levels (F = 7.70, p = 0.0000) during both years and locations (F = 10.22, p = 0.0022) (Table 2). The P-K levels, particularly T_6 (80–40 kg ha^{-1}), improved fodder yield by 40% compared to control (Table 3). Locations (F = 69.13, p = 0.0000), P-K levels (F = 32.68, p = 0.0000) and their interaction locations × P-K levels (F = 3.31, p = 0.0045) dominated on fodder yield. Higher dry matter yield (11%) was recorded at Okara than Multan. Similarly, higher fodder yield was produced (15.6%) at Okara than Multan. Moreover, T_6 (80–40 kg ha^{-1}) produced 40% higher fodder yield than T_0 (Table 3). Linear regression equation of dry matter yield and fodder yield (dependent) with yield attributes were calculated during both years, location and P-K levels. In case of regression equation (Table 4), dry matter yield showed highly significant dependence (90% and 91%) on plant height, while had non-significant effect on germination. Fodder yield showed significant dependence (83% and 84%) on plant height (Table 4).

Table 2. ANOVA table of growth and yield traits of sorghum during both years and locations.

S. O. V.	DF	Plant Height (cm)		Dry Matter Yield (Mg ha^{-1})		Fodder Yield (Mg ha^{-1})	
		MS	F-Value	MS	F-Value	MS	F-Value
Replication	2						
Years	1	11,305 **	1431.8	0.01 NS	0.01	6.90 NS	0.68
Locations	1	105,457 **	13,356.0	16.43 **	8.29	697.25 **	69.13
P-K levels	7	20,043 **	2538.43	15.25 **	7.70	329.59 **	32.68
Years × Locations	1	1866 **	236.39	20.25 **	10.22	5.44 NS	0.54
Years × P-K levels	7	345 **	43.67	0.16 NS	0.08	9.65 NS	0.96
Locations × P-K levels	7	2992 **	378.94	0.96 NS	0.48	33.39 **	3.31
Years × Locations × P-K levels	7	420 **	53.20	0.46 NS	0.23	10.61 NS	1.05
Error mean square		7.89		1.98		10.08	
General average		227.83		7.28		31.86	
C.V. %		1.23		19.33		9.97	

S.O.V., source of variation; MS, mean squares, NS, non-significant; ** = highly significant; C.V., coefficient of variation.

Table 3. Influence of P-K levels on sorghum growth, yield and yield traits during both years and locations.

Treatments	Plant Height (cm)	Dry Matter Yield (Mg ha^{-1})	Fodder Yield (Mg ha^{-1})
	Years		
2015	238.7 A ± 2.5	7.3 ± 0.7	31.6 ± 1.8
2016	217.0 B ± 3.2	7.3 ± 0.7	32.1 ± 1.8
HSD 5%	1.14	NS	NS
	Locations (Loc)		
Multan	194.7 B ± 2.7	6.8 B ± 0.7	29.2 B ± 1.8
Okara	261.0 A ± 2.9	7.7 A ± 0.7	34.5 A ± 1.8
HSD 5%	1.14	0.57	1.29
	P-K levels (kg ha^{-1})		
T_0 (0–0)	176.3 G ± 2.6	5.4 D ± 0.7	23.0 E ± 1.8
T_1 (40–0)	196.5 F ± 2.8	6.2 CD ± 0.8	28.0 D ± 1.8

Table 3. *Cont.*

Treatments	Plant Height (cm)	Dry Matter Yield (Mg ha^{-1})	Fodder Yield (Mg ha^{-1})
T$_2$ (80–0)	195.0 F ± 3.2	7.0 B–D ± 0.8	30.0 CD ± 1.8
T$_3$ (0–40)	212.8 E ± 2.5	7.5 A–C ± 0.7	31.9 B–D ± 1.8
T$_4$ (0–60)	228.7 D ± 2.9	7.3 A–C ± 0.8	32.4 BC ± 1.8
T$_5$ (40–40)	248.6 C ± 2.7	8.2 AB ± 0.7	33.9 BC ± 1.8
T$_6$ (80–40)	298.0 A ± 2.8	9.1 A ± 0.7	40.9 A ± 1.8
T$_7$ (60–80)	265.8 B ± 3.2	7.6 A–C ± 0.8	34.7 B ± 1.8
HSD 5%	3.59	1.80	4.06
Interactions			
Years × Locations	**	**	NS
Years × P-K levels	**	NS	NS
Locations × P-K levels	**	NS	**
Years × Locations × P-K levels	**	NS	NS

NS, non-significant; ** = highly significant. Different letters in each column shows significant difference at 95% probability (HSD).

Table 4. Multiple linear regression equation of different sorghum yield traits on dry matter yield and fodder yield as affected by years, locations, and P-K levels during 2015 and 2016.

Regression Equation	Adj. (R^2)	R^2	GER	PLH
DMY = −1.958 + 0.004 × PLH + 0.276 × PAL − 0.059 × PAW + 0.028 × NPP + 1.478 × WET	90.6% ***	91.2% ***	NS	***
FDY = −4.940 + 0.288 × PLH − 1.054 × PAL − 0.052 × PAW + 0.207 × NPP + 3.824 × WET	82.9%	84%	NS	***

Dry matter yield, DMY; fodder yield, FDY; plant height, PLH; germination, GER. Significance codes: *** = significant; NS = non-significant.

Different quality traits of sorghum fodder like crude protein and fiber, ash, ether, acid and neutral detergent fiber and cellulose and hemicellulose contents were significantly affected by P-K levels (Table 5). Crude protein was significantly affected by P-K levels (F = 7.50, p = 0.0000) and varied in both years (F = 7.08, p = 0.0098). The P-K levels, particularly T$_6$ (80–40 kg ha^{-1}), increased crude protein contents by 26% in sorghum plants compared to T$_0$, while T$_7$ (60–80 kg ha^{-1}) and T$_5$ (40–40 kg ha^{-1}) were statistically similar with T$_6$ (80–40 kg ha^{-1}) (Table 6). Crude fiber was significantly influenced by locations (F = 4.43, p = 0.0392) and different P-K levels (F = 11.72, p = 0.000). At Multan, 2.98% higher crude fiber contents were recorded than Okara. Significantly higher (17.3%) crude fiber contents were recorded in control when plants were not fertilized as compared to T$_7$ (60–80 kg ha^{-1}), T$_6$ (80–40 kg ha^{-1}) and T$_5$ (40–40 kg ha^{-1}) (Table 6). Ash % was significantly affected by P-K levels (F = 7.44, p = 0.000). Significantly higher ash % (25%) contents were recorded when sorghum plants were fertilized with 80–40 kg ha^{-1} (T$_6$) compared to no fertilization treatment (T$_0$: 0–0 kg ha^{-1}) (Table 6). Significant effect of years (F = 23.92, p = 0.0000), locations (F = 8.11, p = 0.0059), P-K levels (F = 38.05, p = 0.0000) and interaction among years × locations (F = 4.48, p = 0.0382) were recorded on ether %. The T$_6$ (80–40 kg ha^{-1}) had 35% higher ether % than control and other fertilizer levels (Table 6). Statistically, 7% and 5% higher ether % was recorded at Okara compared to Multan during 2015 and 2016, respectively. Significant effect of both acid detergent fiber and neutral detergent fiber was recorded for locations (F = 16.04, p = 0.0002), (F = 20.82, p = 0.0000) and P-K levels (F = 0.22.48, p = 0.9079), (F = 7.90, p = 0.0000). Higher acid and neutral detergent fiber were produced (19% and 8%, respectively) in control, while in T_7 (60–80 kg ha^{-1}) sorghum plants had the lowest acid and neutral detergent fiber (Table 6). Overall, Multan location had higher acid and neutral detergent fiber than Okara (4% and 3%) during 2015 and 2016, respectively. Cellulose and hemicellulose contents were significantly influenced by locations (F = 17.24, p = 0.0001), (F = 10.10, p = 0.0023) and P-K levels (F = 20.25, p = 0.0000), (F = 3.48, p = 0.0032). Sorghum plants fertilized with T_6 (80–40 kg ha^{-1}) recorded 13% and 8% higher cellulose and hemicellulose contents than T_0 (0–0 kg ha^{-1}) (Table 6). Multan location recorded 3% and 4% higher cellulose and hemicellulose contents than Okara, respectively.

Multiple linear regression equation for dry matter and fodder yields (dependent) with fodder quality traits were calculated during both years, location, and P-K levels included in the study. In case of regression equation (Table 7), dry matter yield showed highly significant dependence (86% and 87%) on crude protein, crude fiber, ash content, ether content, acid detergent fiber, neutral detergent fiber, while significant effect on hemicellulose and non-significant on cellulose content. Fodder yield showed highly significant dependence (88% and 89%) on crude fiber, ash content, ether contents, while significant effect on crude protein and cellulose content and non-significant effect on acid detergent fiber, neutral detergent fiber and hemicellulose content (Table 7).

Table 5. ANOVA table of fodder quality traits of sorghum during both years and locations.

S. O. V.	DF	Crude Protein (%)		Crude Fiber (%)		Ash (%)		Ether (%)		Acid Detergent Fiber (%)		Neutral Detergent Fiber (%)		Cellulose Content (%)		Hemicellulose Content (%)	
		MS	F-Value	MS	F-Value	MS	F-Value	MS	F-Value	MS	F-Value	MS	F-Value	MS	F-Value	MS	F-Value
Replication	2																
Years	1	7.40 **	7.08	0.80 NS	0.21	2.43 NS	2.44	0.37 **	23.92	0.66 NS	0.42	0.77 NS	0.63	0.042 NS	0.03	0.05 NS	0.02
Locations	1	0.02 NS	0.03	17.16 *	4.43	0.96 NS	0.96	0.12 **	8.11	25.27 **	16.04	25.89 **	20.82	24.03 **	17.24	24.60 **	10.10
P-K levels	7	7.85 **	7.50	45.38 **	11.72	7.44 **	7.44	0.60 **	38.05	35.42 **	22.48	9.82 **	7.90	28.22 **	20.25	8.46 **	3.48
Years × Locations	1	0.37 NS	0.35	0.006 NS	0.00	0.54 NS	0.55	0.07 *	4.48	0.0001 NS	0.00	0.003 NS	0.00	0.002 NS	0.00	0.0009 NS	0.00
Years × P-K levels	7	0.44 NS	0.43	0.034 NS	0.01	0.068 NS	0.07	0.006 NS	0.38	0.005 NS	0.00	0.017 NS	0.01	0.0002 NS	0.00	0.0007 NS	0.00
Locations × P-K levels	7	0.21 NS	0.21	0.408 NS	0.11	0.205 NS	0.20	0.01 NS	0.86	0.012 NS	0.01	0.005 NS	0.00	0.001 NS	0.00	0.003 NS	0.00
Years × Locations × P-K levels	7	0.17 NS	0.17	0.01 NS	0.00	0.163 NS	0.16	0.002 NS	0.17	0.009 NS	0.01	0.004 NS	0.00	0.001	0.00	0.0004 NS	0.00
Error mean square		1.04		3.87		0.99		0.01		1.57		1.24		1.39		2.43	
General average		8.08		28.00		8.69		1.55		24.86		37.26		29.84		24.32	
C.V. %		12.65		7.03		11.51		8.10		5.05		2.99		3.96		6.41	

NS, non-significant; * = significant; ** = highly significant.

Table 6. Influence of P-K levels on sorghum quality traits during both years and locations.

Treatments	Crude Protein (%)	Crude Fiber (%)	Ash (%)	Ether (%)	Acid Detergent fiber (%)	Neutral Detergent Fiber (%)	Cellulose Content (%)	Hemicellulose Content (%)
Years								
2015	8.4 A ± 0.5	28.1 ± 1.0	8.8 ± 0.5	1.6 A ± 0.07	24.9 ± 0.6	37.4 ± 0.5	29.9 ± 0.6	24.4 ± 0.8
2016	7.8 B ± 0.3	27.9 ± 0.9	8.5 ± 0.6	1.5 B ± 0.08	24.8 ± 0.6	37.2 ± 0.6	29.8 ± 0.6	24.3 ± 0.9
HSD 5%	0.41	NS	NS	0.05	NS	NS	NS	NS
Locations								
Multan	8.1 A ± 0.7	28.4 A ± 0.9	8.8 ± 0.5	1.5 B ± 0.07	25.4 A ± 0.6	37.8 A ± 0.5	30.3 A ± 0.6	24.8 A ± 0.8
Okara	8.1 A ± 0.7	27.6 B ± 0.9	8.6 ± 0.6	1.6 A ± 0.08	24.4 B ± 0.6	36.7 B ± 0.6	29.3 B ± 0.6	23.8 B ± 0.9
HSD 5%	NS	0.80	NS	0.05	0.51	0.45	0.48	0.63
P-K levels (kg ha^{-1})								
T_0 (0–0)	7.0 D ± 0.7	30.8 A ± 1.1	7.6 C ± 0.6	1.2 F ± 0.04	28.0 A ± 0.5	38.8 A ± 0.6	27.7 E ± 0.6	23.3 B ± 0.8
T_1 (40–0)	7.6 CD ± 0.8	30.2 A ± 0.8	8.1 BC ± 0.5	1.3 EF ± 0.05	26.3 B ± 0.6	38.1 AB ± 0.5	27.8 DE ± 0.6	23.4 AB ± 0.8
T_2 (80–0)	7.8 B–D ± 0.7	29.7 AB ± 0.9	8.6 BC ± 0.5	1.5 DE ± 0.1	24.8 BC ± 0.6	37.5 A–C ± 0.6	29.3 CD ± 0.6	23.7 AB ± 0.8
T_3 (0–40)	7.5 CD ± 0.7	26.9 C ± 0.9	8.3 BC ± 0.5	1.5 DE ± 0.7	24.7 C ± 0.6	37.1 B–D ± 0.5	30.3 BC ± 0.6	24.4 AB ± 0.8
T_4 (0–60)	7.9 B–D ± 0.7	27.5 BC ± 1.1	8.5 BC ± 0.6	1.6 CD ± 0.1	25.4 BC ± 0.7	37.2 B–D ± 0.5	29.7 C ± 0.6	24.1 AB ± 0.9
T_5 (40–40)	8.5 A–C ± 0.7	27.0 C ± 0.8	9.2 AB ± 0.6	1.7 BC ± 0.1	23.9 CD ± 0.7	36.9 B–D ± 0.6	30.8 A–C ± 0.6	25.0 AB ± 0.9
T_6 (80–40)	9.4 A ± 0.6	26.6 C ± 0.9	10.1 A ± 0.6	1.9 A ± 0.07	23.1 D ± 0.6	36.6 CD ± 0.6	31.9 A ± 0.6	25.4 A ± 0.9
T_7 (60–80)	9.0 AB ± 0.6	25.4 C ± 0.9	9.1 AB ± 0.6	1.8 AB ± 0.5	22.8 D ± 0.6	35.9 D ± 0.6	31.3 AB ± 0.6	25.3 A ± 0.8
HSD 5%	1.30	2.51	1.27	0.16	1.60	1.42	1.50	1.99
Interactions								
Years × Locations	NS	NS	NS	*	NS	NS	NS	NS
Years × P-K levels	NS	NS	NS	NS	NS	NS	NS	NS
Locations × P-K levels	NS	NS	NS	NS	NS	NS	NS	NS
Years × Locations × P-K levels	NS	NS	NS	NS	NS	NS	NS	NS

NS, non-significant; * = significant. Different letters in each column shows significant difference at 95% probability (HSD).

Table 7. Multiple linear regression equation of different sorghum quality traits on dry matter yield and fodder yield as affected by years, locations and P-K levels during 2015 and 2016.

Regression Equation	Adj. (R^2)	R^2	CPR	CRF	ASH	ETH	ADF	NDF	CLC	HCC
DMY = 25.530 + 0.755 × CRP − 0.778 × CRF + 0.793 × ASH−4.138 × ETH−0.749 × ADF + 1.079 × NDF − 0.194×CLC − 0.771 × HCC	86.2% ***	87.4% ***	***	***	**	***	**	**	NS	.
FDY = 205.490 + 1.335 × CRP − 2.282 × CRF + 5.987 × ASH − 12.792 × ETH − 1.020 × ADF − 0.799 × NDF − 1.790 × CLC − 1.811 × HCC	87.6% ***	88.7% ***	.	***	***	**	NS	NS	.	NS

Dry matter yield, DMY; fodder yield, FDY; crude protein, CRP; crude fiber, CRF; ash content, ASH; ether content, ETH; acid detergent fiber, ADF; neutral detergent fiber, NDF; cellulose content, CLC; hemicellulose content, HCC. Significance codes: 0 = "***"; 0.001 = "**"; 0.05 = "."; 0.1 "NS".

4. Discussion

Application of P-K in Aridisol significantly improved fodder yield and quality of sorghum (Tables 2–7). The substantial improvement (44%) in sorghum plant height with the application of 80–40 kg ha^{-1} P-K (Table 3) revealed the balanced availability of nutrients to the plants [35–37]. Moreover, better and efficient utilization of nutrients might bring variation within P-K levels, leading to improved plant height, which contributed towards higher fodder yield [38]. Optimum P-K level (i.e., 80–40 kg ha^{-1}) resulted in the improvement of physiological activities that resulted in improved plant height [23]. Furthermore, P-K have significant role in crop growth; therefore, biomass or plant size was increased [39]. Higher production of amino acids and growth-promoting chemicals within plants leads towards improvement in meristematic activities such as cell division, enlargement, and elongation, which resulted in taller plant height [40]. Suitable application of K leads to the synthesis of proteins, opening and closing of stomata and osmotic adjustments [26]. The shorter plant height in control treatment might be due to nutrient deficiency leading to weak plant metabolism without fertilizers [41].

Availability of balanced nutrients improves meristematic and physiological activities [36,42–44]. This resulted in increased resource use efficiency and more dry matter accumulation per unit area/time as well as higher yield. Being a C$_4$ plant, sorghum has deep and extensive root system, which helped in more uptakes of P and K leading to more plant growth [45,46]. Soil application of 80–40 kg P-K ha^{-1} increased 40% and 44% dry mater yield and fodder yield, respectively (Table 3). Balanced nutrient application might promote enzyme activities which triggered sorghum growth and development, consequently improving the yield of sorghum [44]. Higher dry matter and fodder yield in T$_6$ (80–40 kg ha^{-1}) is a result of better soil nutrients suitable for nutrient uptake and accessibility, which resulted in accelerated cell division, enlargement and elongation. Sufficient nutrients availability flourished growth and resulted in higher fodder yield [45]. This supports our hypothesis that balanced fertilization and favorable weather conditions improved availability and uptake of plant nutrients, particularly in silty-clay-loam soil, which enhanced meristematic and physiological activities [36]. High temperature during 2016 might impacted the quantity and quality traits of sorghum, as well as large-scale fodder production pattern (Table 1). Our results corroborate the findings of Mathur and Jajaoo [46], that higher temperature had adverse effects in metabolism stability and reactions within cells which disturb metabolism in plants, such as crude protein, crude fiber, ash %, ether %, acid and neutral detergent fiber and cellulose and hemicellulose contents. Moreover, high temperature and change in sowing dates decreased nutrient availability and protein concentrations that leads to lower nutritional quality of edible portions of food and forage crops [35]. Furthermore, high temperature can decrease nutritional quality of crops through a reduction in protein, K and calcium levels [47] (Table 6). Our results showed that less rainfall and high temperature (Table 1) significantly reduced forage quality and crude protein content (Table 6) [48].

Forage quality is mainly determined by crude protein and fiber [49]. Soil application of 80–40 kg ha^{-1} P-K improved crude protein (26%) compared to the rest of the treatments (Table 6). In our results, significant improvements were recorded in crude protein that might be linked with the availability of P that played an important role in development of structural component like DNA and RNA that has dominant role in protein synthesis [50]. Phosphorus is main constituent of ATP that is required in many metabolic processes that leads to photosynthetic and protein activity [50]. Moreover, higher crude protein-containing fodder is considered a good quality fodder that is due to improvement in the proximate compositions at different developmental stages [51]. The weak plants had lower ability to develop potential and adapt mechanisms, which enhanced the P and K uptake under nutrients starvation [52]. Our study further supported the results of Jégo et al. [53], indicating that high temperature along with lower nutrients' availability reduces plants growth and contents of crude protein and increase crude fiber and acid and neutral detergent fiber contents. Soil applied 80–40 kg ha^{-1} P-K showed 14% lower crude fiber than control and other treatments that have lower P-K levels (Table 6). It is proven that fodder having lower crude fiber is superior in quality and vice versa and most of the studies supports our hypothesis that lower and control treatments had higher crude fiber and lower crude protein content and vice versa [51]. However, contradictory findings were reported by Eltelib and Eltom [54] that higher nutrients' levels increased crude fiber, whereas current results showed that higher crude fiber was measured at control and lower nutrients levels (Table 5). Soil applied 80–40 kg ha^{-1} P-K significantly improved ash % and ether % in sorghum than control and other treatments (Table 6). Our findings are further supported by earlier results [55] indicating maximum total ash and ether contents content at higher rate of P and K fertilizer levels (80–40 kg ha^{-1}) (Table 6). Acid detergent fiber (ADF) and neutral detergent fiber (NDF) were lowered by 18% and 8% at 80–40 kg ha^{-1} P-K application than control and other treatments (Table 6) which might be due to higher lignin contents. These results supported our hypothesis that balanced nutrients application significantly influences forage quality [11]. Moreover, plant cell walls constituents (i.e., ADF, NDF cellulose, hemicelluloses, lignin, and tannins), which represent crude fiber content in forage, have a large influence on fodder digestibility.

Several studies reported that sorghum plants grown under balanced fertilization and harvested at mature stage have higher crude fiber than plants harvested at the booting stage [15,56] (Table 6). On the contrary, Atis et al. [57] revealed that lignin content tended to increase with prolonged maturity from booting stage to physiological maturity stage. The increase in fodder crude fiber with development of growing stage may be due to increased concentration of cell wall constituents within stem and leaves as well as decreased soluble proportion of the cell [56]. This could be due to lignin accumulation and synthesis during secondary cell wall development [11]. Furthermore, balanced fertilization resulted in lower concentration of acid and neutral detergent fiber and increased in lignin, which reduced the digestibility of the plant [58].

5. Conclusions

Soil incorporated 80–40 kg ha^{-1} P-K noticeably improved the growth and fodder yield of better quality of sorghum cultivated at both locations (i.e., Multan and Okara) in Aridisol in both years. The results reveal that sufficient nutrient availability is necessary for improving the yield and fodder quality of sorghum. However, the nutrients' requirement may vary according to soil type and fertilizer levels must be selected based on the fertility status of the soils. For Aridisol, 80–40 P-K kg ha^{-1} significantly improved the yield and quality of fodder sorghum. Thus, this is recommended for such soils. Nevertheless, long-term studies are needed to obtain more reliable results regarding yield and soil health.

Author Contributions: Conceptualization, A.-u.-R., M.S.A., R.A.-Y. and M.H.; formal analysis, A.-u.-R., R.Q. and M.M.A.; funding acquisition, A.-u.-R., M.S.A. and R.A.-Y.; methodology, A.-u.-R. and R.Q.; project administration, A.-u.-R.; validation, R.Q.; visualization, R.Q.; writing—original draft, A.-u.-R. and M.M.A.; writing—review and editing, R.Q., M.M.A., M.S.A., R.A.-Y. and M.H. All authors have read and agreed to the published version of the manuscript.

Funding: This research was supported by Bahauddin Zakariya University Multan. The authors extend their appreciation to the Researchers supporting project number (RSP-2021/173), King Saud University, Riyadh, Saudi Arabia.

Institutional Review Board Statement: Not applicable.

Informed Consent Statement: Not applicable.

Data Availability Statement: All data are within the manuscript.

Acknowledgments: We acknowledge Bahauddin Zakariya University, Multan, Pakistan for financial assistance during the study. The authors extend their appreciation to the Researchers supporting project number (RSP-2021/173), King Saud University, Riyadh, Saudi Arabia.

Conflicts of Interest: The authors declare no conflict of interest. The funders had no role in the design of the study; in the collection, analyses, or interpretation of data; in the writing of the manuscript; or in the decision to publish the results.

References

1. Amandeep, S. Forage quality of sorghum (*Sorghum bicolor* L.) as influenced by irrigation, N levels and harvesting stage. *Indian J. Sci. Res.* **2012**, *3*, 67–72.
2. Government of Pakistan. *Economic Survey of Pakistan, 2019–2020*; Government of Pakistan, Ministry of Food, Agriculture and Livestock, Agriculture & Live-stock Division (Economic Wing): Islamabad, Pakistan, 2020; p. 121.
3. Kumar, P.S.; Tomer, S.K.; Sengar, S.S. Effect of wheat straw based complete diets on feed intake and nutrient utilization in buffaloes. *Indian J. Anim. Nutr.* **2000**, *17*, 341–343.
4. Khan, S.; Bhatti, M.B.; Thorp, T.K. *Improved Fodders and Seed Production in the Northern Areas*; NARC: Islamabad, Pakistan, 2003.
5. Maiti, R.K. *Sorghum Science*; Science Publishers: Lebanon, NH, USA, 1996.
6. Warner, R.M.; Erwin, J.E. Naturally occurring variation in high temperature induced floral bud abortion across Arabidopsis thaliana accessions. *Plant Cell Environ.* **2005**, *28*, 1255–1266. [CrossRef]
7. Stocker, T.F.; Qin, D.; Plattner, G.K.; Tignor, M. *IPCC: Workshop Report of the Intergovernmental Panel on Climate Change Workshop on Regional Climate Projections and Their Use in Impacts and Risk Analysis Studies*; IPCC Working Group I Technical Support Unit; University of Bern: Bern, Switzerland, 2015.
8. Gray, S.B.; Brady, S.M. Plant developmental responses to climate change. *Dev. Biol.* **2016**, *419*, 64–77. [CrossRef] [PubMed]
9. Fanzo, J.; Davis, C.; McLaren, R.; Choufani, J. The effect of climate change across food systems: Implications for nutrition outcomes. *Glob. Food Sec.* **2018**, *18*, 12–19. [CrossRef]
10. Pretty, J.; Benton, T.G.; Bharucha, Z.P.; Dicks, L.V.; Flora, C.B.; Godfray, H.C.J.; Wratten, S. Global assessment of agricultural system redesign for sustainable intensification. *Nat. Sustain.* **2018**, *8*, 441–446. [CrossRef]
11. Carmi, A.; Umiel, N.; Hagiladi, A.; Yosef, E.; Ben-Ghedalia, D.; Miron, J. Field performance and nutritive value of a new forage sorghum variety 'Pnina' recently developed in Israel. *J. Sci. Food Agric.* **2005**, *85*, 2567–2573. [CrossRef]
12. Glamoclija, D.; Jankovic, S.; Rakic, S.; Maletic, R.; Ikanovic, J.; Lakic, Z. Effects of nitrogen and harvesting time on chemical composition of biomass of Sudan grass, fodder sorghum, and their hybrid. *Turk. J. Agric. For.* **2011**, *35*, 127–138.
13. Azam, M.E.A.; Waraich, A.; Pervaiz, F. Response of a newly developed fodder sorghum (*Bicolor* L. Monech) variety (f-9917) to NPK application. *Pak. J. Life Soc. Sci.* **2010**, *8*, 117–120.
14. Ball, D.M.; Collins, M.; Lacefield, G.D.; Martin, N.P.; Mertens, D.A.; Olson, K.E.; Putnam, D.H.; Understander, D.J.; Wolf, M.W. *Understanding Forage Quality*; American Farm Bureau Federation Publication: Park Ridge, IL, USA, 2001; pp. 1–10.
15. Tariq, M.; Ayub, M.; Elahi, M.; Ahmad, A.H.; Chaudhary, M.N.; Nadeem, M.A. Forage yield and some quality attributes of millet (*Pennisetum americannum* L.) hybrid under various regimes of nitrogen fertilization and harvesting dates. *Afr. J. Agric. Res.* **2011**, *6*, 3883–3890.
16. Meena, B.S.; Nepalia, V. Nutrient management for higher fodder production of sorghum: A review. *Int. J. Chem. Stud.* **2018**, *6*, 1524–1528.
17. Wandruszka, R.V. Phosphorus Retention in Calcareous Soils and the Effect of Organic Matter on Its Mobility. 2006. Available online: http://www.geochemicaltransactions.com/content/7/1/6 (accessed on 12 January 2022).
18. Wortmann, C.S.; Mamo, M. Starter fertilizer and row cleaning did not affect yield of early planted, no-till grain sorghum. *Crop Manag.* **2006**, *5*, 1–5. [CrossRef]
19. Lehman, R.M.; Taheri, W.I.; Osborne, S.L. Fall cover cropping can increase arbuscular mycorrhizae in soils supporting intensive agricultural production. *Appl. Soil. Ecol.* **2012**, *61*, 300–304. [CrossRef]

20. Chattha, M.B.; Iqbal, A.; Chattha, M.U.; Hassan, M.U.; Khan, I.; Ashraf, I.; Faisal, M.; Usman, M. PGPR inoculated-seed increases the productivity of forage sorghum under fertilized conditions. *J. Basic Appl. Sci.* **2017**, *13*, 150–153. [CrossRef]

21. NFDC. *The Impact of Deregulation on the Fertilizer Sector and Crop Productivity in Pakistan*; NFDC (National Fertilizer Development Centre): Islamabad, Pakistan, 2001.

22. Plaxton, W.C.; Tran, H.T. Metabolic adaptations of phosphate-starved plants. *Plant Physiol.* **2011**, *156*, 1006–1015. [CrossRef]

23. Roy, P.R.S.; Khandaker, Z.H. Effect of phosphorus fertilizer on yield and nutritional value of sorghum (*Sorghum bicolor*) fodder at tree cuttings. *Bang. J. Anim. Sci.* **2010**, *39*, 106–115. [CrossRef]

24. Wang, X.; Hoogmoed, W.B.; Cai, D.; Perdok, U.D.; Oenema, O. Crop residue, manure and fertilizer in dryland maize under reduced tillage in northern China: Nutrient balances and soil fertility. *Nutr. Cycl. Agroecosyst.* **2007**, *79*, 17–34. [CrossRef]

25. Lal, K.; Swarup, A.; Singh, K.N. Potassium balance and release kinetics of non-exchangeable K in a typic natrustalf as influenced by long-term fertilizer use in rice-wheat cropping system. *Agrochimica* **2007**, *51*, 95–104.

26. Oosterhuis, D.M.; Loka, D.A.; Kawakami, E.M.; Pettigrew, W.T. Chapter III the Physiology of Potassium in Crop Production. *Adv. Agron.* **2014**, *126*, 203–233.

27. Tonitto, C.; Ricker-Gilbert, J.E. Nutrient management in African sorghum cropping systems: Applying meta-analysis to assess yield and profitability. *Agron. Sustain. Dev.* **2016**, *36*, 10. [CrossRef]

28. Akram, M.S.; Ashraf, M.; Shahbaz, M.; Akram, N.A. Role of foliar applied potassium from different sources on physio-biochemical attributes of sunflower (*Helianthus annuus* L.) under NaCl stress. *J. Plant Nutr. Soil Sci.* **2009**, *172*, 884–893.

29. Helrich, K. *Official Methods of Analysis*, 15th ed.; Association of Official Analytical Chemists International (AOAC): Washington, DC, USA, 1990; pp. 1028–1039.

30. Harris, L.E. *Nutrition Research Techniques for Domestic and Wild Animals*; Utah State University: Logan, UT, USA, 1970.

31. Van Soest, P.J.; Robertson, J.B. *Analysis of Forages and Fibrous Foods*; Cornell University: New York, NY, USA, 1985.

32. Soest, P.V.; Wine, R.H. Use of detergents in the analysis of fibrous feeds. IV. Determination of plant cell-wall constituents. *J. AOAC Int.* **1967**, *50*, 50–55. [CrossRef]

33. SAS Institute. *SAS/STAT 9.1 User's Guide the Regular Procedure, (Book Excerpt)*; SAS Institute: Cary, NC, USA, 2008.

34. Steel, R.G.D.; Torrie, J.H.; Dickey, D.A. *Principles and Procedures of Statistic: A Biometrical Approach*, 3rd ed.; Mc Graw Hill Book Co., Inc.: New York, NY, USA, 1997; pp. 400–428.

35. Rehman, A.; Aurangzeb; Qamar, R.; Rehman, A.; Shoaib, M.; Qamar, J.; Hassan, F. Quality response of maize fodder cultivars to harvest time. *Asian J. Agri. Biol.* **2017**, *5*, 165–172.

36. Akongwubel, A.O.; Ewa, U.B.; Prince, A.; Jude, O.; Martins, A.; Simon, O.; Nicholas, O. Evaluation of agronomic performance of maize (*Zea mays* L.) under different rates of poultry manure application in an ultisol of Obubra, cross river state, Nigeria. *Int. J. Res. Agric. For.* **2012**, *2*, 138–144.

37. Yang, W. Effect of nitrogen, phosphorus and potassium fertilizer on growth and seed germination of (*Capsella bursa-pastoris* L.) medikus. *J. Plant Nutr.* **2018**, *41*, 636–644. [CrossRef]

38. Niamatullah, M.; Khan, M.; Khan, M.Q.; Sadiq, M.; Zaman, K.U.; Hayat, C.S.; Rehman, S. Impact of NPK applications on the number of productive tillers and cost benefit analysis of wheat in hill-torrent irrigated area of D.I. Khan Division, Khyber Pakhtoonkhwa. *J. Anim. Plant Sci.* **2011**, *21*, 211–214.

39. Bothe, D.T.; Sable, R.N.; Raundal, P.U. Effect of Phosphorus, plant population and P solubilizer on soybean fenugreek cropping system. *J. Maharastra Agri. Univ.* **2000**, *25*, 310–311.

40. Mubeena, P.; Halepyati, A.S.; Chittapur, B.M. Effect of date of sowing and nutrient management on nutrient uptake and yield of foxtail millet (*Setaria italica* L.). *Int. J. Bio-Resour. Stress* **2019**, *10*, 92–95. [CrossRef]

41. Singh, S.; Shukla, G.P. Genetic diversity in the nutritive value of dual-purpose sorghum hybrids. *Anim. Nutr. Feed Technol.* **2010**, *10*, 93–100.

42. Buah, S.S.; Polito, T.A.; Killorn, R. No-tillage soybean response to banded and broadcast and direct and residual fertilizer phosphorus and potassium applications. *Agron. J.* **2000**, *92*, 657–662. [CrossRef]

43. Alvarez, V.H.; Novais, R.F.; Dias, L.E.; Oliveira, J.A. Determination and use of residual phosphorus. *Bol. Inf. Soc. Bras. Cienc. Do Solo* **2000**, *25*, 27–32.

44. Sweeney, D.W.; Moyer, J.L.; Jardine, D.J.; Whitney, D.A. Nitrogen, phosphorus, and potassium effects on grain sorghum production and stalk rot following alfalfa and birds foot trefoil. *J. Plant Nutr.* **2011**, *34*, 1330–1340. [CrossRef]

45. Sumeriya, H.K.; Singh, P. Productivity of sorghum (*Sorghum bicolor*) genotypes as influenced by different fertility levels and their residual effect on succeeding wheat. *Ann. Biol.* **2014**, *30*, 266–275.

46. Mathur, S.; Jajoo, A. Effects of heat stress on growth and crop yield of wheat (*Triticum aestivum*). In *Physiological Mechanisms and Adaptation Strategies in Plants under Changing Environment*; Springer: Berlin, Germany, 2014; pp. 163–191.

47. Moretti, C.L.; Mattos, L.M.; Calbo, A.G.; Sargent, S.A. Climate changes and potential impacts on postharvest quality of fruit and vegetable crops: A review. *Food Res. Int.* **2010**, *43*, 1824–1832. [CrossRef]

48. Xu, Z.Z.; Zhou, G.S. Combined effects of water stress and high temperature on photosynthesis, nitrogen metabolism and lipid peroxidation of a perennial grass *Leymus chinensis*. *Planta* **2006**, *224*, 1080–1090. [CrossRef] [PubMed]

49. Rashid, M.; Iqbal, M. Response of sorghum (*Sorghum bicolor* L.) fodder to phosphorous fertilizer on Torripsamment soil. *J. Anim. Plant Sci.* **2011**, *21*, 220–225.

50. Abd El-Lattief, E.A. Growth and fodder yield of forage pearl millet in newly cultivated land as affected by date of planting and integrated use of mineral and organic fertilizers. *Asian J. Crop Sci.* **2011**, *3*, 35–42. [CrossRef]
51. Shobha, V.; Kasturiba, B.; Naik, R.K.; Yenagi, N. Nutritive value and quality characteristics of sorghum genotypes. *Karnataka J. Agric. Sci.* **2008**, *20*, 586–588.
52. Wissuwa, M.; Gamat, G.; Ismail, A.M. Is root growth under phosphorus deficiency affected by source or sink limitations? *J. Exp. Bot.* **2005**, *56*, 1943–1950. [CrossRef]
53. Jégo, G.; Bélanger, G.; Tremblay, G.F.; Jing, Q.; Baron, V.S. Calibration and performance evaluation of the STICS crop model for simulating timothy growth and nutritive value. *Field Crops Res.* **2013**, *151*, 65–77. [CrossRef]
54. Eltelib, H.A.; Eltom, E.A. Effect of time of nitrogen application on growth, yield and quality of four forage sorghum cultivars. *Agric. J.* **2006**, *1*, 59–63.
55. Roy, D.C.; Das, H.; Pakhira, M.C.; Bera, S. Growth, green forage yield and quality of sorghum (*Sorghum bicolor*) fodder as influenced by application of phosphorus fertilizer. *Biosci. Trends* **2015**, *8*, 1712–1718.
56. Qu, H.; Liu, X.B.; Dong, C.F.; Lu, X.Y.; Shen, Y.X. Field performance and nutritive value of sweet sorghum in eastern China. *Field Crops Res.* **2014**, *157*, 84–88. [CrossRef]
57. Atis, I.; Konuskan, O.; Duru, M.; Gozubenli, H.; Yilmaz, S. Effect of harvesting time on yield, composition and forage quality of some forage sorghum cultivars. *Int. J. Agric. Biol.* **2012**, *14*, 879–886.
58. Huang, H.; Faulkner, D.B.; Singh, V.; Danao, M.C.; Eckhoff, S.R. Effect of harvest date on yield, composition and nutritive value of corn stover and DDGS. *Trans. ASABE* **2012**, *55*, 1859–1864. [CrossRef]

Article

Osmopriming Combined with Boron-Tolerant Bacteria (*Bacillus* sp. MN54) Improved the Productivity of Desi Chickpea under Rainfed and Irrigated Conditions

Noman Mehboob [1], Tauqeer Ahmad Yasir [2], Shahid Hussain [3], Shahid Farooq [4], Muhammad Naveed [5] and Mubshar Hussain [1,6,*]

1. Department of Agronomy, Bahauddin Zakariya University, Multan 60800, Pakistan
2. College of Agriculture, Bahadur Sub-Campus, Bahauddin Zakariya University, Layyah 32200, Pakistan
3. Department of Soil Science, Bahauddin Zakariya University, Multan 60800, Pakistan
4. Department of Plant Protection, Faculty of Agriculture, Harran University, Sanliurfa 63050, Turkey
5. Institute of Soil and Environmental Sciences, University of Agriculture Faisalabad, Faisalabad 38000, Pakistan
6. School of Veterinary and Life Sciences, Murdoch University, 90 South Street, Murdoch, WA 6150, Australia
* Correspondence: mubashir.hussain@bzu.edu.pk; Tel.: +92-301-6174879

Citation: Mehboob, N.; Yasir, T.A.; Hussain, S.; Farooq, S.; Naveed, M.; Hussain, M. Osmopriming Combined with Boron-Tolerant Bacteria (*Bacillus* sp. MN54) Improved the Productivity of Desi Chickpea under Rainfed and Irrigated Conditions. *Agriculture* 2022, 12, 1269. https://doi.org/10.3390/agriculture12081269

Academic Editor: Riccardo Scotti

Received: 20 June 2022
Accepted: 17 August 2022
Published: 20 August 2022

Publisher's Note: MDPI stays neutral with regard to jurisdictional claims in published maps and institutional affiliations.

Abstract: Chickpeas are rich source of protein and predominantly grown in boron (B)-deficient sandy-loam soils in Pakistan. Boron-tolerant bacteria (BTB) could tolerate higher B levels in soil and increase B availability to the plants. Field trials were conducted under irrigated (district Layyah) and rainfed (district Chakwal) conditions to evaluate the interactive effects of pre-optimized B application methods and BTB (*Bacillus* sp. MN54) on the nodule's population, grain quality, productivity, and grain-B concentration in desi chickpea during 2019–2020 and 2020–2021. Boron was applied as soil application (1 kg B ha^{-1}), foliar application (0.025% B), osmopriming (0.001% B), and seed coating (1.5 g B kg^{-1} seed) with or without BTB inoculation. Untreated seeds receiving no B through any of the methods were regarded as control. The individual and interactive effects (up to three-way interaction of location × BTB inoculation × B application methods) of year, location, B application methods and BTB inoculation significantly altered the growth and yield-related traits of desi chickpea. The four-way interaction of year × location × BTB inoculation × B application methods was non-significant for all recorded growth and yield-related traits. Regarding individual effects, the higher values of growth and yield-related traits were noted for 2020–2021, rainfed location, BTB inoculation and B application through seed priming. Similarly, in two-way interactions 2020–2021 with rainfed location and BTB inoculation, rainfed location with BTB inoculation and osmopriming and osmopriming with BTB inoculation recorded higher values of the growth and yield-related traits. Osmopriming combined with BTB inoculation significantly improved dry matter accumulation and leaf area index in both locations. Boron application through all the methods significantly improved grain quality, yield grain B concentration. The highest grain and biological yields, and nodules' population were recorded with osmopriming followed by soil application of B combined with BTB inoculation. The highest plant B concentration (75.05%) was recorded with foliar application of B followed by osmopriming (68.73%) combined with BTB inoculation. Moreover, the highest economic returns (USD 2068.5 ha^{-1}) and benefit–cost ratio (3.7%) were recorded with osmopriming + BTB inoculation in 2020–2021 under rainfed conditions. Overall, B application through osmopriming and soil application combined with BTB inoculation could be used to increase productivity and profitability of desi chickpea, whereas foliar application is a better method to enhance grain and plant B concentration.

Keywords: osmopriming; grain yield; grain B concentration; boron-tolerant bacteria

1. Introduction

Balanced nutrition is essential for optimum growth and quality seed production in crop plants. Micro- and macronutrients are equally important for better crop growth and productivity [1]. The farmers seldom use recommended quantities of micronutrients and mostly focus on primitive methods of macronutrients' application which result in nutritional imbalance [2]. Micronutrients are less mobile within plants; therefore, must be supplied for better growth and productivity [3,4]. Boron is a micronutrient and its intake in daily diet is permissible up to 20 mg day^{-1} in adults and 1.5–3 mg day^{-1} in children under the age of three years. Boron controls many diseases, including memory loss, and low testosterone level.

Boron is required in minute quantity; thus, its toxicity and deficiency both are detrimental for plants. Soils are mostly deficient in B as compared to other micronutrients, as B-deficiency is observed in ~80 countries across the globe, including Pakistan. It has been observed that 65% of Pakistani soils are B-deficient on which 132 plant species are being cultivated [5–7].

Like other plant species, B is important for chickpea (*Cicer arietinum* L.) as it plays significant role in cell wall synthesis, cell division, pod formation, seed setting, nodulation and improving seed quality [8–10]. Insufficient B supply restricts seedling growth, diminishes fruit quality, decreases number of grains per spike and increases anthesis which alters seed setting [11,12]. Excessive B supply reduces chlorophyll contents and chloroplast in leaves, and results in water loss during stomatal limitations, ultimately lowering photosynthetic capacity [13–16]. Chickpea is a leguminous crop and Pakistan ranks 4th in term of production in the world. Chickpea is cultivated on both irrigated and rainfed lands in Pakistan [17]. Two types of chickpeas (i.e., desi and kabuli) are cultivated in the world. Desi type has small size and angular shape, which is cultivated on ~80–85% of the area in the Indo-Pak subcontinent [18].

Biofortification is the process of improving micronutrients' density in crop plants through improved agronomic practices, conventional breeding, and modern biotechnological approaches. Agronomic biofortification is gaining importance, in which nutrient status of crop plants can be enhanced by using different application methods such as seed treatment (priming and coating), foliar and soil application [19]. Plants can easily uptake nutrients through soil application, and it is one of the primitive and beneficial methods of biofortification. It has been observed that soil-applied B along with *Bacillus* sp. strain MN54 considerably increased roots' growth, nodulation, grain production and nutritional quality of chickpea [20]. Foliar feeding is the direct application of nutrients on the plant canopy, especially leaves, to improve grain yield and nutritional value of crop [21,22]. It has been reported foliar spray of B at booting stage produced higher grain yield in several cereal crops [12,23].

Seed coating is the most economical method of nutrient application directly to seeds which involves adhering the required nutrients to the seed surface [24,25]. Boron seed coating in chickpea improved grain yield by 25%, whereas B seed coating in tandem with inoculation of boron-tolerant bacteria (BTB) MN54 increased grain yield by 37% compared to untreated seeds [26]. Seed priming is a controlled hydration technique that involves soaking of seeds in liquid osmotica for a specific time and then seeds are re-dried to their original weight [27–29]. Combined application of B seed priming and bacterial inoculation has been found helpful to improve early stand establishment, seedling emergence, grains' productivity and grain biofortification of different crops [30–32].

Microbial inoculation has been extensively used to improve the mobility of micronutrients in the soil and their availability to the plants, which correspondingly avoid the excessive use of chemical fertilizers [33]. The BTB has the ability to survive under harsh conditions, tolerate B-toxicity and deficiency, which promote crop growth and grain production [34]. Recent studies clearly showed that combined application of *Bacillus* spp. MN-54 and B through seed coating [26], soil application [20] and seed priming [32] improved B uptake and productivity of chickpea. The BTB enables plants to survive under higher B level by controlling its dynamics. Under harsh environmental and nutritional conditions, survival rate of normal bacteria is very low; however, BTB could survive under these harsh conditions and lower toxic effects of B.

Several studies have been conducted to evaluate the positive effects of B fertilization along with bacterial strains' inoculation on different crops; however, information regarding combined application of pre-optimized B levels for different application methods and BTB strains on the productivity of desi chickpea has not reported yet. We optimized the B application levels for different methods in pot experiments in earlier studies and then tested these optimized levels under field conditions in the current study. Thus, the main objectives of this study were to assess the most suitable B application method along with BTB inoculation to improve growth, grain productivity, quality and biofortification of desi chickpea under rainfed and irrigated conditions. It was hypothesized that combined application of BTB and B through different methods would increase the growth, nodulation, grain quality, productivity and biofortification of chickpea. It was further hypothesized that osmopriming combined with BTB inoculation will improve productivity, whereas foliar application of B coupled with BTB inoculation would improve plant and grain B concentration.

2. Materials and Methods

2.1. Experimental Site

Field experiments were conducted at irrigated (Bahauddin Zakariya University, Bahadur Sub Campus, Layyah, 30°57′36.0″ N, 70°55′48.0″ E) and rainfed (Research Farm Chakwal, 32°55′48.0″ N, 72°51′36.0″ E) conditions during 2019–2020 and 2020–2021. The B levels used in different methods in the current study were selected based on earlier studies [20,26,32]. Before sowing, the soils of the experimental sites were analyzed for physical and chemical properties which are given in Table 1.

Table 1. Soil physical and chemical properties of experiment locations prior to the initiation of experiments.

Soil Properties	2019–2020		2020–2021	
	Rainfed Area	**Irrigated Area**	**Rainfed Area**	**Irrigated Area**
Soil texture	Sandy loam	Sandy loam	Sandy loam	Sandy loam
EC	$1.12\ \mathrm{dS\ m^{-1}}$	$3.52\ \mathrm{dS\ m^{-1}}$	$1.14\ \mathrm{dS\ m^{-1}}$	$3.48\ \mathrm{dS\ m^{-1}}$
pH	7.35	8.00	7.50	8.20
Saturation	36.00%	30.00%	36.00%	32.00%
Organic matter	0.41%	0.55%	0.35%	0.53%
Available phosphorous	$3.23\ \mathrm{mg\ kg^{-1}}$	$5.43\ \mathrm{mg\ kg^{-1}}$	$3.32\ \mathrm{mg\ kg^{-1}}$	$4.92\ \mathrm{mg\ kg^{-1}}$
Available potassium	$78\ \mathrm{mg\ kg^{-1}}$	$132\ \mathrm{mg\ kg^{-1}}$	$74\ \mathrm{mg\ kg^{-1}}$	$134\ \mathrm{mg\ kg^{-1}}$
Available nitrogen	0.032%	0.053%	0.041%	0.049%
Available boron	$0.32\ \mathrm{mg\ kg^{-1}}$	$0.45\ \mathrm{mg\ kg^{-1}}$	$0.35\ \mathrm{mg\ kg^{-1}}$	$0.42\ \mathrm{mg\ kg^{-1}}$

2.2. Planting Material and Treatment Details

The best performing desi chickpea variety 'Punjab-2008' based on an earlier optimization study was used in the current study due to its better performance. All possible combinations of two inoculation levels (control and inoculated) and B application methods (seed priming, seed coating, foliar and soil application) were tested under two locations (irrigated and rain-fed). Seeds were coated with $1.5\ \mathrm{g\ B\ kg^{-1}}$ by using Arabic gum as the adhering agent.

For soil application, $1\ \mathrm{kg\ B\ ha^{-1}}$ was applied in each plot through side dressing. For osmopriming, seeds were soaked in aerated B solution (0.001% B), while seeds were soaked in distilled water only for hydropriming. In foliar application, 0.025% B solution was sprayed 55 days after sowing at N-node stage (unfolded leaf and flat leaflets before flowering) on plant leaves. Water spray with distilled water, untreated dry seeds and hydro-priming were taken as control to compare with B application treatments. Boric acid (CAS No. 10043-35-3, a product of Merck, Darmstadt, Germany) was used as B source. The seeds in all treatments were either B-inoculated or non-inoculated and sown. The experiment was laid out according to randomized complete block design with factorial arrangements. All treatments had three replications. The row-to-row and plant-to-plant distance was 45 and 10 cm, respectively. The net plot size was 5×3 m.

Optical density bacterial strain *Bacillus* sp. MN54 (Accession no. KT375574) of inoculum was adjusted to 109 cfu per mL before seed inoculation. This strain has already been used for enhancing growth and production of different crops [35–37].

2.3. Crop Husbandry

The seeds were sown on well-prepared seedbeds at both locations. Before sowing, pre-soaking irrigation of 10 cm was applied to the irrigated experiment. When soil reached a workable condition, two cultivations followed by planking were completed. Sowing was undertaken on 24th and 26th of October during 2019, and 20th and 23rd October during 2020 in irrigated and rainfed locations, respectively. Sowing was undertaken with the help of a hand drill by using a seed rate of 80 kg ha^{-1}. Before sowing 40 kg ha^{-1} nitrogen and 80 kg ha^{-1} phosphorous was applied using urea and tipple super phosphate to fulfill nutrient requirements. In the irrigated location, four irrigations were applied to avoid moisture stress, whereas rainwater was the only moisture source under rainfed conditions. Weeds were manually controlled at both locations. Mature crop was harvested on 18 April and 4 May 2020, and 1 and 18 April 2021 in irrigated and rainfed locations, respectively.

2.4. Observations

2.4.1. Nodule Population

Three plants were uprooted randomly form each treatment before flowering and the number of nodules was counted and averaged to record the nodule population.

2.4.2. Allometric Traits

Chlorophyll index was measured by taking the Soil Plant Analysis Development (SPAD) value (SPAD-502, Minolta, Konica Minolta Inc., Osaka, Japan) of five randomly selected leaves of different plants from each treatment.

Three plants from each treatment were taken and their leaves were separated from the stem. The area of the detached leaves was determined by leaf area meter (DT Area Meter, Model MK2, Delta T Devices, Cambridge, UK). Afterwards, leaf area index (LAI) was calculated by using following formula given by Watson [38].

$$LAI = \frac{\text{Leaf area}}{\text{Ground area}} \tag{1}$$

Crop growth rate (CGR) was determined by harvesting the plants from an area of 1 m^2. Sampling was completed from 35 to 95 DAS with an interval of 20 days. All the samples were weighed to take fresh weight and then sun dried in open air. Furthermore, sun-dried samples were placed in a hot air oven for 48 h at 70 °C for drying. Oven-dried samples were weighed and CGR was calculated by using the following formula given by Hunt [39].

$$CGR = \frac{W2 - W1}{t2 - t1} \tag{2}$$

In the equation, W_2 = dry weight per unit land area (g m^{-2}) at second harvest, W_1 = dry weight per unit land area (g m^{-2}) at first harvest, t_2 = time corresponding to second harvest and t_1 = time corresponding to first harvest.

2.4.3. Yield and Related Traits

Number of pods per plant were counted by picking pods from 10 randomly selected plants from each plot. Number of grains per pod were counted from manually threshed 20 pods from each plot and averaged. Number of grains per plant was calculated by using the formula given by Hussain et al. [26].

Number of grains per plant = Number of pods per plant × Number of grains per pod

For estimating 1000-grain weight, three samples of 100 grains from each plot were taken, weighed and multiplied by 10 to record 1000-grain weight. Plants were harvested from an area of 1 m^2 from each experimental unit, sun dried and weighed to get biological yield, followed by threshing to account for grain yield. Then, biological yield and grain yield of 1 m^2 were converted into hectares by using the unitary method. Harvest index was calculated as:

$$\text{Harvest index} = \frac{\text{Grain yield}}{\text{Biological yield}} \tag{3}$$

2.4.4. Grain Boron Analysis

Grain and leaf samples were oven dried at 70 °C for B analysis. After drying, 1 g plant and grain sample from each treatment was taken in a porcelain crucible and incinerated in a muffle furnace at 550 °C for 6 h [40]. Extraction of samples was completed by adding 10 mL of 0.36 N H_2SO_4 for 1 h. Extracted samples were then filtered with Whatman no. 1 filter paper and transferred to 50 mL transparent plastic bottles and the volume was raised to 50 mL by adding distilled water. Buffer solution was prepared by adding 250 mL ammonium acetate, 15 mL Ethylenediamine tetraacetic acid (EDTA) and 125 mL acetic acid gently into 400 mL distilled water. The azomethine solution was formed by adding 0.45 g azomethine-H and 1 g L-ascorbic acid into 100 mL of distilled water. Afterwards, 1 mL solution was taken from filtered extracted solution and mixed with 2 mL of each buffer solution and azomethine-H solution. Prepared samples were kept at room temperature for 45 min to develop color. After that, samples were analyzed for grain and plant B concentrations by using a spectrophotometer (double beam product of Bristol Myers Squibb, Madrid, Spain) at 420 nm wavelength.

2.5. Statistical and Economic Analysis

The recorded data were tested for normality and the variables with non-normal distribution were normalized using Arcsine transformation technique to meet the normality assumption of analysis of variance (ANOVA). A four-way ANOVA (year × B application methods × BTB inoculation × location) was used to test the significance in the dataset. Least significant difference at 95% probability level was used to compare the means where ANOVA denoted significant differences [41]. Economic analysis was undertaken to test the economic efficiency of desi chickpea grown with various B application methods combined with BTB inoculation. For economic analysis, production costs (cost required for land rent, seedbed preparation, sowing and harvesting of crop, and purchase of inputs such as seed, fertilizer pesticides and irrigation) were computed. Net income was attained by excluding all expenses from the gross income. Gross income was divided production cost to compute benefit–cost ratio (BCR) [42]. The significant interactions were presented and interpreted in the manuscript.

3. Results

3.1. Nodules' Population

Nodules' population was significantly altered by individual effects of year, locations, BTB inoculation and B application methods. The significant two-way interactions were year × location, year × BTB inoculation and location × BTB inoculation, while the remaining possible interactions were non-significant (Table 2).

Regarding individual effects, 2020–2021, rainfed location, BTB inoculation and osmo-priming recorded higher root nodulation compared to the remaining individual effects of the study (Table 3). Rainfed location and BTB inoculation interaction with 2020–2021 and rainfed location by BTB inoculation interaction recorded higher root nodulation compared to the other possible interactions (Table 4). The three-way and four-way interactions remained non-significant.

Table 2. Analysis of variance (*p* values) of individual and interactive effects of year, location, BTB (*Bacillus* sp. MN54) inoculation and boron nutrition for growth and yield-related traits of desi chickpea.

Variables	Y	S	B	I	Y × S	Y × B	Y × I	S × B	S × I	B × I	Y × S × B	Y × S × I	Y × B × I	S × B × I	Y × S × B × I
Nodule population	0.0000 *	0.0009 *	0.0000 *	0.0000 *	0.0024 *	0.0740 NS	0.0248 *	0.1118 NS	0.0358 *	0.1770 NS	0.5980 NS	0.2101 NS	0.5498 NS	0.9538 NS	0.8909 NS
Number of pods per plant	0.0000 *	0.0000 *	0.0000 *	0.0000 *	0.0084 *	0.0448 *	0.0040 *	0.0000 *	0.0019 *	0.0001 *	0.0448 *	0.9040 NS	0.9917 NS	0.0000 *	0.9917 NS
Number of grains per pod	0.0000 *	0.0000 *	0.0000 *	0.0000 *	0.4902 NS	0.6013 NS	0.9738 NS	0.6926 NS	0.5763 NS	0.1286 NS	0.9422 NS	0.9214 NS	0.9499 NS	0.9388 NS	0.9769 NS
Number of grains per plant	0.0000 *	0.0000 *	0.0000 *	0.0000 *	0.0000 *	0.1407 NS	0.7561 NS	0.0001 *	0.4804 NS	0.0159 *	0.0001 *	0.2725 NS	0.9262 NS	0.0000 *	0.7403 NS
1000-grain weight (g)	0.0000 *	0.0000 *	0.0000 *	0.0000 *	0.0000 *	0.8217 NS	0.4896 NS	0.0000 *	0.2835 NS	0.0218 *	0.8217 NS	0.4896 NS	0.8217 NS	0.0000 *	0.8217 NS
Grain yield (t ha^{-1})	0.0000 *	0.0000 *	0.0000 *	0.0000 *	0.0000 *	0.1392 NS	0.0124 *	0.0000 *	0.0911 NS	0.0006 *	0.1392 NS	0.9762 NS	0.1392 NS	0.0000 *	0.1392 NS
Biological yield (t ha^{-1})	0.0000 *	0.0000 *	0.0000 *	0.0000 *	0.0000 *	0.8715 NS	0.9936 NS	0.0000 *	0.0000 *	0.0000 *	0.8715 NS	0.9936 NS	0.9629 NS	0.0000 *	0.9629 NS
Harvest index (%)	0.2466 NS	0.0000 *	0.0001 *	0.0035 *	0.0000 *	0.2601 NS	0.2476 NS	0.0000 *	0.0015 *	0.0000 *	0.6734 NS	0.8382 NS	0.9735 NS	0.0000 *	0.4769 NS
Grain B concentration (mg)	0.0000 *	0.0000 *	0.0000 *	0.0000 *	0.0000 *	0.0399 *	0.0003 *	0.1302 NS	0.0091 *	0.2548 NS	0.6130 NS	0.0801 NS	0.2105 NS	0.0001 *	0.064 NS
Plant B concentration (mg)	0.0000 *	0.0000 *	0.0000 *	0.0000 *	0.0000 *	0.0615 NS	0.0011 *	0.0024 *	0.0166 *	0.3670 NS	0.0003 *	0.1120 NS	0.0320 *	0.7781 NS	0.1054 NS

Here: Y = year; S = locations (i.e., irrigated or rainfed); B = boron application methods; I = BTB (*Bacillus* sp. MN54) inoculation; * indicates that the relevant individual or interactive effect is significant for the respective trait, where NS indicates that the individual or interactive effect is non-significant for the respective trait.

Table 3. The impact of year, location, BTB (*Bacillus* sp. MN54) inoculation and boron nutrition on growth and yield-related traits of desi chickpea.

	Nodule Population	Number of Pods per Plant	Number of Grains per Pod	Number of Grains per Plant	1000-Grain Weight (g)	Grain Yield (t ha^{-1})	Biological Yield (t ha^{-1})	Harvest Index (%)	Grain B Concentration (mg)	Plant B Concentration (mg)
					Years					
Y$_1$	20.31 b	31.94 b	2.28 b	75.48 b	227.63 b	2.41 b	5.27 b	-	40.24 b	43.11 b
Y$_2$	25.93 a	34.31 a	2.49 a	95.84 a	236.55 a	2.60 a	5.71 a	-	48.39 a	50.30 a
LSD 0.05	1.53	0.71	0.07	2.55	3.45	0.11	0.06	-	1.33	1.62
					Locations					
S$_1$	24.44 a	34.36 a	2.46 a	94.94 a	236.09 a	2.46 b	6.33 a	39.17 b	42.68 b	48.82 a
S$_2$	21.80 b	31.89 b	2.31 b	76.38 b	228.09 b	2.55 a	4.64 b	54.88 a	45.95 a	44.60 b
LSD 0.05	1.54	0.75	0.10	2.87	4.21	0.07	0.12	0.54	1.76	2.13
					Bacterial inoculation					
I$_1$	26.21 a	37.84 a	2.54 a	101.29 a	239.24 a	2.66 a	5.80 a	47.43 a	47.81 a	49.93 a
I$_2$	20.02 b	28.41 b	2.23 b	70.03 b	224.94 b	2.35 b	5.18 b	46.62 b	40.82 b	43.48 b
LSD 0.05	1.56	0.77	0.08	3.01	4.33	0.12	0.17	0.48	2.45	2.93
					Boron application methods					
B$_1$	15.38 e	23.85 f	1.90 e	53.19 f	216.28 g	2.21 f	4.73 e	47.39 a	20.37 e	22.37 e
B$_2$	15.71 e	26.79 e	2.07 d	62.92 e	223.04 e	2.34 e	4.99 d	47.95 a	19.95 e	23.08 e
B$_3$	23.25 c	33.83 c	2.51 b	88.15 c	235.33 c	2.57 c	5.57 c	47.45 a	69.80 a	71.08 a
B$_4$	29.46 b	38.19 b	2.63 b	103.48 b	243.89 b	2.71 b	5.96 b	47.05 ab	56.56 c	59.00 c
B$_5$	18.92 d	29.94 d	2.25 c	73.81 d	220.11 f	2.35 e	5.03 d	47.52 a	23.47 d	26.88 d
B$_6$	34.54 a	45.51 a	2.79 a	129.78 a	255.72 a	2.88 a	6.50 a	46.25 bc	63.49 b	65.52 b
B$_7$	24.58 c	33.77 c	2.55 b	88.29 c	230.26 d	2.48 d	5.62 c	45.59 c	56.56 c	59.00 c
LSD 0.05	2.87	1.32	0.13	4.77	3.90	0.06	0.11	1.01	2.49	3.04

Here: Y$_1$ = 2019–2020; Y$_2$ = 2020–2021; S$_1$ = rainfed; S$_2$ = irrigated; I$_1$ = BTB inoculation; I$_2$ = no BTB inoculation; B$_1$ = control (0.00); B$_2$ = water spray; B$_3$ = foliar spray at 0.05% B; B$_4$ = soil application; B$_5$ = hydropriming; B$_6$ = osmopriming 0.001% B; and B$_7$ = seed coating 1.5/kg seed. LSD = least significant difference. Any two means sharing one letter in common within a column are statistically similar at 95% probability level. - = the relevant parameter was not significantly altered by the respective explanatory variable.

Table 4. The impact of year by location, year by BTB (*Bacillus* sp. MN54) inoculation and location by BTB (*Bacillus* sp. MN54) inoculation interaction on growth and yield-related traits of desi chickpea.

	Nodule Population	Number of Pods per Plant	Number of Grains per Plant	1000-Grain Weight (g)	Grain Yield (t ha^{-1})	Biological Yield (t ha^{-1})	Harvest Index (%)	Grain B Concentration (mg)	Plant B Concentration (mg)
				Year × Location					
Y_1S_1	20.43 c	32.86 b	80.62 b	230.09 b	2.39 d	6.01 b	40.15 c	40.36 c	43.21 c
Y_1S_2	20.19 c	31.02 c	70.33 c	225.18 c	2.42 c	4.53 d	53.59 b	40.12 c	43.00 c
Y_2S_1	28.45 a	35.86 a	109.26 a	242.09 a	2.53 b	6.66 a	38.20 d	44.99 b	54.42 a
Y_2S_2	23.41 b	32.77 b	82.42 b	231.00 b	2.67 a	4.76 c	56.17 a	51.78 a	46.19 b
LSD 0.05	2.16	1.04	3.61	4.44	0.02	0.18	0.76	1.88	2.30
				Year × BTB inoculation					
Y_1I_1	24.29 b	36.67 b	-	-	2.55 b	-	-	42.47 b	44.96 b
Y_1I_2	16.33 c	27.20 d	-	-	2.26 d	-	-	38.02 c	41.25 c
Y_2I_1	28.14 a	39.01 a	-	-	2.77 a	-	-	53.16 a	54.91 a
Y_2I_2	23.71 b	29.62 c	-	-	2.44 c	-	-	43.62 b	45.70 b
LSD 0.05	2.18	1.09			0.19			2.91	2.54
				Location × BTB inoculation					
S_1I_1	27.67 a	39.64 a	-	-	-	6.72 a	39.14 c	45.28 b	53.04 a
S_1I_2	24.76 b	29.07 c	-	-	-	5.94 b	39.21 c	40.07 c	44.59 bc
S_2I_1	21.21 c	36.04 b	-	-	-	4.88 c	55.73 a	50.34 a	46.83 b
S_2I_2	18.83 d	27.75 d	-	-	-	4.41 d	54.03 b	41.56 c	42.37 c
LSD 0.05	2.20	1.43				0.21	0.78	2.21	2.33

Here: Y_1 = 2019–2020; Y_2 = 2020–2021; S_1 = rainfed; S_2 = irrigated; I_1 = BTB inoculation; and I_2 = no BTB inoculation. LSD = least significant difference. Any two means sharing one letter in common within a column are statistically similar at 95% probability level. - = the relevant parameter was not significantly altered by the respective explanatory variable.

3.2. Allometric Traits

Chlorophyll index, LAI and CGR increased from 35 to 95 days after sowing (DAS) under all B application methods and BTB inoculation in irrigated and rainfed conditions (Figures 1–3). However, chlorophyll index suddenly declined after 75 DAS during 2020–2021 at both locations (Figure 1). Regarding interaction, osmopriming combined with BTB inoculation resulted in the highest chlorophyll index, LAI and CGR compared to the rest of the treatments under rainfed condition, whereas lowest values were noted for the control treatment under irrigated conditions during both years (Figures 1–3).

3.3. Yield and Related Traits

The individual and interactive effects of year, different B application methods, BTB inoculation and locations had significant effects on all yield-related traits except four-way interactions (Table 2). Number of pods per plant was significantly affected by all individual, two-way, and three-way (except for year × location × BTB inoculation and year × B application methods × BTB inoculation) interactions of year, different B application methods, BTB inoculation and locations (Table 2). Number of grains per pod were only altered by individual effects of year, B application methods, BTB inoculation and locations, whereas their all two-, three- and four-way interactions remained non-significant in this regard. Similarly, individual effects of all studied factors, two-way interaction among year and location, location, and B application methods and BTB inoculation and B application methods, and three-way interaction among year, location and B application methods and location, BTB inoculation and B application methods had a significant impact on number of grains per plant, whereas the remaining possible interactions were non-significant (Table 2).

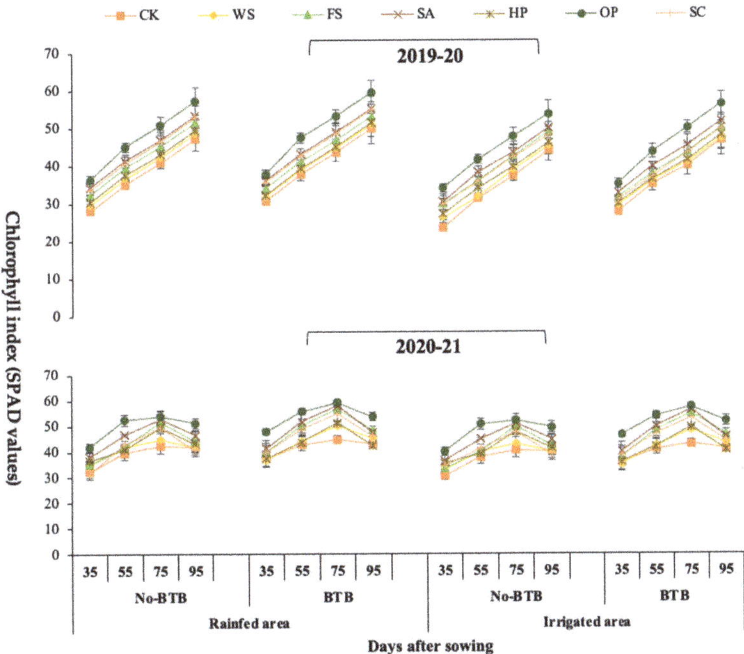

Figure 1. The impact of different boron application methods and boron-tolerant bacteria (*Bacillus* sp. MN54) on chlorophyll index (SPAD value) of desi chickpea grown under rainfed and irrigated conditions. Here: CK = control; WS = water spray; FS = foliar spray of B; SA = soil application of B; HP = hydropriming; OP = osmopriming of B; SC = seed coating of B; and BTB = boron-tolerant bacteria.

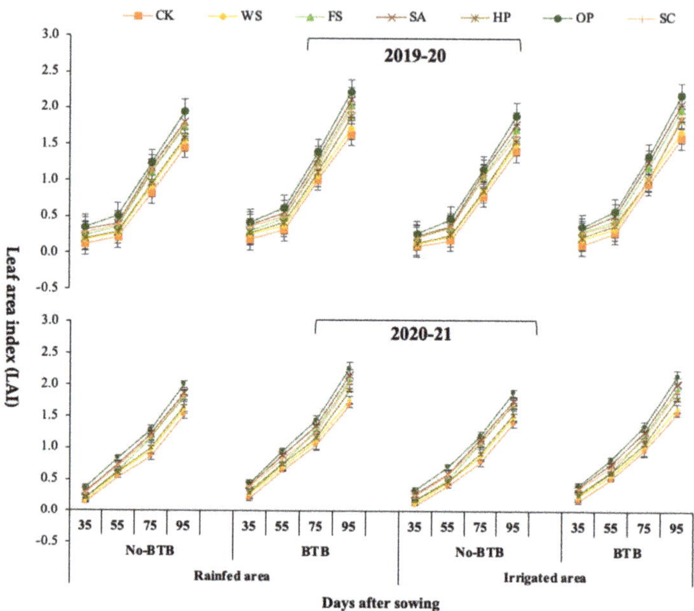

Figure 2. The impact of different boron application methods and boron-tolerant bacteria (*Bacillus* sp. MN54) on leaf area index of desi chickpea grown under rainfed and irrigated conditions. Here: CK = control; WS = water spray; FS = foliar spray of B; SA = soil application of B; HP = hydropriming; P = osmopriming of B; SC = seed coating of B; and BTB = boron-tolerant bacteria.

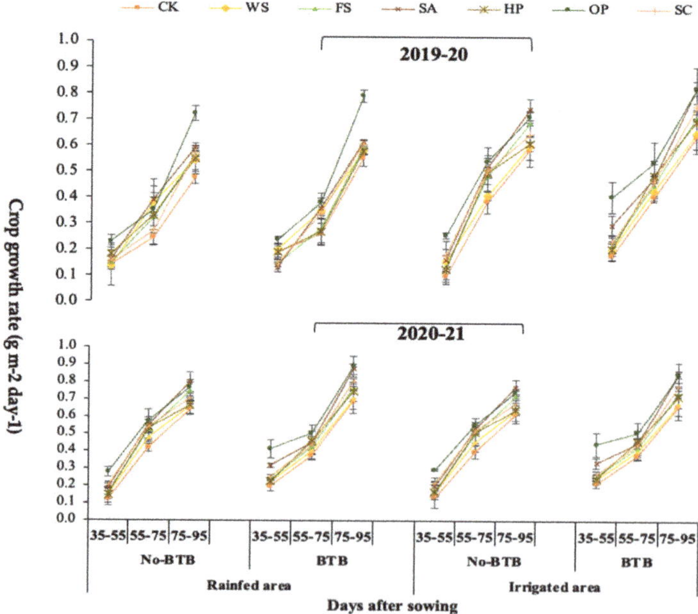

Figure 3. The impact of different boron application methods and boron-tolerant bacteria (*Bacillus* sp. MN54) on crop growth rate of desi chickpea grown under rainfed and irrigated conditions. Here: CK = control; WS = water spray; FS = foliar spray of B; SA = soil application of B; HP = hydropriming; OP = osmopriming of B; SC = seed coating of B; and BTB = boron-tolerant bacteria.

The individual effects of all factors significantly altered (except for the non-significant effect of year on harvest index) 1000-grain weight, grain and biological yields, harvest index and plant and grain B concentration. The two-way interaction among year and location, location, and B application methods and BTB inoculation and B application methods and three-way interaction among location, BTB inoculation and B application methods had a significant effect on 1000-grain weight, while the rest of the interactions remained non-significant. All possible two-way interactions (except year × B application methods and location × BTB inoculation) and three-way interaction among location, BTB inoculation and B application methods significantly affected grain yield, whereas the rest of the interactions were non-significant. Likewise, biological yield and harvest index were significantly altered by two-way interactions of year and location, B application methods with location and BTB inoculation and locations and BTB inoculation and three-way interaction of location, BTB inoculation and B application methods, whereas the remaining possible interactions proved non-significant. All two-way interactions (except B applications methods with location and BTB inoculation) and three-way interaction of location, BTB inoculation and B application methods had a significant effect on plant B concentration, while the remaining possible interactions remained non-significant. Similarly, grain B concentration was significantly altered by all two-way interactions (except B applications methods with year and BTB inoculation) and three-way interaction among year × B application methods × BTB inoculation and year × B application methods × BTB inoculation. However, the remaining interactions remained non-significant for grain B concentration (Table 2).

The higher values of yield and related traits were generally noted for 2020–2021, rainfed location, osmopriming and BTB inoculation. The year 2020–2021 recorded 9.21%, 26.97%, 3.92%, 7.88%, 8.35%, 20.25% and 16.68% higher number of pods per plant, number of grains per pod, number of grains per plant, 1000-grain weight, grain yield, biological yield, grain B concentration and plant B concentration, respectively, compared to 2019–2020. Similarly, BTB inoculation improved number of pods per plant, number of grains per pod, number of grains per plant, 1000-grain weight, grain yield, biological yield, harvest index, grain B concentration and plant B concentration by 24.92%, 12.20%, 30.86%, 5.98%, 11.65%, 10.69%, 1.71%, 14.62% and 12.92%, respectively (Table 3). Likewise, 90.82%, 46.84%, 143.99%, 18.24%, 30.32%, 37.42%, 211.68% and 192.89% improvement was recorded in number of pods per plant, number of grains per pod, number of grains per plant, 1000-grain weight, grain yield, biological yield, grain B concentration and plant B concentration, respectively, with osmopriming compared to control treatment of the study. The highest improvement in grain B concentration (242.66%) and plant B concentration (217.75%) was noted for foliar application of B compared to the control treatment of the study (Table 3).

Regarding the two-way interaction among year and location, rainfed location during 2020–2021 recorded higher values for number of pods per plant, number of grains per pod, number of grains per plant, 1000-grain weight, biological yield, and plant B concentration, whereas the irrigated location during 2020–2021 noted higher values for grain yield, harvest index and grain B concentration, whereas the irrigated location during 2019–2020 recorded the lowest values of all these traits (Table 4). Similarly, BTB inoculation during 2020–2021 recorded the highest values for number of pods per plant, grain yield, grain B concentration and plant B concentration, while no BTB inoculation during 2019–2020 recorded the lowest values of these traits. In the same way, rainfed location with BTB inoculation recorded higher values for number of pods per plant, biological yield and plant B concentration and irrigated location with BTB inoculation noted higher values for harvest index and grain B concentration, while irrigated location with no BTB inoculation recorded the lowest values of these traits (Table 4).

The highest values for number of pods and grains per plant, 1000-grain weight and grain and biological yields were noted for rainfed location with osmopriming, whereas rainfed location with foliar application of B recorded higher plant B concentration (Table 5). Similarly, osmopriming combined with BTB inoculation recorded the highest values of all yield-related traits, whereas control treatment with no BTB inoculation noted the lowest values for all yield-related traits (Table 5).

Table 5. The impact of location and BTB inoculation interactions with boron application methods on growth and yield-related traits of desi chickpea.

	Pods Plant^{-1}	Grains Plant^{-1}	1000-Grain Weight (g)	Grain Yield (t ha^{-1})	Biological Yield (t ha^{-1})	Harvest Index (%)	Plant B Concentration (mg)
			Location $\times$ B application methods				
S_1B_1	26.17 h	66.82 i	222.07 f	2.13 j	5.20 e	41.04 c	26.83 ef
S_1B_2	28.67 g	76.10 h	229.30 e	2.24 i	5.64 d	39.71 cd	26.94 de
S_1B_3	36.67 d	99.77 d	238.60 c	2.53 e	6.49 c	39.08 d	72.97 a
S_1B_4	39.83 c	112.71 c	249.60 b	2.71 c	6.99 b	38.99 d	58.99 c
S_1B_5	29.17 fg	82.02 gh	224.70 f	2.32 h	5.70 d	40.88 c	31.20 d
S_1B_6	47.00 a	136.50 a	260.80 a	2.91 a	7.80 a	37.42 e	65.84 b
S_1B_7	33.00 e	90.66 ef	227.55 e	2.41 fg	6.50 c	37.09 e	58.95 c
S_2B_1	21.53 i	39.55 k	210.49 h	2.29 h	4.25 h	53.74 b	17.91 h
S_2B_2	24.91 h	49.73 j	216.78 g	2.45 f	4.35 h	56.19 a	19.23 gh
S_2B_3	31.00 f	76.54 h	232.06 d	2.60 d	4.65 g	55.82 a	69.19 ab
S_2B_4	36.55 d	94.24 de	238.18 c	2.71 c	4.93 f	55.11 ab	59.00 c
S_2B_5	30.33 fg	65.61 i	215.53 g	2.37 g	4.37 h	54.15 b	22.57 fg
S_2B_6	44.02 b	123.05 b	250.65 b	2.86 b	5.21 e	55.07 ab	65.20 b
S_2B_7	34.53 e	85.92 fg	232.96 d	2.56 de	4.74 g	54.08 b	59.06 c
LSD 0.05	1.88	6.75	4.76	0.12	0.17	1.43	4.30
			BTB inoculation $\times$ B application methods				
I_1B_1	27.03 i	65.64 g	223.24 g	2.40 g	4.91 h	50.01 a	-
I_1B_2	30.33 g	75.30 f	229.64 f	2.48 f	5.16 g	49.27 ab	-
I_1B_3	38.38 d	102.73 d	242.62 d	2.73 c	5.92 cd	47.59 cde	-
I_1B_4	43.42 b	118.60 b	252.51 b	2.84 b	6.47 b	45.29 fgh	-
I_1B_5	36.03 e	93.61 e	228.16 f	2.52 f	5.34 ef	48.34 bcd	-
I_1B_6	50.26 a	148.07 a	263.09 a	3.03 a	7.01 a	45.17 gh	-
I_1B_7	39.44 cd	105.08 cd	235.44 e	2.63 d	5.78 d	46.35 efg	-
I_2B_1	20.67 k	40.73 i	209.33 j	2.02 j	4.55 j	44.76 h	-
I_2B_2	23.24 j	50.53 h	216.44 h	2.20 i	4.82 hi	46.63 ef	-
I_2B_3	29.29 gh	73.58 f	228.04 f	2.40 g	5.23 fg	47.30 de	-
I_2B_4	32.97 f	88.35 e	235.27 e	2.58 e	5.45 e	48.81 abc	-
I_2B_5	23.85 j	54.01 h	212.07 i	2.17 i	4.73 i	46.69 ef	-
I_2B_6	40.76 c	111.48 c	248.36 c	2.73 c	6.00 c	47.33 de	-
I_2B_7	28.09 hi	71.50 fg	225.07 g	2.34 h	5.46 e	44.82 h	-
LSD 0.05	1.93	5.98	3.67	0.16	0.15	1.40	

Here: S_1 = rainfed; S_2 = irrigated; I_1 = BTB inoculation; I_2 = no BTB inoculation; B_1 = control (0.00); B_2 = water spray; B_3 = foliar spray at 0.05% B; B_4 = soil application; B_5 = hydropriming; B_6 = osmopriming 0.001% B; and B_7 = seed coating 1.5/kg seed. LSD = least significant difference. Any two means sharing one letter in common within a column are statistically similar at 95% probability level. - = the relevant parameter was not significantly altered by the respective explanatory variable.

The year $\times$ location $\times$ boron application methods' interaction was only significant for number of pods and grains per plant and plant B concentration (Table 6). Rainfed location during 2020–2021 with osmopriming recorded the highest values for number of pods and grains per plant, while rainfed location with foliar application of B during 2020–2021 recorded the highest plant B concentration. The lowest values of these traits were recorded for the control treatment in the irrigated location during 2019–2020 (Table 6).

Table 6. The impact of year $\times$ location $\times$ boron application methods' interaction on number of pods and grains per plant and plant boron concentration of desi chickpea.

B Application Methods	2019–2020		2020–2021	
	Rainfed	**Irrigated**	**Rainfed**	**Irrigated**
Number of pods per plant				
Control (0.00)	24.67 kl	21.07 n	27.67 ij	22.00 mn
Water spray	27.17 jk	24.32 lm	30.17 hi	25.50 jkl
Foliar spray at 0.05% B	35.17 fg	30.66 h	38.17 de	31.33 h
Soil application	38.33 de	35.93 efg	41.33 c	37.17 ef
Hydropriming	27.67 ij	30.42 h	30.67 h	31.00 h
Osmopriming 0.001% B	45.50 b	39.98 cd	48.50 a	48.05 ab
Seed coating 1.5/kg seed	31.50 h	34.73 fg	34.50 g	34.33 g
LSD 0.05			9.55	
Number of grains per plant				
Control (0.00)	45.38 no	34.70 p	88.27 ghi	44.40 o
Water spray	57.18 lm	45.11 no	95.02 efg	54.35 mn
Foliar spray at 0.05% B	88.28 ghi	71.43 jk	111.25 c	81.65 hi
Soil application	101.13 de	87.67 ghi	124.28 b	100.82 de
Hydropriming	65.05 kl	64.27 kl	98.98 def	66.95 k
Osmopriming 0.001% B	127.05 b	107.88 cd	145.95 a	138.22 a
Seed coating 1.5/kg seed	80.28 ij	81.26 hi	101.04 de	90.58 fgh
LSD 0.05			10.12	
Plant B concentration				
Control (0.00)	15.67 i	17.66 hi	38.00 g	18.17 hi
Water spray	19.45 hi	19.26 hi	34.43g	19.19 hi
Foliar spray at 0.05% B	71.91 ab	65.92 bc	74.02 a	72.47 a
Soil application	54.96 f	56.53 ef	63.03 cd	61.48 cde
Hydropriming	23.61 h	22.73 h	38.79 g	22.41 h
Osmopriming 0.001% B	60.09 cdef	60.37 cdef	71.59 ab	70.04 ab
Seed coating 1.5/kg seed	56.79 ef	58.57 def	61.10 cde	59.55 def
LSD 0.05			6.09	

LSD = least significant difference. Any two means sharing one letter in common within a column are statistically similar at the 95% probability level.

Regarding three-way interactions among location, BTB inoculation and B application methods, osmopriming at rainfed location with BTB inoculation recorded the highest values of number of pods and grains per plant. Similarly, foliar application of B at rainfed location with BTB inoculation noted higher plant B concentration (Table 7). The lowest values of these traits were noted for the control treatment with no BTB inoculation at the irrigated location (Table 7).

Table 7. The impact of location × BTB inoculation × boron application methods' interaction on number of pods and grains per plant and plant boron concentration of desi chickpea.

	Rainfed		Irrigated	
	BTB	**No BTB**	**BTB**	**No BTB**
Number of pods per plant				
Control (0.00)	31.17 h	21.17 mn	22.90 lm	20.17 n
Water spray	34.83 g	22.50 lmn	25.83 jk	23.98 kl
Foliar spray at 0.05% B	42.17 bc	31.17 h	34.58 g	27.41 ij
Soil application	43.83 b	35.83 fg	43.00 b	30.10 h
Hydropriming	35.50 g	22.83 lm	36.55 efg	24.87 jkl
Osmopriming 0.001% B	50.83 a	43.17 b	49.68 a	38.35 def
Seed coating 1.5/kg seed	39.17 de	26.83 ij	39.72 cd	29.35 hi
LSD 0.05	9.63			
Number of grains per plant				
Control (0.00)	84.73 fg	56.81 l	48.92 kl	32.55 m
Water spray	93.78 ef	89.87 jk	58.42 jk	42.65 l
Foliar spray at 0.05% B	115.58 bc	114.77 fg	83.95 g	63.20 ij
Soil application	122.43 b	84.50 bc	102.98 de	73.72 h
Hydropriming	102.73 de	145.41 fg	61.30 ij	46.72 l
Osmopriming 0.001% B	150.73 a	102.96 de	122.27 b	100.69 de
Seed coating 1.5/kg seed	107.20 cd	56.81 de	74.12 h	68.88 hi
LSD 0.05	9.55			
Plant B concentration				
Control (0.00)	32.91 g	20.76 hij	20.53 hij	15.29 j
Water spray	32.57 g	21.31 hij	22.15 hi	16.30 ij
Foliar spray at 0.05% B	75.05 a	70.88 ab	71.12 ab	67.26 bcd
Soil application	63.78 cd	54.20 f	63.15 cd	54.85 f
Hydropriming	36.07 g	26.33 h	22.94 h	22.20 hi
Osmopriming 0.001% B	68.37 bc	63.31 cd	66.29 bcd	64.11 cd
Seed coating 1.5/kg seed	62.56 cde	55.33 f	61.58 de	56.54 ef
LSD 0.05	6.21			

LSD = least significant difference. Any two means sharing one letter in common within a column are statistically similar at the 95% probability level.

Similarly, osmopriming with BTB inoculation at the rainfed location recorded the highest values of 1000-grain weight, and grain and biological yields, whereas control treatment with no BTB inoculation recorded the lowest values of these traits (Table 8).

Table 8. The impact of location $\times$ BTB inoculation $\times$ boron application methods' interaction on yield and related traits of desi chickpea.

Treatments	Rainfed		Irrigated		Rainfed		Irrigated	
	BTB	No BTB	BTB	No BTB	BTB	No BTB	BTB	No BTB
	1000-Grain Weight (g)				Grain Yield (t ha^{-1})			
Control (0.00)	228.67 i	215.47 m	217.81 lm	203.18 o	2.25 k	2.01 n	2.55 h	2.03 n
Water spray	234.90 fg	223.70 k	224.38 jk	209.18 n	2.34 j	2.13 m	2.62 fg	2.28 k
Foliar spray at 0.05% B	245.73 d	231.47 ghi	239.51 e	224.61 jk	2.72 de	2.35 j	2.73 d	2.46 i
Soil application	258.43 bc	240.77 e	246.58 d	229.78 hi	2.84 c	2.58 gh	2.84 c	2.59 gh
Hydropriming	228.13 ij	221.27 kl	228.18 ij	202.88 o	2.48 i	2.17 lm	2.56 gh	2.18 lm
Osmopriming 0.001% B	266.77 a	254.83 c	259.41 b	241.88 e	3.07 a	2.75 d	3.00 b	2.72 de
Seed coating 1.5/kg seed	238.10 ef	217.00 m	232.78 gh	233.15 gh	2.59 gh	2.23 kl	2.67 ef	2.45 i
LSD 0.05	5.12				0.06			
	Biological yield (t ha^{-1})				Harvest index (%)			
Control (0.00)	5.48 gh	4.92 i	4.33 klm	4.17 m	41.09 hi	40.99 hij	58.93 a	48.54 g
Water spray	5.87 f	5.41 gh	4.46 jkl	4.24 lm	39.92 hijk	39.49 hijk	58.61 a	53.76 def
Foliar spray at 0.05% B	6.96 c	6.03 ef	4.88 i	4.43 jkl	39.17 ijkl	38.99 jkl	56.02 bc	55.61 bcd
Soil application	7.61 b	6.37 d	5.33 h	4.54 jk	37.34 lm	40.63 hij	53.24 ef	56.99 ab
Hydropriming	6.13 e	5.27 h	4.55 j	4.18 m	40.48 hijk	41.29 h	56.21 b	52.09 f
Osmopriming 0.001% B	8.46 a	7.13 c	5.55 g	4.86 i	36.28 mn	38.56 kl	54.06 cdef	56.09 b
Seed coating 1.5/kg seed	6.54 d	6.47 d	5.03 i	4.45 jll	39.68 hijk	34.51 n	53.03 f	55.13 bcde
LSD 0.05	0.22				2.01			

LSD = least significant difference. Any two means sharing one letter in common within a column are statistically similar at the 95% probability level.

3.4. Economic Analysis

The economic analysis described that B application methods combined with BTB inoculation positively improved the net benefit and benefit–cost ratio during both years of study under rainfed and irrigated conditions (Table 9). The highest net benefit (USD 2068.5 ha^{-1}) and benefit–cost ratio (3.7%) were obtained under osmopriming combined with BTB inoculation followed by soil application (Table 9). The BTB inoculation significantly improved net benefits and benefit–cost ratio compared with no-BTB inoculation (Table 9).

Table 9. Effects of B application methods along with BTB strain on economic analysis of desi chickpea during 2019–2020 and 2020–2021.

Treatments		Rainfed Condition							Irrigated Condition							
		2019–2020				2020–2021				2019–2020				2020–2021		
		Total Cost (USD ha^{-1})	Gross Income (USD ha^{-1})	Net Income (USD ha^{-1})	BCR %	Total Cost (USD ha^{-1})	Gross Income (USD ha^{-1})	Net Benefit (USD ha^{-1})	BCR%	Total Cost (USD ha^{-1})	Gross Income (USD ha^{-1})	Net Benefit (USD ha^{-1})	BCR%	Total Cost (USD ha^{-1})	Gross Income (USD ha^{-1})	Net Benefit (USD ha^{-1})	BCR%
No-BTB	Control (0.00)	695.40	1696.1	1000.7	2.4	722.66	1801.1	1078.4	2.5	710.8	1661.3	950.5	2.3	746.5	1861.4	1114.9	2.5
	Water spray	703.11	1799.1	1095.9	2.6	739.60	1903.4	1163.8	2.6	727.8	1878.8	1151.0	2.6	763.5	2078.8	1315.4	2.7
	Foliar spray	738.54	1987.5	1249.0	2.7	823.25	2091.9	1268.6	2.5	811.4	2041.2	1229.8	2.5	847.1	2241.2	1394.1	2.6
	Soil application	744.71	2194.8	1450.1	2.9	872.55	2299.2	1426.6	2.6	860.7	2148.4	1287.7	2.5	896.4	2348.5	1452.1	2.6
	Hydropriming	704.65	1835.3	1130.6	2.6	731.90	1939.7	1207.8	2.7	720.1	1791.8	1071.8	2.5	755.8	1991.9	1236.1	2.6
	Osmopriming	728.20	2336.3	1608.1	3.2	772.40	2440.7	1668.3	3.2	760.6	2267.3	1506.7	3.0	796.3	2467.4	1671.1	3.1
	Seed coating	703.11	1885.7	1182.6	2.7	780.11	1990.1	1210.0	2.6	768.3	2032.5	1264.2	2.6	804.0	2232.5	1428.5	2.8
BTB	Control (0.00)	701.64	1902.0	1200.3	2.7	728.89	2006.4	1277.5	2.8	717.0	2102.0	1385.0	2.9	752.8	2336.9	1584.1	3.1
	Water spray	709.34	1981.7	1272.4	2.8	745.84	2086.1	1340.3	2.8	734.0	2157.1	1423.1	2.9	769.7	2392.0	1622.3	3.1
	Foliar spray	744.78	2313.7	1568.9	3.1	829.48	2418.1	1588.6	2.9	817.6	2258.6	1441.0	2.8	853.4	2493.5	1640.1	2.9
	Soil application	750.94	2415.2	1664.2	3.2	878.78	2519.5	1640.8	2.9	866.9	2348.5	1481.5	2.7	902.7	2583.3	1680.7	2.9
	Hydropriming	710.88	2100.6	1389.7	3.0	738.14	2205.0	1466.8	3.0	726.3	2107.8	1381.5	2.9	762.0	2342.7	1580.7	3.1
	Osmopriming	734.44	2493.5	1759.0	3.4	778.64	2847.2	2068.5	3.7	766.8	2490.6	1723.8	3.2	802.5	2725.4	1922.9	3.4
	Seed coating	709.34	2199.2	1489.8	3.1	786.34	2303.5	1517.2	2.9	774.5	2200.6	1426.1	2.8	810.2	2435.5	1625.2	3.0

Here: BCR = benefit–cost ratio; ha = hectare.

4. Discussion

This two-year study revealed that B application through different methods and BTB inoculation under irrigated and rainfed conditions significantly improved root nodulation, productivity, quality and biofortification of desi chickpea (Figures 1–3; Tables 3–9). These findings are supported by Khanam et al. [43] who reported that chickpea grain yield was significantly increased by B application. We further noticed that B application through osmopriming and seed coating (seed treatment), soil application and foliar application performed better, when combined with BTB inoculation and significantly improved morphological parameters, grain yield and grain B concentration. In another study, B application through seed priming and seed coating improved early seedling growth of rice [44]. Osmopriming proved better in improving growth and yield-related traits. Osmopriming enables delayed, osmotic fluid rehydration of seeds with the goal of triggering early germination-related metabolic processes without radicle protrusion. The quicker germination and better growth in osmopriming probably enabled plants to utilize resources more efficiently, which resulted in higher productivity. The 2nd year of the study noted higher values of all traits. Since the experiment was repeated on the same experimental field, B availability and higher BTB density during second year resulted in higher values of yield and related traits. Similarly, rainfed location resulted in better productivity due to wider adaptability of chickpea to rainfed conditions in the country. The better yield and related traits in BTB inoculation are owed to the bacterial activity which improved B supply compared to no BTB inoculation.

Increased grain yield under different B application methods might be linked to higher root nodulation (Table 3). Improvement in root nodules was higher with BTB inoculation in rainfed conditions compared with untreated seeds (Table 3). Soil microbes improve nutrient supply by releasing several organic acids, dissolving fixed minerals and making them available to plants [32]. The higher nodulation in BTB inoculation might be the reason for improved chickpea performance. The number of nodules per plant increased with the increasing rates of B application at normal amount (1.5 kg B ha^{-1}) [45]. The same pattern was noted in our study that different B application methods and BTB inoculation significantly improved matter production and grain productivity (Figure 3; Tables 3–8). Farooq et al. [46] revealed that osmopriming enhanced fresh weight and root elongation in rice. Soil applied B at 0.25 mg kg^{-1} soil coupled with BTB strain resulted in the highest dry weight and plant height of chickpea [20].

Chickpeas are mostly grown in desert areas, where soils are mostly deficient in micronutrients, particularly B. Boron is required by plants in minute quantity; however, metabolic activities are significantly altered by B-deficiency. Boron stabilizes plant cell wall, membrane integrity and sugar transport, and improves utilization of calcium and nitrogen [47,48]. In our study, chlorophyll index, leaf area index and dry matter accumulation are increased by B application combined with BTB inoculation. Chlorophyll index and leaf area index directly linked with photosynthesis. Higher chlorophyll index and leaf area index resulted in more green parts which enhanced the production potential of the crop. Yamori et al. [49] reported that leaf photosynthesis was the most significant factor for attaining higher grain yield and biomass. Ali et al. [50] reported that there was an association between LAI and total grain yield, and that higher LAI produced higher yield. Lower chlorophyll index, poor growth rate and dry matter accumulation was observed for no B application (Figures 1–3). Boron deficiency during reproductive stage lowers plant height, reduces dry matter production and interrupts photosynthesis [51]. The same results were obtained in our study, where growth, allometric and yield related parameters were reduced in the treatments receiving no B application (Figures 1–3, Tables 3–8).

Improvement in grain yield was higher under B application with osmopriming as compared to no B application (Table 6). This might be due to improved yield-related parameters, i.e., number of pods per plant, grains per pod and 1000-grain weight. Osmopriming positively improved plant height compared with untreated seeds (Table 3). Bangar et al. [52] also reported increase in plant height in soybean (*Glycine max* L.) with B application. Number of pods per plant, grains per pod and 1000-grain weight increased under 0.001% B priming compared with control [32]. Chickpea grain yield under rainfed condition was higher than irrigated areas. Pod formation and grain quality was improved under B application because it takes part in assimilated partitioning. Assimilates are stored in the stem and leaves (vegetative portion) of plants in a large amount and then translocate into reproductive parts [53].

The higher number of grains per pod, pods per plant and 1000-grain weight was observed under rainfed condition with B application through osmopriming, which ultimately increased grain yield (Tables 5 and 6). Boron is an essential micronutrient which plays a key role in sugar transport from source to sink, helps in carbohydrate metabolism, improves pollen fertility and flower life; thus, improves yield-related traits. Similarly, higher chickpea yield was recorded in this study with B application. According to Hussain et al. [26], seed coating with 1.5 g B kg^{-1} seed positively enhanced number of seed per plant which further improved grain production. Boron application methods (seed priming, coating, soil, and foliar application) proved helpful in improving chickpea production and grain quality. The highest B grain concentration was recorded under foliar application compared with other levels (Table 3). The BTB inoculation combined with B application increased chickpea production compared to no BTB inoculation. The BTB inoculation exerted significantly improved 1000-grain weight, grain yield and biological yield compared to no BTB inoculation. These results confirmed the findings of Mehboob et al. [20] who stated that number of branches plant^{-1}, number of grains pod^{-1} and 1000-grain weight of chickpea showed improvement with the application of BTB strain MN54. According to Ullah et al. [54], Zn application through soil application or seed coating coupled with plant growth promoting bacteria positively improved productivity, profitability and grain quality of desi chickpea. Boron nutrition as osmopriming in chickpea resulted in higher grain yield as compared to control with an average yield difference of 30%. Another study conducted by Dar [55] reported that chickpea yield was increased by 54% with the application of B at the rate of 1 kg ha^{-1}. Samreen et al. [37] reported that BTB strain MN-54 improved nutrient supply in canola (*Brassica napus*) leading to its improved growth. Therefore, higher productivity of chickpea due to B application combined with BTB application in this study was linked with continuous and higher B supply.

Boron deficiency and toxicity are injurious for plant growth as well as human health since it is required in a lower amount. It was confirmed by this study that chickpea growth and productivity are affected by B-deficiency. These results agree with previous studies reporting both increased and reduced yield responses by various chickpea genotypes to B-deficiency [56].

5. Conclusions

Boron application by different methods along with inoculation of boron-tolerant bacteria *Bacillus* sp. strain MN-54 considerably enhanced nodule population, dry matter accumulation, grain yield, quality and biofortification of desi chickpea under rainfed conditions. Similarly, osmopriming with 0.001% B solution combined with boron-tolerant bacteria inoculation improved chickpea grain production. However, higher grain B concentration was observed under foliar application of 0.025% B combined with boron-tolerant bacteria inoculation. Hence, B application through osmopriming combined with BTB inoculation seemed a pragmatic option to boost production, higher economic returns, and grain B concentration of desi chickpea under irrigated and rainfed conditions.

Author Contributions: Conceptualization, M.H., S.F. and T.A.Y.; methodology, M.H.; software, N.M.; validation, S.F., S.H. and M.H.; formal analysis, N.M.; investigation, N.M.; resources, M.H., M.N.; data curation, N.M.; writing—original draft preparation, N.M.; writing—review and editing, M.H., S.F., S.H., T.A.Y.; visualization, N.M.; supervision, M.H.; project administration, M.H.; funding acquisition, M.H. All authors have read and agreed to the published version of the manuscript.

Funding: The corresponding author (Mubshar Hussain) is highly obliged to the Higher Education Commission of Pakistan for financial assistance to accomplish this study under NRPU grant no. 6998/Punjab/NRPU/R&D/HEC/2017.

Institutional Review Board Statement: Not applicable.

Informed Consent Statement: Not applicable.

Data Availability Statement: All data are within the manuscript.

Conflicts of Interest: The authors declare no conflict of interest.

References

1. Kalsoom, M.; Rehman, F.U.; Shafique, T.; Junaid, S.; Khalid, N.; Adnan, M.; Zafar, I.; Tariq, M.A.; Raza, M.A.; Zahra, A. Biological importance of microbes in agriculture, food and pharmaceutical industry: A review. *Innovare J. Life Sci.* **2020**, *8*, 1–4. [CrossRef]
2. Ghatas, M.; Khan, M.; Gorgey, A. Skeletal muscle stiffness as measured by magnetic resonance elastography after chronic spinal cord injury: A cross-sectional pilot study. *Neural Regen. Res.* **2021**, *16*, 2486. [CrossRef]
3. Mekdad, A.A.A.; Rady, M.M. Response of *Beta vulgaris* L. to nitrogen and micronutrients in dry environment. *Plant Soil Environ.* **2016**, *62*, 23–29. [CrossRef]
4. Nadeem, F.; Azhar, M.; Anwar-ul-Haq, M.; Sabir, M.; Samreen, T.; Tufail, A.; Awan, H.U.M.; Juan, W. Comparative Response of Two Rice (*Oryza sativa* L.) Cultivars to Applied Zinc and Manganese for Mitigation of Salt Stress. *J. Soil Sci. Plant Nutr.* **2020**, *20*, 2059–2072. [CrossRef]
5. Zia, M.H.; Ahmad, R.; Khaliq, I.; Ahmad, A.; Irshad, M. Micronutrients status and management in orchards soils: Applied aspects. *Soil Environ.* **2006**, *25*, 6–16.
6. Rashid, A.; Rafique, E. Boron deficiency diagnosis and management in field crops in calcareous soils of Pakistan: A mini review. *Bor Derg.* **2017**, *2*, 142–152.
7. Goli, E.; Hiemstra, T.; Rahnemaie, R. Interaction of boron with humic acid and natural organic matter: Experiments and modeling. *Chem. Geol.* **2019**, *515*, 1–8. [CrossRef]
8. Wu, X.; Riaz, M.; Yan, L.; Du, C.; Liu, Y.; Jiang, C. Boron Deficiency in Trifoliate Orange Induces Changes in Pectin Composition and Architecture of Components in Root Cell Walls. *Front. Plant Sci.* **2017**, *8*, 1882. [CrossRef]
9. Vera, A.; Moreno, J.L.; García, C.; Morais, D.; Bastida, F. Boron in soil: The impacts on the biomass, composition and activity of the soil microbial community. *Sci. Total Environ.* **2019**, *685*, 564–573. [CrossRef]
10. Souri, M.K.; Hatamian, M. Aminochelates in plant nutrition: A review. *J. Plant Nutr.* **2019**, *42*, 67–78. [CrossRef]
11. Furlani, Â.M.C.; Carvalho, C.P.; de Freitas, J.G.; Verdial, M.F. Wheat cultivar tolerance to boron deficiency and toxicity in nutrient solution. *Sci. Agric.* **2003**, *60*, 359–370. [CrossRef]
12. Tahir, M.; Tanveer, A.; Shah, T.H.; Fiaz, N.; Wasaya, A. Yield response of wheat (*Triticum aestivum* L.) to boron application at different growth stages. *Pak. J. Life Soc. Sci* **2009**, *7*, 39–42.
13. Landi, M.; Margaritopoulou, T.; Papadakis, I.E.; Araniti, F. Boron toxicity in higher plants: An update. *Planta* **2019**, *250*, 1011–1032. [CrossRef]
14. Landi, M.; Remorini, D.; Pardossi, A.; Guidi, L. Boron excess affects photosynthesis and antioxidant apparatus of greenhouse Cucurbita pepo and Cucumis sativus. *J. Plant Res.* **2013**, *126*, 775–786. [CrossRef]
15. Shah, A.; Wu, X.; Ullah, A.; Fahad, S.; Muhammad, R.; Yan, L.; Jiang, C. Deficiency and toxicity of boron: Alterations in growth, oxidative damage and uptake by citrange orange plants. *Ecotoxicol. Environ. Saf.* **2017**, *145*, 575–582. [CrossRef]
16. Papadakis, I.E.; Tsiantas, P.I.; Tsaniklidis, G.; Landi, M.; Psychoyou, M.; Fasseas, C. Changes in sugar metabolism associated to stem bark thickening partially assist young tissues of Eriobotrya japonica seedlings under boron stress. *J. Plant Physiol.* **2018**, *231*, 337–345. [CrossRef]
17. GOP. *Economic Survey of Pakistan*; Economic Advisory Wing: Islamabad, Pakistan, 2021.
18. Gaur, P.M.; Tripathi, S.; Gowda, C.L.L.; Ranga Rao, G.V.; Sharma, H.C.; Pande, S.; Sharma, M. *Chickpea Seed Production Manual*; ICRISAT: Hyderabad, India, 2010.
19. Sadeghzadeh, B. A review of zinc nutrition and plant breeding. *J. Soil Sci. Plant Nutr.* **2013**, *13*, 905–927. [CrossRef]
20. Mehboob, N.; Hussain, M.; Minhas, W.A.; Yasir, T.A.; Naveed, M.; Farooq, S.; Alfarraj, S.; Zuan, A.T.K. Soil-applied boron combined with boron-tolerant bacteria (*Bacillus* sp. mn54) improve root proliferation and nodulation, yield and agronomic grain biofortification of chickpea (*Cicer arietinum* L.). *Sustainability* **2021**, *13*, 9811. [CrossRef]
21. Fageria, N.K.; Filho, M.P.B.; Moreira, A.; Guimarães, C.M. Foliar Fertilization of Crop Plants. *J. Plant Nutr.* **2009**, *32*, 1044–1064. [CrossRef]
22. Sarakhsi, H.S.; Yarnia, M.; Amirniya, R. Effect of nitrogen foliar application in different concentration and growth stage of corn (Hybrid 704). *Adv. Environ. Biol.* **2010**, 291–299.
23. Saleem, M.A.; Tahir, M.; Ahmad, T.; Tahir, M.N. Foliar application of boron improved the yield and quality of wheat (*Triticum aestivum* L.) in a calcareous field. *Soil Environ.* **2020**, *39*, 59–66. [CrossRef]
24. Farooq, M.; Wahid, A.; Siddique, K.H.M. Micronutrient application through seed treatments: A review. *J. Soil Sci. Plant Nutr.* **2012**, *12*, 125–142. [CrossRef]

25. Rehman, A.; Farooq, M. Zinc seed coating improves the growth, grain yield and grain biofortification of bread wheat. *Acta Physiol. Plant.* **2016**, *38*, 238. [CrossRef]
26. Hussain, M.; Mehboob, N.; Naveed, M.; Shehzadi, K.; Yasir, T.A. Optimizing boron seed coating level and boron-tolerant bacteria for improving yield and biofortification of chickpea. *J. Soil Sci. Plant Nutr.* **2020**, *20*, 2471–2478. [CrossRef]
27. Rehman, A.; Farooq, M.; Ahmad, R.; Basra, S.M.A. Seed priming with zinc improves the germination and early seedling growth of wheat. *Seed Sci. Technol.* **2015**, *43*, 262–268. [CrossRef]
28. Mehboob, N.; Minhas, W.A.; Nawaz, A.; Shahzad, M.; Ahmad, F.; Hussain, M. Surface Drying after Seed Priming Improves the Stand Establishment and Productivity of Maize than Seed Re-Drying. *Int. J. Agric. Biol.* **2018**, *20*, 1283–1288.
29. Farooq, M.; Hussain, M.; Imran, M.; Ahmad, I.; Atif, M.; Alghamdi, S.S. Improving the productivity and profitability of late sown chickpea by seed priming. *Int. J. Plant Prod.* **2019**, *13*, 129–139. [CrossRef]
30. Farooq, M.; Gogoi, N.; Hussain, M.; Barthakur, S.; Paul, S.; Bharadwaj, N.; Migdadi, H.M.; Alghamdi, S.S.; Siddique, K.H.M. Effects, tolerance mechanisms and management of salt stress in grain legumes. *Plant Physiol. Biochem.* **2017**, *118*, 199–217. [CrossRef]
31. Farooq, M.; Hussain, M.; Habib, M.M.; Khan, M.S.; Ahmad, I.; Farooq, S.; Siddique, K.H.M. Influence of seed priming techniques on grain yield and economic returns of bread wheat planted at different spacings. *Crop Pasture Sci.* **2020**, *71*, 725–738. [CrossRef]
32. Mehboob, N.; Minhas, W.A.; Naeem, M.; Yasir, T.A.; Naveed, M.; Farooq, S.; Hussain, M. Seed priming with boron and. *Crop Pasture Sci.* **2022**, *73*, 494–502. [CrossRef]
33. Khan, M.I.; Afzal, M.J.; Bashir, S.; Naveed, M.; Anum, S.; Cheema, S.A.; Wakeel, A.; Sanaullah, M.; Ali, M.H.; Chen, Z. Improving nutrient uptake, growth, yield and protein content in chickpea by the co-addition of phosphorus fertilizers, organic manures, and bacillus sp. Mn-54. *Agronomy* **2021**, *11*, 436. [CrossRef]
34. Abubakar, M.; Naveed, M.; Ahmad, Z.; Cheema, S.A.; Khan, A.S.; Park, H.Y.; Kwon, C.H. Ameliorative Effect of Bacillus sp. MN-54 and Organic Amendments Combination on Maize Plants Growth and Physiology Under Chromium Toxicity. *J. Agric. Sci.* **2020**, *12*, 39. [CrossRef]
35. Naveed, M.; Mitter, B.; Yousaf, S.; Pastar, M.; Afzal, M.; Sessitsch, A. The endophyte Enterobacter sp. FD17: A maize growth enhancer selected based on rigorous testing of plant beneficial traits and colonization characteristics. *Biol. Fertil. soils* **2014**, *50*, 249–262. [CrossRef]
36. Yang, A.; Akhtar, S.S.; Iqbal, S.; Amjad, M.; Naveed, M.; Zahir, Z.A.; Jacobsen, S.-E. Enhancing salt tolerance in quinoa by halotolerant bacterial inoculation. *Funct. Plant Biol.* **2016**, *43*, 632. [CrossRef]
37. Samreen, T.; Zahir, Z.A.; Naveed, M.; Asghar, M. Boron tolerant phosphorus solubilizing Bacillus spp. MN-54 improved canola growth in alkaline calcareous soils. *Int. J. Agric. Biol.* **2019**, *21*, 538–546. [CrossRef]
38. Watson, D.J. Comparative physiological studies on the growth of field crops: I. Variation in net assimilation rate and leaf area between species and varieties, and within and between years. *Ann. Bot.* **1947**, *11*, 41–76. [CrossRef]
39. Hunt, R. *Plant Growth Analysis*; Institute of Terrestrial Ecology: Penicuik, UK, 1982; Volume 4, ISBN 090428266X.
40. Chapman, O.L. Spectrometric Identification of Organic Compounds. *J. Am. Chem. Soc.* **1963**, *85*, 3316. [CrossRef]
41. Steel, R.; Torrei, J.; Dickey, D. *Principles and Procedures of Statistics A Biometrical Approach*; McGraw-Hill: New York, NY, USA, 1997.
42. Shahzad, M.; Hussain, M.; Farooq, M.; Farooq, S.; Jabran, K.; Nawaz, A. Economic assessment of conventional and conservation tillage practices in different wheat-based cropping systems of Punjab, Pakistan. *Environ. Sci. Pollut. Res.* **2017**, *24*, 24634–24643. [CrossRef]
43. Khanam, R.; Arefin, M.S.; Haque, M.A.; Islam, M.R.; Jahiruddin, M. Effects of magnesium, boron, and molybdenum on the growth, yield and protein content of chickpea and lentil. *Progress. Agric.* **2000**, *11*, 77–80.
44. Farooq, M.; Atique-ur-Rehman; Aziz, T.; Habib, M. Boron nutripriming improves the germination and early seedling growth of rice (*Oryza sativa* L.). *J. Plant Nutr.* **2011**, *34*, 1507–1515. [CrossRef]
45. Quddus, M.; Hossain, M.; Naser, H.; Naher, N.; Khatun, F. Response of chickpea varieties to boron application in calcareous and terrace soils of Bangladesh. *Bangladesh J. Agric. Res.* **2018**, *43*, 543–556. [CrossRef]
46. Farooq, M.; Barsa, S.M.A.; Wahid, A. Priming of field-sown rice seed enhances germination, seedling establishment, allometry and yield. *Plant Growth Regul.* **2006**, *49*, 285–294. [CrossRef]
47. Wimmer, M.A.; Lochnit, G.; Bassil, E.; Mühling, K.H.; Goldbach, H.E. Membrane-associated, boron-interacting proteins isolated by boronate affinity chromatography. *Plant cell Physiol.* **2009**, *50*, 1292–1304. [CrossRef]
48. Flores, R.A.; da Silva, R.G.; da Cunha, P.P.; Damin, V.; Abdala, K. de O.; Arruda, E.M.; Rodrigues, R.A.; Maranhão, D.D.C. Economic viability of Phaseolus vulgaris (BRS Estilo) production in irrigated system in a function of application of leaf boron. *Acta Agric. Scand. Sect. B Soil Plant Sci.* **2017**, *67*, 697–704. [CrossRef]
49. Yamori, W.; Kondo, E.; Sugiura, D.; Terashima, I.; Suzuki, Y.; Makino, A. Enhanced leaf photosynthesis as a target to increase grain yield: Insights from transgenic rice lines with variable Rieske FeS protein content in the cytochrome b_6/f complex. *Plant. Cell Environ.* **2016**, *39*, 80–87. [CrossRef]
50. Ali, J.; Revilleza, J.E.; Frangi, N.J.; Acero, B. Leaf area index and grain yield of elite green super rice under stress conditions. *Philipp. J. Crop Sci.* **2016**, *42*, 22–23.
51. Zhao, D.; Oosterhuis, D.M. Cotton growth and physiological responses to boron deficiency. *J. Plant Nutr.* **2003**, *26*, 855–867. [CrossRef]

52. Bangarwa, S.K.; Norsworthy, J.K. Brassicaceae Cover-Crop Effects on Weed Management in Plasticulture Tomato. *J. Crop Improv.* **2014**, *28*, 145–158. [CrossRef]
53. ur Rehman, H.; Iqbal, Q.; Farooq, M.; Wahid, A.; Afzal, I.; Basra, S.M.A. Sulphur application improves the growth, seed yield and oil quality of canola. *Acta Physiol. Plant.* **2013**, *35*, 2999–3006. [CrossRef]
54. Ullah, A.; Farooq, M.; Hussain, M. Improving the productivity, profitability and grain quality of kabuli chickpea with co-application of zinc and endophyte bacteria Enterobacter sp. MN17. *Arch. Agron. Soil Sci.* **2020**, *66*, 897–912. [CrossRef]
55. Dar, W.D. Macro-benefits from micronutrients for grey to green revolution in agriculture. In Proceedings of the IFA International Symposium on Micronutrients, New Delhi, India, 23–25 February 2004; pp. 1–13.
56. Saxena, N.P.; Saxena, M.C.; Ruckenbauer, P.; Rana, R.S.; El-Fouly, M.M.; Shabana, R. Screening techniques and sources of tolerance to salinity and mineral nutrient imbalances in cool season food legumes. *Euphytica* **1993**, *73*, 85–93. [CrossRef]

Article

Response of Natural Enemies toward Selective Chemical Insecticides; Used for the Integrated Management of Insect Pests in Cotton Field Plots

Amir Nadeem [1], Hafiz Muhammad Tahir [1], Azhar Abbas Khan [2], Atif Idrees [3,*], Muhammad Faisal Shahzad [4], Ziyad Abdul Qadir [5,6], Naveed Akhtar [1], Arif Muhammad Khan [7], Ayesha Afzal [3,8] and Jun Li [3,*]

[1] Department of Zoology, Government College University Lahore, Lahore 54000, Pakistan
[2] College of Agriculture, Bahauddin Zakariya University Multan, Bahadur Campus, Layyah 31200, Pakistan
[3] Guangdong Key Laboratory of Animal Conservation and Resource Utilization, Guangdong Public Laboratory of Wild Animal Conservation and Utilization, Institute of Zoology, Guangdong Academy of Sciences, Guangzhou 510260, China
[4] Department of Entomology, Gomal University, Dera Ismail Khan 29220, Pakistan
[5] Honeybee Research Institute, National Agricultural Research Centre, Park Road, Islamabad 45500, Pakistan
[6] Department of Entomology and Wildlife Ecology, University of Delaware, Newark, DE 19716, USA
[7] Department of Biotechnology, University of Sargodha, Sargodha 40100, Pakistan
[8] Institute of Molecular Biology and Biotechnology, The University of Lahore, 1-Km Defense Road, Lahore 54000, Pakistan
* Correspondence: atif_entomologist@giabr.gd.cn (A.I.); junl@giabr.gd.cn (J.L.)

Citation: Nadeem, A.; Tahir, H.M.; Khan, A.A.; Idrees, A.; Shahzad, M.F.; Qadir, Z.A.; Akhtar, N.; Khan, A.M.; Afzal, A.; Li, J. Response of Natural Enemies toward Selective Chemical Insecticides; Used for the Integrated Management of Insect Pests in Cotton Field Plots. *Agriculture* **2022**, *12*, 1341. https://doi.org/10.3390/agriculture12091341

Academic Editors: Mubshar Hussain, Sami Ul-Allah, Shahid Farooq and Stanislav Trdan

Received: 12 July 2022
Accepted: 26 August 2022
Published: 30 August 2022

Publisher's Note: MDPI stays neutral with regard to jurisdictional claims in published maps and institutional affiliations.

Abstract: Sucking pests of cotton (*Gossypium hirsutum* L.), such as thrips, or *Thrips tabaci* Lindeman, and jassid, or *Amrasca biguttula* Ishida, are among the most threatening insect pests to young cotton plants in Pakistan. New chemical insecticides have been trialed to control their damage in commercial fields. Formulations that show good suppression of these pest's populations, while sparing bio-controlling agents, are always preferred for obtaining better crop yield. Six different commercially available insecticides, namely Fountain® (fipronil and imidacloprid), Movento Energy® (spirotetramat and imidacloprid), Oshin® (dinotefuran), Concept Plus® (pyriproxyfen, fenpyroximate, and acephate), Maximal® (nitenpyram), and Radiant® (spinetoram) were evaluated in the present study to shortlist the best available insecticide against targeted pests. Harmful impacts of selected insecticides were also evaluated against naturally occurring predators, such as spiders and green lacewings (*Chrysoperla carnea*). Radiant® (spinetoram) and Movento Energy®, respectively, were best at controlling thrips (with 61% and 56% mortality, respectively) and jassid (62% and 57% mortality, respectively) populations during 2018 and 2019. Radiant® proved itself as the best option and showed minimal harmful effects on both major arthropod predators of cotton fields i.e., spiders (with 8–9% mortality) and green lacewings (with 12–16% mortality). Movento Energy® also showed comparatively less harmful effects (with 15–18% mortality) towards natural predatory fauna of cotton crops, as compared to other selective insecticides used in the study. The findings of current study suggest that the judicious use of target-oriented insecticides can be an efficient and predator-friendly management module in cotton fields. However, the impact of these chemicals is also depended on their timely application, keeping in consideration the ETL of pests and the population of beneficial arthropods.

Keywords: biological control; sustainable; natural predation; habitat management; sucking pests

1. Introduction

Currently, *Bt* cotton, *Gossypium hirsutum* L., is officially authorized in Pakistan since 2010 due to its better yield and high profitability [1], which are why it is named the "golden cash crop" of Pakistan [2]. In Pakistan, the cotton crop is attacked by a variety of insect pests which are reported to be as many as ~162 species [3]. The major insect pests of *Bt* cotton include sap-feeding pests, of which jassid, or *Amrasca biguttula* Ishida (Hemiptera: Cicadellidae), and thrips, or *Thrips tabaci* Lindeman (Thysanoptera: Thripidae), are more common. As such, they have emerged as a serious threat to the crop yield for cotton growers [4].

Most of these sap-feeding pests attack the crop at its early phenological growth stage [5]. The insect pests of cotton usually cause a 5 to 10% percent yield loss [6,7], but in favorable environmental conditions, they may pull down crop yield by 35 to 50% [7]. They stunt plant growth by sipping sap from soft tissues and also from the undersides of young leaves of growing plants; they may also deface them [8]. Thrips are the first among different sucking pests which invade cotton fields at the seedling stage of the crop and, hence, cause an adverse effect on the overall yield of the crop. Similarly, jassids are known as the most critical pest at the growing stage of cotton plants. They not only suck the plant's sap, but they also inject poisonous saliva into the plant tissues during its feeding. Early attack by jassid reduces the photosynthetic area of plants by affecting their young leaves [9]. This may cause a 23.67% reduction in the overall cotton yield if goes unchecked [10].

Two different strategies can be used to avoid major cotton yield losses i.e., biological and chemical control to keep these pest species below economic threshold level (ETL). In the early phenological stages when the plants are too young to face these severely damaging pest species, cotton growers mainly rely upon different types of insecticides, which give them quick relief against sudden pest outbreaks. For this purpose, pesticides worth 10 billion rupees are imported and, of these, almost 70–80% are applied to cotton crops alone [11]. Such massive use of these pesticides results in different types of health hazards for farmers, including environmental and soil pollution. Furthermore, insects are developing resistance to these insecticides [12].

Resistance is developed in insect pests against generally used insecticides, which become less effective over time. To overcome this issue, the discovery of some already existing molecules and the invention of new synthetic chemicals is an ongoing process. Spirotetramat (Movento Energy®) is a wide-ranging insecticide suitable for all crops. Newly developed insecticides exhibit excellent results in controlling not only thrips but other sucking pests as well [13]. Thus, the usage of innovative chemicals molecule, such as pyridine carboxamide and neonicotinoids is increasing. These synthetic molecules are required in lower quantities and are also cost-effective for the regulation of sucking pests of the cotton crop [14]. Neonicotinoids are usually applied as foliar sprays and are also broadly used for seed treatments in *Bt* cotton to reduce pest attacks [15]. As with neonicotinoids, fipronil and some other pesticides are also used as foliar sprays to control various pests which appear at different phenological stages of the crop [16].

This study aimed to evaluate the efficacy of two comparatively newer insecticides (Radiant® and Movento Energy®) and conventionally used formulations (Fountain®, Concept Plus®, Oshin®, and Maximal®) against two of the major sucking insect pests, as well as their impact on major natural predatory fauna. The experiments were conducted in cotton fields under natural agricultural conditions for the evaluation of six selective insecticides as a foliar application at their recommended field rates. Their effects on naturally occurring bio-control agents i.e., spiders and green lacewings were also evaluated.

2. Materials and Methods

The present study was conducted at the Agricultural Research Farms of Bahauddin Zakariya University, Bahadur Sub-Campus, located in a semi-arid zone in the Layyah District (70.98401° N, 31.17979° E), Punjab Pakistan. The area has an elevation of 148 m and the soil texture is sandy loam. The present experiments were conducted during the summers of 2018 and 2019. Randomized complete block design (RCBD) was used for the applications of selected insecticides in experimental plots. A total of 10 blocks (10 replications) were selected for the experiment. The block size was 42.5 m × 30 m. Each block was further divided into seven plots in which six different commercial pesticides were evaluated, while the seventh plot was left untreated as a control. Every treatment plot (12 square meters) was separated from the next plot by a buffer zone area of 9 square meters. Plant to plant and row to row distances of 23 cm and 75 cm were maintained, respectively. Each experimental block was divided into a cultivated area, water channels for irrigation purposes, and the paths of the experiment. A main water channel of a 2 m width was made along one side of the field plot. Four sub-water channels (1 m wide), linked to the main water channel, were established to irrigate every treatment plot. Three paths, each of 1 m width, were also made. A single cotton cultivar (*Bt* FH-142) was selected for evaluating the efficacy of the short-listed insecticides (Table 1). The pest and predator data were taken before and after the application of commercial insecticides on the designed dates. The seeds were collected from the Food Machinery Chemical Corporation (FMC), Pakistan. Delinted seeds of *Bt* cotton were used at the amount of 19.77 kg per hectare.

Table 1. List of insecticides used in experimental field plots and their rates of application to the cotton crop.

Product Name	Active Ingredient (IRAC Class)	Application Rate (per Acre)	Company Name
(Fountain® 80% WDG)	Fipronil (GABA-gated chloride channel blockers) + imidacloprid (nicotinic acetylcholine receptor competitive modulators)	50 g	Four Brothers Chemicals (Pvt.) Ltd. Lahore-Pakistan
(Movento Energy® 48% SC)	Spirotetramat 12% (inhibitor of acetyl CoA carboxylase) + imidacloprid 36% (nicotinic acetylcholine receptor competitive modulators)	150 mL + 250 mL adjuvant	Bayer Crop Science, Karachi-Pakistan
(Oshin® 20SG)	Dinotefuran (nicotinic acetylcholine receptor competitive modulators)	100 g	Arysta Life Sciences Pvt. Ltd. Karachi-Pakistan
(Concept Plus® 35% EC)	Pyriproxyfen (juvenile hormone mimics) + fenpyroximate (mitochondrial complex 1 electron transport inhibitors) + acephate (acetylcholinesterase inhibitors)	750 mL	Kanzo AG Multan-Pakistan
(Maximal®)	Nitenpyram (nicotinic acetylcholine receptor competitive modulators)	150 g	FMC Karachi-Pakistan
(Radiant® SC)	Spinetoram (nicotinic acetylcholine receptor allosteric modulators—Site 1)	100 mL	Dow Agro Sciences, Karachi-Pakistan
(Untreated)	Control	-	-

Cotton sowing was carried out on May 20th and May 15th in the years 2018 and 2019, respectively, on ridges created through proper plowing of field plots. The recommended fertilizers i.e., DAP, nitrogen, phosphorous, and potash were applied according to the cotton cultivation technology [16]. Irrigation was carried out weekly or when it was required according to field moisture conditions, but not extended to the interval of more than two weeks. No herbicides were applied and, instead, handpicking or plowing was performed for weed control.

Manual counting of seedlings was carried out after 15 days of cotton sowing in the field plots. To keep the recommended plant × plant (6–9 inches) and row × row (2.5 feet) distance, thinning was performed accordingly [17]. The treatments were applied when thrips or jassid populations reached their economic threshold level (ETL) (the ETL of jassid is 1 adult or nymph per leaf, while the ETL of thrips is 8–10 adults or nymphs per leaf), after taking pre-application data. All insecticides were liquefied in water according to the label directions before their application. For applications of treatments, a hand-operated knapsack sprayer (UK Registered Design No. 2025702) was used. The control plot was left untreated. Pests and bio-control agents were monitored regularly on a weekly basis by making direct observations, and pre-application data was recorded a day before the application of the insecticides. During each visit, adult pest populations were recorded from three leaves of five randomly selected plants. From every randomly selected plant, the first leaf was taken from slightly above the ground level, the second leaf from the middle portion of it, and the third leaf from its top canopy. Similarly, data on arthropod predators were also recorded [18].

The normality and distribution of data were checked before the statistical analysis. A generalized linear model (GLM) under a one-way analysis of variance (ANOVA) was performed, leading to the LSD to discern the means at $p < 0.05$. Correlations of thrips and jassid populations with each other and with temperature and humidity at different dates were also calculated using Spearman's correlation. The statistical package SPSS[®] version 16 was used for the analyses of data.

3. Results

Before the application of selected insecticides, significant differences were observed in the populations of *T. tabaci* in all experimental plots during both study years ($F_{6,63} = 2.55$; $p = 0.0283$ for 2018 and $F_{6,63} = 3.39$; $p = 0.0059$ for 2019). A significant decline in the *T. tabaci* populations of all treated groups was recorded as compared to the control for both years after first application (Table 2). Radiant[®] caused a maximum reduction in thrip population (0.65 ± 0.08) two days after treatment (2 DAT) ($F_{6,63} = 8.08$; $p < 0.001$), as compared to the pre-treatment during 2018. The same treatment showed the maximum population reduction (0.50 ± 0.08) on 2 DAT ($F_{6,63} = 8.62$; $p < 0.001$), as compared to pre-treatment in 2019.

Table 2. Effect of selective insecticides on number (mean $\pm$ SEM) of adult thrips per leaf on cotton plants during 2018 and 2019. The data were recorded pre-24 h and 2, 9, and 16 days after treatment (DAT).

Treatments	Pre-24 h		2 DAT		9 DAT		16 DAT	
	2018	2019	2018	2019	2018	2019	2018	2019
	The first application (1 July 2018, and 26 June 2019)							
Fountain[®]	2.08 ± 0.25 [ab]	1.78 ± 0.21 [ab]	1.10 ± 0.13 [bc]	0.94 ± 0.11 [bc]	1.33 ± 0.16 [ab]	1.14 ± 0.13 [ab]	1.83 ± 0.23 [ab]	1.57 ± 0.19 [ab]
Movento Energy[®]	2.66 ± 0.43 [a]	2.61 ± 0.40 [a]	1.15 ± 0.19 [bc]	1.13 ± 0.17 [bc]	1.46 ± 0.24 [ab]	1.43 ± 0.22 [ab]	2.13 ± 0.35 [ab]	2.09 ± 0.32 [a]
Oshin[®]	1.87 ± 0.20 [ab]	1.72 ± 0.19 [ab]	1.19 ± 0.13 [bc]	1.09 ± 0.12 [bc]	1.38 ± 0.15 [ab]	1.28 ± 0.14 [ab]	1.75 ± 0.19 [ab]	1.62 ± 0.17 [ab]
Concept Plus[®]	1.27 ± 0.24 [b]	1.18 ± 0.22 [b]	0.61 ± 0.11 [c]	0.56 ± 0.56 [c]	0.76 ± 0.14 [b]	0.71 ± 0.13 [b]	1.05 ± 0.20 [b]	0.98 ± 0.18 [b]
Maximal[®]	2.61 ± 0.47 [ab]	2.39 ± 0.44 [ab]	1.50 ± 0.26 [ab]	1.37 ± 0.24 [ab]	1.80 ± 0.33 [a]	1.65 ± 0.30 [a]	2.35 ± 0.43 [a]	2.16 ± 0.39 [a]
Radiant[®]	1.70 ± 0.22 [ab]	1.32 ± 0.22 [b]	0.65 ± 0.08 [c]	0.50 ± 0.08 [c]	0.88 ± 0.12 [b]	0.68 ± 0.11 [b]	1.27 ± 0.17 [ab]	0.99 ± 0.16 [b]
Control	2.35 ± 0.28 [ab]	2.14 ± 0.26 [ab]	2.07 ± 0.24 [a]	1.89 ± 0.23 [a]	2.15 ± 0.25 [a]	1.96 ± 0.24 [a]	2.30 ± 0.27 [a]	2.09 ± 0.26 [a]
	The second application (20 July 2018, and 15 July 2019)							
Fountain[®]	1.71 ± 0.24 [bc]	1.72 ± 0.24 [bc]	0.90 ± 0.12 [bc]	0.91 ± 0.12 [bc]	1.09 ± 0.15 [bc]	1.10 ± 0.15 [bc]	1.51 ± 0.21 [bc]	1.52 ± 0.21 [bc]
Movento Energy[®]	2.31 ± 0.24 [bc]	2.33 ± 0.24 [bc]	1.00 ± 0.10 [bc]	1.01 ± 0.10 [bc]	1.27 ± 0.13 [bc]	1.28 ± 0.13 [bc]	1.86 ± 0.20 [bc]	1.87 ± 0.20 [bc]
Oshin[®]	1.99 ± 0.40 [bc]	2.00 ± 0.40 [bc]	1.26 ± 0.25 [bc]	1.26 ± 0.25 [bc]	1.47 ± 0.30 [bc]	1.49 ± 0.30 [bc]	1.87 ± 0.38 [bc]	1.88 ± 0.38 [bc]
Concept Plus[®]	1.56 ± 0.22 [bc]	1.57 ± 0.22 [bc]	0.75 ± 0.11 [bc]	0.75 ± 0.11 [bc]	0.94 ± 0.13 [bc]	0.95 ± 0.13 [bc]	1.30 ± 0.19 [c]	1.31 ± 0.19 [c]
Maximal[®]	2.85 ± 0.37 [ab]	2.87 ± 0.37 [ab]	1.64 ± 0.20 [b]	1.65 ± 0.21 [b]	1.97 ± 0.25 [b]	1.98 ± 0.25 [b]	2.57 ± 0.33 [b]	2.59 ± 0.34 [b]
Radiant[®]	1.35 ± 0.07 [c]	1.36 ± 0.07 [c]	0.52 ± 0.02 [c]	0.52 ± 0.03 [c]	0.70 ± 0.03 [c]	0.71 ± 0.03 [c]	1.01 ± 0.05 [c]	1.02 ± 0.05 [c]
Control	4.16 ± 0.48 [a]	4.19 ± 0.48 [a]	3.67 ± 0.43 [a]	3.70 ± 0.43 [a]	3.80 ± 0.44 [a]	3.83 ± 0.44 [a]	4.07 ± 0.49 [a]	4.10 ± 0.49 [a]

Note: The values in the columns with different superscripts are significantly different when compared by LSD.

The minimum population reduction was recorded by the Oshin[®] on 2 DAT (1.19 ± 0.13 and 1.09 ± 0.12 for 2018 and 2019, respectively) as compared to the population before treatment (1.87 ± 0.20 and 1.72 ± 0.19 for 2018 and 2019, respectively). A similar trend was also recorded against the second application of insecticide treatments on 20 July 2018, and 15 July 2019. The differences in the populations of *T. tabaci* at 9 DAT ($F_{6,63} = 5.30$; $p = 0.0002$) and 16 DAT ($F_{6,63} = 3.26$; $p = 0.0074$) of 2018 and 9 DAT ($F_{6,63} = 5.92$; $p = 0.001$) and 16 DAT ($F_{6,63} = 4.04$; $p = 0.0017$) of 2019 were also statistically significant.

After the application of various insecticides, a significant reduction in the jassid population was observed in all treated plots. The results given in Table 3 showed mean population reduction in jassid by various insecticide treatments in both applications during the years 2018 and 2019. The data indicated that maximum mortality (0.87 ± 0.09 and 0.56 ± 0.05) was recorded with the Radiant[®] at 2 DAT ($F_{6,63} = 20.4$; $p < 0.001$) in 2018. A similar trend was also recorded at 2 DAT ($F_{6,63} = 21.6$; $p < 0.001$) for the year 2019. In the second application, the difference at 9 DAT ($F_{6,63} = 5.71$; $p < 0.0001$) and 16 DAT ($F_{6,63} = 3.38$; $p = 0.0059$) of 2018, and 9 DAT ($F_{6,63} = 5.58$; $p = 0.0001$) and 16 DAT ($F_{6,63} = 3.29$; $p = 0.0071$) of 2019 were also significant.

Table 3. Effect of selective insecticides on number (mean $\pm$ SEM) of adult jassids per leaf on cotton plants during 2018 and 2019. The data was recorded pre-24 h and 2, 9, and 16 days after treatment (DAT).

Treatments	Pre-24 h		2 DAT		9 DAT		16 DAT	
	2018	2019	2018	2019	2018	2019	2018	2019
	The first application (1 July 2018, and 26 June 2019)							
Fountain®	1.48 ± 0.17 [c]	1.53 ± 0.18 [c]	0.79 ± 0.09 [c]	0.81 ± 0.09 [c]	0.95 ± 0.11 [c]	0.98 ± 0.11 [c]	1.30 ± 0.15 [c]	1.35 ± 0.15 [cd]
Movento Energy®	2.85 ± 0.40 [bc]	2.94 ± 0.41 [bc]	1.24 ± 0.18 [bc]	1.28 ± 0.18 [bc]	1.57 ± 0.22 [bc]	1.62 ± 0.23 [bc]	2.29 ± 0.33 [bc]	2.36 ± 0.33 [bcd]
Oshin®	3.39 ± 0.49 [ab]	3.49 ± 0.50 [ab]	2.15 ± 0.31 [b]	2.21 ± 0.32 [b]	2.51 ± 0.36 [b]	2.59 ± 0.37 [b]	3.18 ± 0.46 [ab]	3.28 ± 0.48 [ab]
Concept Plus®	1.52 ± 0.19 [c]	1.58 ± 0.20 [c]	0.73 ± 0.09 [c]	0.76 ± 0.10 [c]	0.91 ± 0.12 [c]	0.95 ± 0.12 [c]	1.26 ± 0.16 [c]	1.32 ± 0.17 [d]
Maximal®	3.13 ± 0.36 [bc]	3.23 ± 0.37 [abc]	1.79 ± 0.20 [bc]	1.85 ± 0.20 [bc]	2.16 ± 0.25 [bc]	2.23 ± 0.25 [bc]	2.82 ± 0.32 [bc]	2.91 ± 0.33 [bc]
Radiant®	2.29 ± 0.24 [bc]	2.36 ± 0.25 [bc]	0.87 ± 0.09 [c]	0.90 ± 0.10 [c]	1.19 ± 0.13 [bc]	1.23 ± 0.13 [c]	1.71 ± 0.18 [bc]	1.77 ± 0.19 [bcd]
Control	4.88 ± 0.69 [a]	4.92 ± 0.65 [a]	4.29 ± 0.60 [a]	4.33 ± 0.57 [a]	4.46 ± 0.63 [a]	4.50 ± 0.60 [a]	4.76 ± 0.67 [a]	4.80 ± 0.63 [a]
	The Second application (20 July 2018, and 15 July 2019)							
Fountain®	1.59 ± 0.16 [ab]	1.75 ± 0.17 [ab]	0.85 ± 0.09 [cd]	0.93 ± 0.09 [bcd]	1.02 ± 0.10 [bc]	1.12 ± 0.11 [bc]	1.40 ± 0.14 [ab]	1.53 ± 0.15 [ab]
Movento Energy®	1.70 ± 0.27 [ab]	1.85 ± 0.29 [ab]	0.73 ± 0.11 [cd]	0.80 ± 0.12 [cd]	0.94 ± 0.15 [c]	1.02 ± 0.16 [c]	1.36 ± 0.21 [b]	1.48 ± 0.23 [b]
Oshin®	1.80 ± 0.20 [ab]	1.95 ± 0.22 [ab]	1.14 ± 0.13 [abc]	1.24 ± 0.14 [abc]	1.33 ± 0.15 [abc]	1.45 ± 0.17 [abc]	1.69 ± 0.19 [ab]	1.83 ± 0.21 [ab]
Concept Plus®	1.83 ± 0.25 [ab]	1.98 ± 0.26 [ab]	0.88 ± 0.11 [bcd]	0.95 ± 0.12 [bcd]	1.10 ± 0.15 [abc]	1.19 ± 0.16 [abc]	1.52 ± 0.21 [ab]	1.65 ± 0.22 [ab]
Maximal®	2.46 ± 0.27 [a]	2.66 ± 0.29 [a]	1.41 ± 0.16 [ab]	1.53 ± 0.17 [ab]	1.70 ± 0.18 [a]	1.84 ± 0.20 [a]	2.22 ± 0.24 [a]	2.40 ± 0.26 [a]
Radiant®	1.47 ± 0.12 [b]	1.60 ± 0.13 [b]	0.56 ± 0.05 [d]	0.61 ± 0.05 [d]	0.76 ± 0.06 [c]	0.84 ± 0.07 [c]	1.10 ± 0.09 [b]	1.20 ± 0.10 [b]
Control	1.80 ± 0.23 [ab]	1.96 ± 0.25 [ab]	1.58 ± 0.20 [a]	1.73 ± 0.22 [a]	1.64 ± 0.21 [ab]	1.79 ± 0.23 [ab]	1.76 ± 0.23 [ab]	1.92 ± 0.26 [ab]

Note: The values in the columns with different superscripts are significantly different when compared by LSD.

The preliminary data analysis showed an almost similar trend of both predator (green lacewings and spiders) densities during both years of the study period and, hence, the data was pooled together for further statistical analysis. The selected insecticides affected both green lacewings and spiders in the following trends i.e., Oshin® > Maximal® > Fountain® > Concept Plus® > Movento Energy® > Radiant®. The results indicated that maximum population reduction in the beneficial arthropod, *C. carnea* was recorded against the application of Oshin® before pre-treatment (0.649), ($F_{6,63}$ = 0.85; p = 0.5348) followed by post-treatment i.e., 2 DAT (0.189), ($F_{6,63}$ = 24.7; p = 0.0000) in 2018, which caused maximum mortality among all treatments. Radiant®, on the other hand, caused minimum mortality of the *C. carnea* (pre-treatment, 0.656 $\pm$ 0.0741 and post-treatment, 0.577 $\pm$ 0.0645) having ANOVA values ($F_{6,63}$ = 0.85; p = 0.5348) and 2 DAT ($F_{6,63}$ = 24.7; p = 0.0000), and spiders (pre-treatment, 1.062 $\pm$ 0.0783 and post-treatment, 0.987 $\pm$ 0.0745) with ANOVA values ($F_{6,63}$ = 3.27; p = 0.0073) and 2 DAT ($F_{6,63}$ = 35.1; p = 0.0000), during the year 2018. The varying effect of these insecticides on these beneficial arthropods was also revealed by post-hoc test results (Figures 1 and 2).

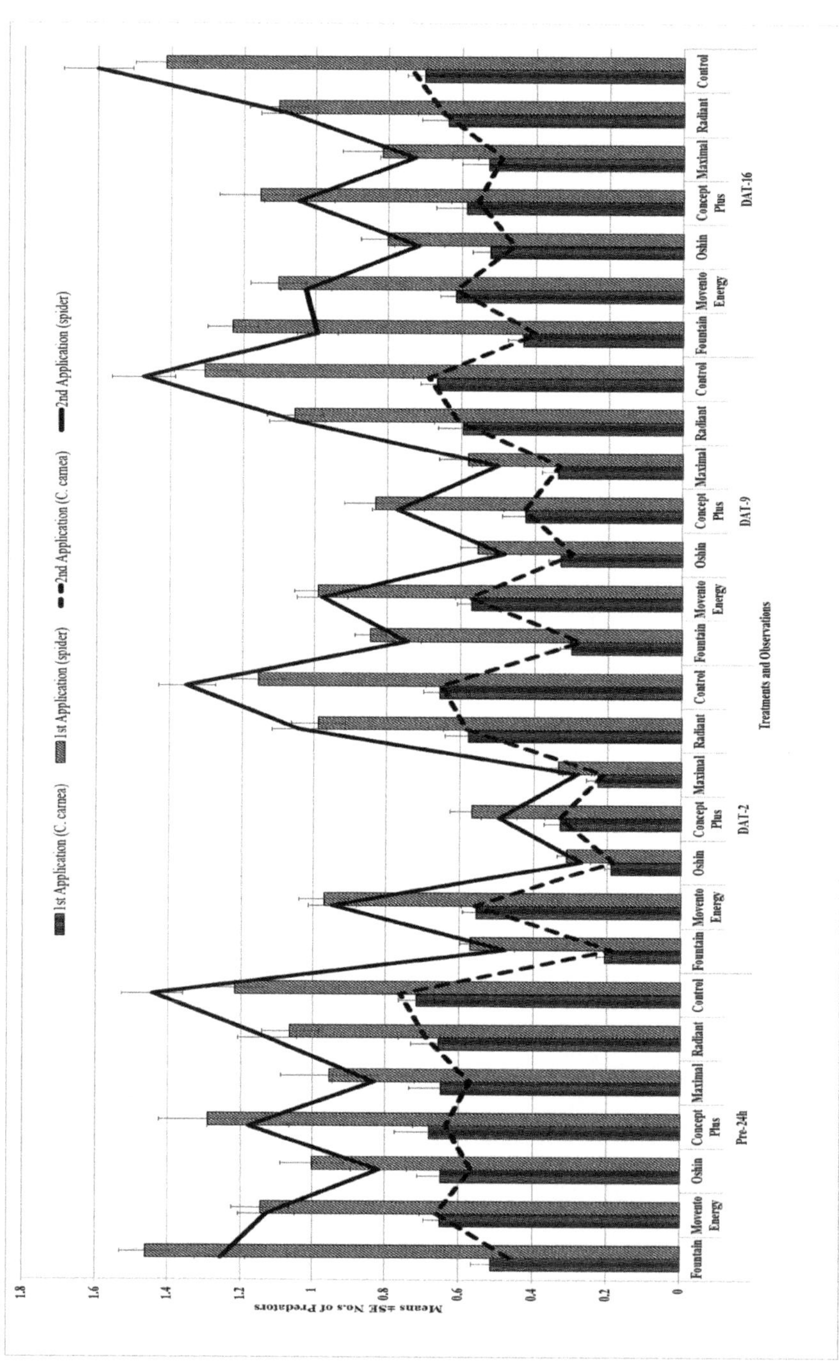

Figure 1. Effect of different insecticides on numbers of *C. carnea* and predatory spiders per plant. Mean (±SEM) in *Bt* cotton field plots on 1 July 2018 and 20 July 2018, observed at pre-24 h, 2 DAT, 9 DAT, and 16 DAT (days after treatment).

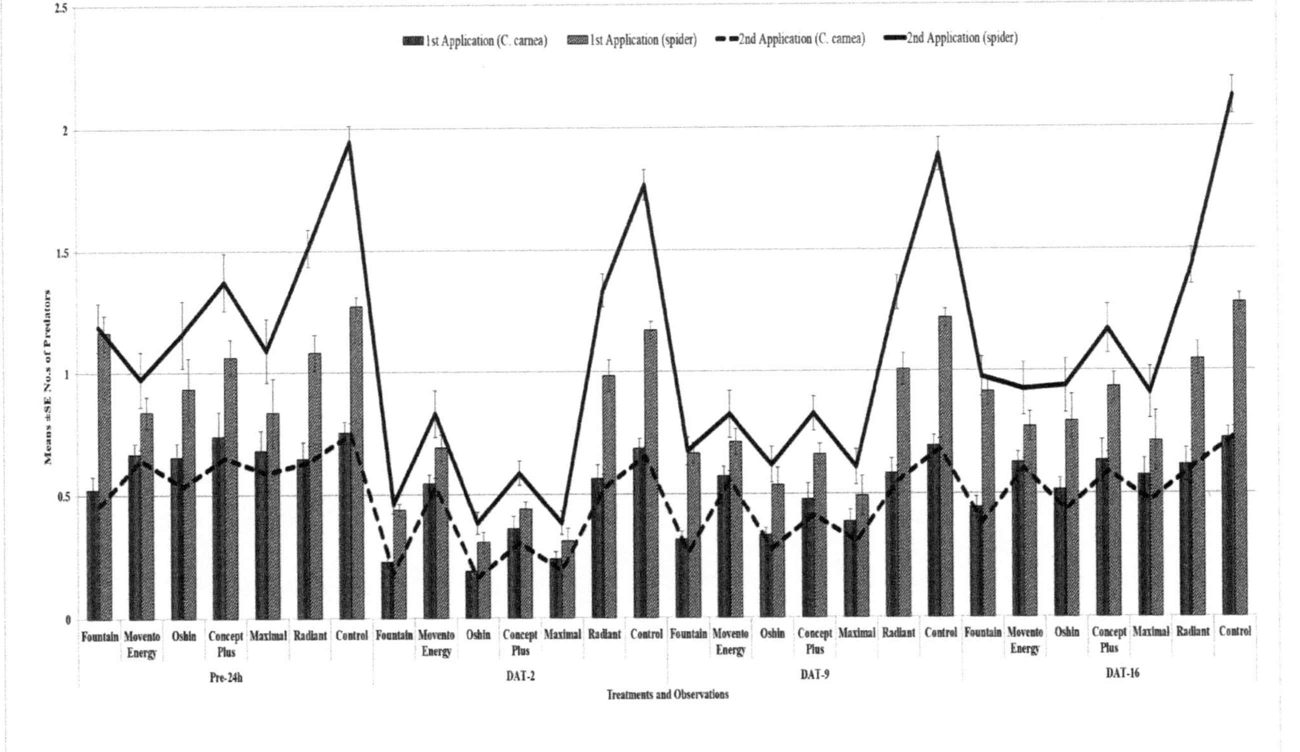

Figure 2. Effect of different insecticides on numbers of *C. carnea* and predatory spiders per plant. Mean (±SEM) in *Bt* cotton field plots on 26 June 2019 and 15 July 2019, observed at pre-24 h, 2 DAT, 9 DAT, and 16 DAT (days after Treatment).

4. Discussion

Cotton growers used a wide range of commercially available insecticides against sucking pests but, due to their regular usage, pests can develop resistance against these insecticides, meaning that they become less effective in controlling pests [19]. Moreover, their impact on beneficial fauna like common predators including beetles (coccinellids), green lacewings, ants, and parasitoids (wasps), along with environmental contamination, are major issues related to the unjust usage of these broad-spectrum insecticides [20]. Therefore, there is a need to evaluate and select those insecticide formulations and active ingredients which are more target-specific in their mode of action against sucking insect pests of cotton and are friendly towards the environment and beneficial predators. Recently, it has been proposed that controlling these pests biologically is eco-friendlier than using synthetic insecticides [21].

Results of the present study showed that all selected insecticides imposed a significant reduction in the populations of two major sucking pests (thrip and jassid) in treated field plots of respective treatments when compared to the pest populations found in untreated control plots, suggesting their efficacy. However, Radiant® proved itself as the most bio-friendly insecticide among all the selected chemical insecticides for the study. All selected treatments work efficiently and showed a reduction in *T. tabaci* populations for both years of study as compared to control (untreated). Radiant®, Movento Energy®, and Concept Plus® showed more of a reduction in *T. tabaci* populations at their first and second applications as compared to other insecticides during the years 2018–2019. Chloridis et al. [22] also reported that spinetoram proved more efficient against many insect pests as compared to the plots sprayed with spinosad formulations.

Maximum mortality (percentage) of thrips was observed in the field plots treated with Radiant® after 2 DAT (61.76% and 61.48%) to 16 DAT (25.29% and 25.18%) after its first and second application, respectively, during the experiment year 2018 (Table 2). Spinetoram was described by Dripps et al. [23] as an active ingredient of a semi-synthetic nature that demonstrated higher levels of efficacy than spinosad, especially against lepidopteran larvae, thrips, and leaf miners on a variety of crops and horticultural plants. Waters and Walsh [24] reported spinetoram as an efficacious insecticide against onion thrips, while spirotetramat provides satisfactory control of the thrip population. Ghelani et al. [25] also evaluated the effectiveness of different insecticides of both chemical and botanical origin and documented that all of the evaluated botanicals and insecticides were effective against thrips and other major sucking insect pests.

All of the six tested insecticides showed significant mortality in jassid populations observed post-treatment for both experimental years 2018–2019. However, two insecticides, namely Radiant® and Movento Energy®, gave the highest mortality ratios of jassid. Gogi et al. [26], also mentioned that, in cotton fields, rapid control of different pest species through chemical insecticides is the best strategy, and it plays an important role in different integrated pest management (IPM) programs. In the present study, all selected chemical insecticides showed control against the major sucking pests of the crops.

All of the selective insecticides reduced the total number of beneficial fauna (non-target) to almost one-half of their total population. However, in the present study, Radiant® proved to be the much more friendly option towards the beneficial fauna of the crops, especially towards green lacewings and spiders. Some new insect growth regulators, such as buprofezin and pyriproxyfen, proved themselves to be quite target-specific and showed minimal impact on beneficial predatory fauna [27].

5. Conclusions

At the vegetative stage of the cotton crop, a repeated spray of chemical insecticides for controlling different sucking pests, especially jassid and thrip, is of great importance. However, to provide a safeguard for the environment, beneficial predatory fauna, as well as cotton growers, safe and more specific insecticides should be developed and tested regularly. The results of our tested chemicals showed that two bio-based chemical insecticides, namely Radiant® and Movento Energy®, showed less harm to beneficial predatory fauna while controlling the sucking pest populations. These insecticides could be selected as the first choice for future IPM strategies and also in places where usage of conventional insecticides is restricted, for example in organic farming land areas. The findings of pests' resistance against commonly practiced broad-spectrum insecticides, such as neonicotinoid and acephate, show the need for their low usage, and also the need to integrate them with some milder insecticides.

Author Contributions: Conceptualization, A.N., H.M.T., A.A.K. and A.I.; Investigation, A.N., H.M.T. and A.A.K.; Methodology A.N., H.M.T. and A.A.K.; Project administration, J.L.; Software, M.F.S., Z.A.Q. and A.A.; Supervision, A.A.K. and H.M.T.; Formal analysis, A.I. and A.M.K.; Writing—original draft, A.N. and A.I.; Writing—review and editing, H.M.T., A.A.K., M.F.S., Z.A.Q., N.A., A.M.K., A.A. and J.L.; Validation, A.M.K. and N.A.; Data curation, A.I. and A.A.; Funding acquisition, A.I. and J.L. All authors have read and agreed to the published version of the manuscript. All the authors have contributed substantially to the work reported.

Funding: This research work was supported by GDAS Special Project of Science and Technology Development (No. 2020GDASYL-20200301003) and GDAS Action Capital Project to build a comprehensive industrial technology innovation center (No. 2022GDASZH-2022010106).

Institutional Review Board Statement: Not applicable.

Informed Consent Statement: Not applicable.

Data Availability Statement: Not applicable.

Acknowledgments: We are thankful to the local farmers for the provision of crop field's area for the current research. We are also grateful to Department of Entomology, College of Agriculture, Bahauddin Zakariya University, Bahadur Campus Layyah-Pakistan for providing entomological equipment and related facilities.

Conflicts of Interest: The authors declare no conflict of interest. The funders had no role in the design of the study; in the collection, analyses, or interpretation of data; in the writing of the manuscript; or in the decision to publish the results.

References

1. Kouser, S.; Qaim, M. Valuing financial, health, and environmental benefits of *Bt* cotton in Pakistan. *Agric. Econ.* **2013**, *44*, 323–335. [CrossRef]
2. Iqbal, J.; Ali, Z.; Aasi, M.S.; Ali, A.; Rasul, A.; Begum, H.A.; Nadeem, M. Evaluation of some new chemistry insecticides against cotton whitefly (*Bemisia tabaci* Genn.) (Hemiptera: Aleyrodidae). *Pak. Entomol.* **2018**, *40*, 19–23.
3. Sarwar, M.; Sattar, M. An Analysis of Comparative Efficacies of Various Insecticides on the Densities of Important Insect Pests and the Natural Enemies of Cotton, *Gossypium hirsutum* L. *Pak. J. Zool.* **2016**, *48*, 131–136.
4. Vennila, S. Pest management for cotton ecosystems or ecosystem management for cotton protection? *Curr. Sci.* **2008**, *94*, 1351–1352.
5. Reed, J.T.; Jackson, C.S. *Thrips on Mississippi Seedling Cotton: Pest Overview and 15-Year Summary of Pesticide Evaluation*; Bulletin 1124; Mississippi State University Extension Service: Washington, DC, USA, 2002. Available online: https://lccn.loc.gov/2004356107 (accessed on 27 October 2019).
6. Central Cotton Research Institute (CCRI). *Annual Report*; CCRI: Multan, Pakistan, 2005; p. 23.
7. Ali, A. Physio-Chemical Factors Affecting Resistance in Cotton against Jassid, Amrasca Devastans (Dist.) and Thrips, Thrips Tabaci (Lind.) in Punjab, Pakistan. Ph.D. Thesis, Department of Entomology, University of Agriculture, Faisalabad, Pakistan, 1992. Volume 1. pp. 1–293.
8. Masood, A.; Arif, M.J.; Hamed, M.; Talpur, M.A. Field performance of *Trichogramma chilonis* against cotton bollworms infestation in different cotton varieties as a sustainable IPM approach. *PJAAEVS* **2011**, *27*, 176–184.
9. Sahito, H.A.; Shah, Z.H.; Kousar, T.; Mangrio, W.M.; Mallah, N.A.; Jatoi, F.A.; Kubar, W.A. Comparative efficacy of novel pesticides against Jassid, *Amrasca biguttula biguttula* (Ishida) on cotton crop under field conditions at Khairpur, Sindh, Pakistan. *Singap. J. Sci. Res.* **2017**, *1*, 1–8. [CrossRef]
10. Razaq, M.; Suhail, A.; Aslam, M.; Arif, M.J.; Saleem, M.A.; Khan, M.H.A. Evaluation of neonicotinoids and conventional insecticides against cotton jassid, *Amrasca devastans* (Dist.) and cotton whitefly, *Bemisia tabaci* (Genn.) on cotton. *Pak. Entomol.* **2005**, *27*, 75–78.
11. Government of Pakistan. *Economic Survey of Pakistan, 2007–2008*; Finance Division: Islamabad, Pakistan, 2008; pp. 17–37.

12. Mahmood, I.; Imadi, S.R.; Shazadi, K.; Gul, A.; Hakeem, K.R. Effects of Pesticides on Environment. In *Plant, Soil and Microbes*; Springer: Cham, Switzerland, 2016; pp. 253–269.

13. Udikeri, S.S.; Patil, S.B.; Hirekurubar, R.B.; Guruprasad, G.S.; Shailah, H.M.; Matti, P.V. Management of sucking pests in cotton with new insecticides. *Karnataka J. Agric. Sci.* **2009**, *22*, 798–802.

14. Gaurkhede, A.S.; Bhalkare, S.K.; Sadawarte, A.K.; Undirwade, D.B. Bio efficacy of new chemistry molecules against sucking pests of *Bt* transgenic cotton. *Int. J. Plant Protec.* **2015**, *8*, 7–12. [CrossRef]

15. Zhang, Z.; Wang, Y.; Zhao, Y.; Li, B.; Lin, J.; Zhang, X.; Liu, F.; Mu, W. Nitenpyram seed treatment effectively controls against the mirid bug *Apolygus lucorum* in cotton seedlings. *Sci. Rep.* **2017**, *7*, 8573. [CrossRef] [PubMed]

16. Bonmatin, J.M.; Giorio, C.; Girolami, V.; Goulson, D.; Kreutzweiser, D.P.; Krupke, C.; Liess, M.; Long, E.; Marzaro, M.; Mitchell, E.A.D.; et al. Environmental fate and exposure; neonicotinoids and fipronil. *Environ. Sci. Pollut. Res.* **2015**, *22*, 35–67. [CrossRef] [PubMed]

17. Government of Punjab. *Agriculture Department, Crop Advisory Wing, Sir Agha Khan Road*; Government of Punjab: Lahore, Pakistan, 2017; pp. 1–27.

18. Nadeem, A.; Tahir, H.M.; Khan, A.A. Plant age, crop stage and surrounding habitats: Their impact on sucking pests and predators complex in cotton (*Gossypium hirsutum* L.) field plots in arid climate at district Layyah, Punjab, Pakistan. *Braz. J. Biol.* **2021**, *82*, e236494. [CrossRef] [PubMed]

19. Luttrell, R.G.; Teague, T.G.; Brewer, M.J. Cotton Insect Pest Management. In *Cotton*, 2nd ed.; Agronomy Monographs 57; ASA, CSSA, and SSSA: Madison, WI, USA, 2015; pp. 509–546. [CrossRef]

20. Simon-Delso, N.; Amaral-Rogers, V.; Belzunces, L.P.; Bonmatin, J.M.; Chagnon, M.; Downs, C.; Goulson, D. Systemic insecticides (neonicotinoids and fipronil): Trends, uses, mode of action and metabolites. *Environ. Sci. Pollut. Res.* **2015**, *22*, 5–34. [CrossRef] [PubMed]

21. Haider, I.; Suhail, A.; Aziz, A. Toxicity of some insecticides against cotton jassid (*Amrasca devastans* Dist.) and its predator (*Chrysoperla carnea* Stephens). *J. Agric. Res.* **2017**, *55*, 311–321.

22. Chloridis, A.; Downard, P.; Dripps, J.E.; Kaneshi, K.; Lee, L.C.; Min, Y.K.; Pavan, L.A. Spinetoram (XDE-175): A new spinosyn. In Proceedings of the XVI International Plant Protection Congress, Scotland, UK, 15–18 October 2007; pp. 44–49.

23. Dripps, J.; Olson, B.; Sparks, T.; Crouse, G. Spinetoram: How artificial intelligence combined natural fermentation with synthetic chemistry to produce a new spinosyn insecticide. *Plant Health Progress* **2008**, *22*. Available online: http://www.plantmanagementnetwork.org/pub/php/perspective/2008/spinetoram/ (accessed on 27 October 2019).

24. Waters, T.D.; Walsh, D.B. Onion Thrips Control in Washington State. 2010. Available online: https://www.unce.unr.edu/ (accessed on 17 November 2019).

25. Ghelani, M.K.; Kabaria, B.B.; Chhodavadia, S.K. Field efficacy of various insecticides against major sucking pests of *Bt* cotton. *J. Biopestic.* **2014**, *7*, 27.

26. Gogi, M.D.; Sarfraz, R.M.; Dosdall, L.M.; Arif, M.J.; Keddie, A.B.; Ashfaq, M. Effectiveness of two insect growth regulators against *Bemisia tabaci* (Gennadius) (Hemiptera: Aleyrodidae) and *Helicoverpa armigera* (Hübner) (Lepidoptera: Noctuidae) and their impact on population densities of arthropod predators in cotton in Pakistan. *Pest Manag. Sci.* **2006**, *62*, 982–990. [CrossRef] [PubMed]

27. Naranjo, S.E.; Ellsworth, P.C.; Hagler, J.R. Conservation of natural enemies in cotton: Role of insect growth regulators in management of *Bemisia tabaci*. *Biol. Control* **2004**, *30*, 52–72. [CrossRef]

Article

Legume Species Alter the Effect of Biochar Application on Microbial Diversity and Functions in the Mixed Cropping System—Based on a Pot Experiment

Akari Kimura [1], Yoshitaka Uchida [2,*] and Yvonne Musavi Madegwa [2]

1   Environmental Biogeochemistry Laboratory, Graduate School of Global Food Resources, Hokkaido University, Sapporo 0600809, Japan
2   Environmental Biogeochemistry Laboratory, Graduate School of Agriculture, Hokkaido University, Sapporo 0608589, Japan
*   Correspondence: uchiday@chem.agr.hokudai.ac.jp; Tel.: +81-11-706-4006

Abstract: Biochar application to legume-based mixed cropping systems may enhance soil microbial diversity and nitrogen (N)-cycling function. This study was conducted to elucidate the effect of biochar application on soil microbial diversity and N-cycling function with a particular focus on legume species. Therefore, we performed a pot experiment consisting of three legume species intercropped with maize: cowpea, velvet bean, and common bean. In addition, one of three fertilizers was applied to each crop: biochar made of chicken manure (CM), a chemical fertilizer, or no fertilizer. Amplicon sequencing for the prokaryotic community and functional prediction with Tax4Fun2 were conducted. Under the CM, Simpson's diversity index was higher in soils with common beans than those in other legume treatments. On the other hand, N-cycling genes for *ammonia oxidation* and *nitrite reductase (NO-forming)* were more abundant in velvet bean/maize treatment, and this is possibly due to the increased abundance of *Thaumarchaeota* (6.7%), *Chloroflexi* (12%), and *Planctomycetes* (11%). Cowpea/maize treatment had the lowest prokaryotes abundances among legume treatments. Our results suggest that the choice of legume species is important for soil microbial diversity and N-cycling functions in CM applied mixed cropping systems.

Keywords: mixed cropping; legume; microbial diversity; biochar; nitrogen cycling

Citation: Kimura, A.; Uchida, Y.; Madegwa, Y.M. Legume Species Alter the Effect of Biochar Application on Microbial Diversity and Functions in the Mixed Cropping System—Based on a Pot Experiment. *Agriculture* **2022**, *12*, 1548. https://doi.org/10.3390/agriculture12101548

Academic Editors: Mubshar Hussain, Sami Ul-Allah and Shahid Farooq

Received: 15 August 2022
Accepted: 21 September 2022
Published: 25 September 2022

Publisher's Note: MDPI stays neutral with regard to jurisdictional claims in published maps and institutional affiliations.

1. Introduction

Modern agricultural practices, including monoculture cropping systems and the intensive use of inorganic fertilizers, have led to soil degradation and a loss of genetic diversity in soil [1,2]. Among the different scales of biodiversity, the diversity of soil microorganisms is especially important for the stability of agricultural ecosystems because soil microbes can be considered the main drivers of the biogeochemical reactions that are critical for soil health and crop productivity [3]. For example, they are especially involved in biogeochemical processes essential for plant health and growth, including nutrient absorption, immune function, pathogen prevention, and stress tolerance [4,5]. Therefore, agricultural management systems that maintain or increase soil microbial diversity and functioning should be established.

Among the various agricultural practices that can potentially promote soil microbial diversity, the use of mixed cropping systems has received increasing attention. Legume-based intercropping systems have been reported to enhance soil microbial diversity and functions, including the mineralization of available phosphorus (P) and nitrogen (N) [6,7]. In particular, sweet maize (*Zea may* L.)/soybean (*Glycine max* L.) intercropping systems have been reported to promote the expression of key genes involved in N-cycling (e.g., ammonium oxidation, nitrite reductase, and nitrous-oxide reductase) [8]. Moreover, mixed cropping systems with field pea varieties have been demonstrated by Horner et al. [9] to build stronger and larger co-occurrence networks in rhizosphere prokaryotic communities compared to monocropping systems. These studies have suggested that crop diversification, including mixed cropping systems, improves soil microbial diversity and nutrient availability in agricultural soil. However, outside of certain successful cases, not all intercropping systems improve soil microbial conditions and crop productivity. Previous research has indicated that microbial community composition and structure are often less affected by intercropping systems [10–12]. In addition, cereal/legume intercropping systems did not improve plant production compared with monocropping systems [13,14]. This inconsistency can be explained by interspecific interactions between legumes and cereals, as plant combinations in intercropping systems significantly alter soil microbial communities composition and relative abundance. In addition, the soil microbes in the rhizosphere of legume crops have a high diversity and are species-specific due to their variable N acquisition abilities through symbiosis with rhizobia [15]. However, few studies have yet examined how different legume combinations in mixed cropping systems may alter soil microbial community composition, structure, and functions (specifically, N-cycling).

In addition to the type of legume species, fertilizer is another common agricultural management technique that plays a major role in determining microbial community composition and functioning. The use of organic fertilizer often enhances the abundance and diversity of prokaryotes, when compared to inorganic fertilizer [16,17]. Among variable organic fertilizers, biochar has been reported by many previous studies to have positive effects on soil microbial diversity and functioning in mixed cropping systems [18,19]. Furthermore, the combined use of biochar and legume-based intercropping systems could alter the expression of microbial N-cycling genes, which is important for plant growth. For example, biochar application stimulates microbial *ammonium oxidation* and reduces gaseous N loss from soil by reducing the denitrification potential [20]. However, the effect of biochar on plant beneficial functions are reported as plant species-specific. For example, a number of studies reviewed by Kochaneck et al. [21] showed that the variability of plant species significantly impacted the soil microbial community in biochar-applied soil. Plant root exudates contain organic acids (citric acid, malic acid, and ethanoic acid) that promote biochar decomposition, with the amount and quality varying based on plant species [22]. Therefore, the difference in legume species used in mixed cropping might be responsible for the variable nutrient availability of biochar, which is important for the soil microbial community and N cycling function. Although most studies have examined the effects of biochar in comparison with monocropping systems, more studies are necessary to address the biochar effect in different intercropping systems.

This study elucidated legume species-specific effects of maize/legume mixed cropping on soil microbial community structure and their predicted functions, with a special focus on the interaction between legume species and fertilizer. The main hypotheses of this study were: (i) soil microbial community compositions are different among the legume species used in the mixed cropping system, and (ii) combined use of specific legumes and biochar promotes microbial diversity and enhances microbial N-cycling rates. To test these hypotheses, a greenhouse legume-maize mixed cropping experiment was set-up using three legume varieties commonly used in mixed cropping systems [23–25] and a chemical or biochar (carbonized chicken manure (CM)) as fertilizer. To test the impact of these factors on soil microbial communities structure, prokaryotic diversity and functional diversity were assessed using a 16S rRNA gene survey.

2. Materials and Methods

2.1. Soil Sampling

The soil used in this experiment was sampled from land that had been abandoned for 30 years to develop a low-nutrient-input system at a university farm located at the Field Science Centre for Northern Biosphere, Hokkaido University, Japan (43°04′ N, 141°20′ E). The properties of the soil are given in Table 1. The soil type was clay loam with 44.6% sand, 21.5% silt, and 33.9% clay.

Table 1. Chemical properties of soil and carbonized chicken manure (CM).

Chemical Properties	Soil	CM
Water content (%)	19.5 ± 0.1	19.1 ± 1.2
C/N	14.7 ± 0.2	10.0 ± 0.4
pH (1:10)	6.3 ± 0.0	10.4 ± 0.3
EC (µs/cm)	78.5 ± 3.5	12740 ± 793
Total nitrogen (%)	0.25 ± 0.01	4.06 ± 0.45
Total carbon (%)	3.9 ± 0.1	40.5 ± 3.4
P g kg^{-1} soil	0.21 ± 0.01	36.8 ± 1.0
Ca g kg^{-1} soil	5.9 ± 0.3	134 ± 5
Mg g kg^{-1} soil	0.53 ± 0.01	15.1 ± 0.3
K g kg^{-1} soil	0.56 ± 0.01	45.3 ± 0.9

Errors represent standard deviation ($n = 3$).

2.2. Experimental Design and Soil Sampling

A pot experiment was performed in a greenhouse at the Graduate School of Hokkaido University. The sampled soil was air-dried and sieved with a 2 mm mesh and subsequently poured into Wagner pots (surface area = 1/5000 a, diameter = 16 cm, and height = 19 cm). Each pot contained 1.8 kg of air-dried soil. The experimental design, which was completely randomized and included three fertilizer treatments × four plant treatments (three mixed cropping treatments and a single maize treatment), was conducted in triplicate. Three common leguminous species used for legume/maize mixed cropping systems, namely cowpea (*Vigna unguiculata* (L.) Walp.), velvet bean (*Mucuna pruriens* (L.) DC.), and common bean (*Phaseolus vulgaris* L.), were selected for our study. One of the four types of plant treatment was then planted into each pot: (1) single maize (*Zea mays* L.; SM), (2) mixture of cowpea and maize (VM), (3) mixture of velvet bean and maize (MM), and (4) mixture of common bean and maize (PM). The pots received one of three fertilizer treatments, namely control ('Ctr') without fertilizer, chemical fertilizer containing P and K ('CF'), or biochar made from CM (50 g pot^{-1} of carbonized CM; 'CM'). The application rate for CF was 30 kg P ha^{-1} and 50 kg K ha^{-1}. Soil and CM chemical properties are described in Table 1. The application amount of CM was designed to optimize P uptakes and the growth rate of plants [26,27]. As well as CF, recommended amounts of P and K were applied according to previous studies [28,29]. Three replicates were performed for 12 treatments (3 fertilizer treatments × 4 crop types); therefore, 36 pots were prepared in total. During these treatments, maize, cowpea, velvet bean, and common bean were sprouted for 2 weeks in small pots filled with vermiculite before they were transplanted to Wagner pots. The temperature was maintained at 25 °C to 30 °C for the duration of the experiment. Plants were grown for 50 days after transplanting. This was in agreement with previous studies that showed that plant N demand was highest after 50 days [30,31].

The soil was sampled from each pot at 0 to 10 cm and passed through a 2 mm sieve to homogenize the sample and remove roots and stones. Then, it was stored at 4 °C or −20 °C for subsequent chemical analysis and DNA extraction, respectively. Legume roots were gently washed with water and then the nodule number per plant was visually counted. Harvested plants were dried in an oven at 65 °C for 3 days to determine their dry weight.

2.3. Chemical Property Analysis

Within one week after soil sampling, pH and extractable NH_4^+ and NO_3^- concentrations were measured with the following method: for soil pH, 6 g of soil was shaken for 30 min with 30 mL of Milli-Q water, and then the pH was measured using a pH sensor (AS800; ASONE Co., Osaka, Japan). To measure soil NH_4^+ and NO_3^-, samples were extracted with a KCl solution (2 mol L^{-1}) and then subjected to a colorimetric analysis using a flow injection analyzer system (ACLA-700; Aqualab Co., Ltd., Osaka, Japan).

2.4. DNA Extraction and 16S rRNA Sequencing

Using the same sampled soils, DNA was extracted with the NucleoSpin® Soil kit (Takara Bio, Inc., Shiga, Japan), following the manufacturer's instructions. The extracted DNA was subsequently amplified by polymerase chain reaction (PCR) targeting the V4 region of the 16S rRNA gene (amplicon size ~250 bp; forward primer 515F: 5′-GTGCCAG CMGCCGCGGTAA-3′ and reverse primer 806R: 5′-GGACTACHVGGGTWTCTAAT-3′). To perform PCR, 10 μL of AmpliTaq Gold® 360 Master Mix (Applied Biosystems, Foster City, CA, USA), 0.4 μL of the forward primer, 0.4 μL of the reverse primer, 8.2 μL of nuclease-free water, and 1 μL of DNA extract were mixed. The PCR cycles were as follows: first, 95 °C for 10 min, then 20 cycles at 95 °C for 30 s, then 57 °C for 30 s and 72 °C for 1 min, and finally 72 °C for 7 min. The PCR products were subsequently purified with Agencourt AMPure XP (Beckman Coulter, Brea, CA, USA) according to the protocol provided. Purified PCR products were quantified with the QuantiFluor® ONE dsDNA system by a Quantus Fluorometer E6150 (Promega, Madison, WI, USA).

An additional PCR was performed on the original PCR products to add Ion-Torrent-specific barcodes. The 515F forward primer with the Ion Xpress Barcode Adapters Kit sequence and the 806R reverse primer attached to the Ion Xpress sequence of the Ion P1 adaptor were used (Thermo Fisher Scientific K.K., Tokyo, Japan). Amplicons from the first PCR were diluted to 2000 ng mL^{-1}, and 1 μL of each PCR product was subsequently mixed with 10 μL of AmpliTaq Gold® 360 Master Mix, 0.4 μL of forward primer, 0.4 μL of reverse primer, and 7.2 μL of nuclease-free water. The second PCR cycle was set to 95 °C for 10 min and then 5 cycles at 95 °C for 30 s, 57 °C for 30 s, and 72 °C for 1 min, followed by 72; °C, for 7 min. Products from the second PCR were purified following the same method outlined previously. The final length and concentration of the amplicons were confirmed using a Bioanalyzer DNA 1000 Kit (Agilent Technologies, Santa Clara, CA, USA(Agilent Technologies, USA). The library was subsequently diluted to 50 pM and loaded onto the Ion 318 chip using Ion Chef Instruments with an Ion PGM Hi-Q Chef Solutions. The samples were sequenced on an Ion PGM Sequencer with Ion PGM Hi-Q View Sequence Solutions (Ion Torrent Life Technologies, Guilford, CT, USA). Sequence data were deposited in the Sequence Read Archive of the National Center for Biotechnology Information (NCBI) under accession number PRJNA743765.

2.5. Sequence Processing

The barcoded 16S rRNA gene sequences were denoised, quality-filtered, and assessed using the DADA2 algorithm implemented in Quantitative Insights Into Microbial Ecology (QIIME2; see Bolyen et al. [32]). Rarefaction was performed with minimal reads among all samples, and sequence data were subsampled to 41,095 sequences per sample. The R package Vegan was used to assess sequencing depth and to generate an alpha rarefaction curve (Figure S1). The rarefaction curve was then evaluated using the interval of a step sample size of 1000. Clustering of operational taxonomic units (OTUs) was performed at 99% identity and was conducted using the SILVA 123 database.

2.6. Measurement of Prokaryotes Abundance

To measure prokaryotes abundance, quantitative PCR (qPCR) was performed on extracted DNA and diluted 50 times with nuclease-free water. The 515F/806R primer pairs described above were used to amplify the V4 region of the 16S rRNA gene. For the standard curve, PCR products from the DNA extracted from the Ctr pots were used, purified with AMPure XP, and further diluted to five different concentrations. Samples were prepared with 10.4 μL of KAPA SYBR Fast qPCR kit (Kapa Biosystems, Woburn, MA, USA), 0.08 μL of the forward primer, 0.08 μL of the reverse primer, and 2 μL of diluted DNA extract. Nuclease-free water was added to achieve a final volume of 20 μL. CFX96 Touch Real (Bio-Rad Laboratories, Inc., Richmond, CA, USA) was used, and the cycling conditions were 95 °C for 30 s, 35 cycles at 95 °C for 30 s, 58 °C for 30 s, and 72 °C for 1 min, all followed by 95 °C for 1 min and then a subsequent ramp from 55 °C to 95 °C by 1 °C increments for 10 s each. Ct values (threshold cycle) were calculated after quantifying the amplification results using qpcR R package.

2.7. Statistical Analysis

To quantify the diversity of soil microbial communities, the Shannon index and the Simpson index, which are used to estimate community α-diversity, were used. For each diversity index, a two-way ANOVA on prokaryotic community structure was performed using fertilizer treatments and plant species followed by a Tukey's Honest Significant Difference test (emmeans R package). In addition, a correlation test using the Pearson method was performed for each diversity index using the prokaryotic community structure as a correlate.

A nonmetric multidimensional scaling (NMDS) analysis of community structure dissimilarity based on the Bray–Curtis index was performed using the metaMDS function in the vegan package in R. The envfit function in the vegan package was used to illustrate significant correlations ($p < 0.05$) between soil chemistry and relative abundances of phylum (>0.1%), with NMDS values presented as vectors on an NMDS ordination. Differences in prokaryotic community structure between treatments were tested by permutational multivariate analysis of variance (PERMANOVA) with the factors 'plant' and 'fertilizer', using the adonis function of the vegan package in R.

Functional profiling of the prokaryotic community was conducted with the Tax4Fun2 R package [33]. The rarefied OTU table was used for Tax4Fun2 searches, and metagenome functional profiles were predicted against the Kyoto Encyclopedia of Gene and Genomes ortholog tables [34]. To evaluate biogeochemical reactions, 36 gene-coding enzymes related to the N metabolism pathway were selected [35,36] (Table S1) and visualized with 'ComplexHeatmap' R package.

For soil chemical property data, a two-way analysis of variance (ANOVA) was performed to investigate the effect of two main treatments (fertilizer and legume treatments) on soil pH, NH_4^+, and NO_3^-.

3. Results

3.1. Alpha Diversity

The impact of the interaction between legume species and fertilizer treatments was visualized using microbial diversity indices ($p < 0.05$). With the PM crop, a significantly higher microbial diversity (Simpson's index) was observed under CM treatment than the Ctr (Figure 1). In contrast, with the MM crop, there was a lower diversity with CM than in the Ctr treatment. There were no significant differences in the microbial diversity between the SM and VM crop.

Correlation tests between diversity indices (Shannon or Simpson) and prokaryote community structure (phylum or class level) showed the contribution of each microbial taxon on overall diversity (Table S2). At the class level, *Betaproteobacteria*, *Gammaproteobacteria*, *Deltaproteobacteria* (each belonging to the phylum *Proteobacteria*), *Acidimicrobiia* (phylum *Actinobacteria*), and *Sphingobacteriia* (*Bacteroidetes*) positively correlated with diversity indices, and they were more abundant in PM and VM in CM-applied soil than MM. In contrast, *TK10* (*Chloroflexi*), the *Soil Crenarchaeotic Group* (*Thaumarchaeota*), and *Spartobacteria* (*Verrucomicrobia*) that were negatively correlated with diversity indices were more abundant in MM than other legume treatments.

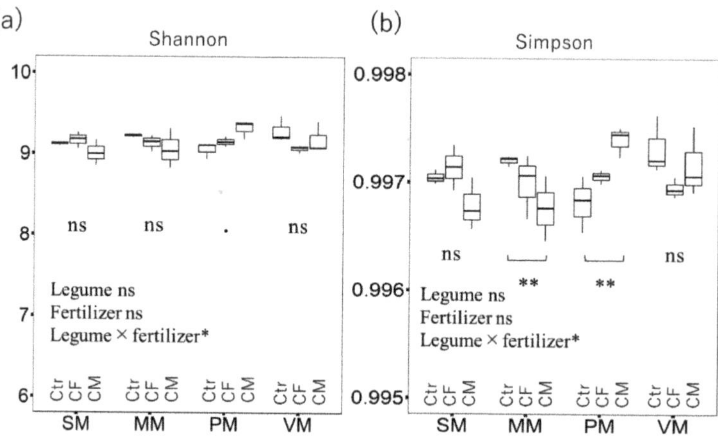

Figure 1. Box plot of bacterial OTUs alpha diversity indicating (**a**) the Shannon's Index and (**b**) the Simpson's index. Two-way ANOVA and Tukey–Kramer's pairwise comparison were performed on the calculated alpha-diversity indices: $p < 0.25$, * $p < 0.05$, and ** $p < 0.01$. The abbreviations of legume treatments were as follows: single maize, SM; maize cropped with velvet bean, MM; maize cropped with common bean, MP; maize cropped with cowpea, VM. Fertilizer treatments were abbreviated as no fertilizer, Ctr; chemical fertilizer, CF; carbonized chicken manure, CM.

3.2. Soil Microbial Community Abundance and Structure

Prokaryotic community structure was significantly affected by legume treatments (PERMANOVA, $p < 0.001$), the fertilizer treatments ($p < 0.001$), and their interaction ($p < 0.01$). NMDS plots based on a Bray–Curtis dissimilarity index showed that CM-treated plots clustered separately from CF and Ctr, which clustered together (Figure 2). For CM-treated soils, the variability among legume treatments was higher than in other fertilizer treatments, whereas for CF-treated pots, microbial beta diversity was highly similar. Moreover, microbial communities under MM treatments clustered separately compared to other legume treatments in the CM treatment ($p < 0.001$, Figure S3). However, the distinct clustering by legume species was not clearly observed within the Ctr and CF treatments. Among the soil chemical properties, pH and $NO_3{}^-$-N concentration were significant factors in the NMDS ordination. Based on a two-way ANOVA with the relative abundances of prokaryotes phyla as a response variable, *Thaumarchaeota*, *Chloroflexi*, *Planctomycetes*, *Verrucomicrobia*, *Proteobacteria*, and *Gemmatimonadetes* contributed the most to changes in community structures, and their relative abundance was influenced by the interactions between legume species and fertilizer treatments (Figure 3; Table 2). Under CM treatment, *Thaumarchaeota*, *Verrucomicrobia*, *Chloroflexi*, and *Planctomycetes* were significantly more abundant in MM than other legume treatments. Comparing fertilizer treatments, *Thaumarchaeota*, *Verrucomicrobia*, *Chloroflexi*, and *Planctomycetes* were more abundant in the CM treatment, whereas *Proteobacteria* was more abundant in the Ctr and CF treatments.

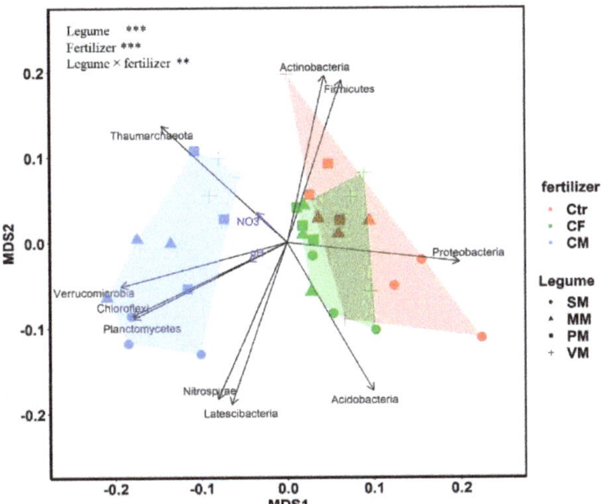

Figure 2. Nonmetric multidimensional scaling plot based on Bray–Curtis dissimilarity (stress = 0.158). The color represents the fertilizer. The shape indicates the legume type. The legend in the upper left shows the significant result of the the Permutational multivariate analyses of variance (PerMANOVA) test for legume and fertilizer treatments. Asterisks show *p*-values (** $p < 0.01$, and *** $p < 0.001$).

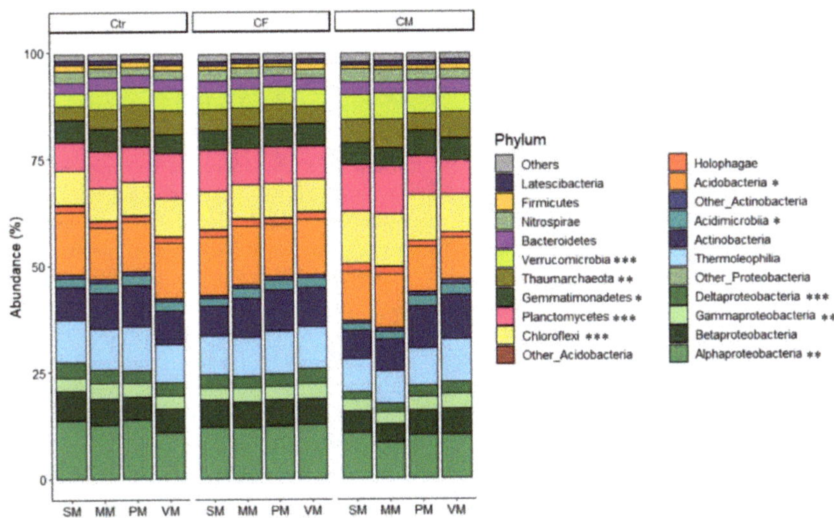

Figure 3. Relative abundance (%) of the main phyla and classes. Phyla with relative abundance > 10% (*Acidobacteria*, *Actinobacteria*, and *Proteobacteria*) are separated into classes. Phyla and classes with relative abundance < 0.1% are subsumed into 'Others'. Asterisk show *p*-value of two-way ANOVA for legume and fertilizer treatments (* $p < 0.05$, ** $p < 0.01$, and *** $p < 0.001$). Legume treatments: SM, MM, PM, and VM; fertilizer treatments: Ctr, CF, and CM.

Table 2. The relative abundance of prokaryotes at the phylum level showed significant interactions between legume species and fertilizer treatments.

Treatment	Relative Abundance (%)					
	Thaumarchaeota	*Gemmatimonadetes*	*Chloroflexi*	*Planctomycetes*	*Verrucomicrobia*	*Proteobacteria*
Ctr						
SM	3.2 [a]	5.2 [a]	7.8 [ab]	6.8 [a]	2.9 [a]	27 [b]
MM	4.7 [ab]	5.3 [a]	7.6 [a]	8.5 [b]	4.4 [b]	26 [ab]
PM	5.2 [ab]	4.7 [a]	7.8 [ab]	8.2 [ab]	4.2 [b]	25 [b]
VM	5.7 [b]	4.4 [a]	8.7 [b]	11 [c]	4.6 [b]	27 [b]
CF						
SM	4.9 [a]	4.6 [a]	8.8 [b]	9.8 [b]	4.1 [a]	24 [a]
MM	4.2 [a]	5.2 [a]	8.0 [ab]	8.5 [ab]	4.4 [a]	24 [a]
PM	4.7 [a]	5.4 [a]	7.9 [ab]	8.7 [ab]	3.9 [a]	24 [a]
VM	4.2 [a]	5.3 [a]	7.7 [a]	7.6 [a]	3.8 [a]	26 [a]
CM						
SM	5.5 [ab]	5.1 [ab]	12 [c]	11 [b]	5.9 [b]	20 [ab]
MM	6.7 [b]	4.3 [a]	12 [c]	11 [b]	6.2 [b]	17 [a]
PM	4.0 [a]	6.0 [b]	11 [b]	9.1 [a]	4.6 [a]	22 [b]
VM	6.2 [b]	5.3 [ab]	8.7 [a]	7.9 [a]	4.5 [a]	23 [b]

The results from multiple pairwise comparisons are shown as different letters, indicating significant differences between treatments ($p < 0.05$).

3.3. 16S rRNA Gene Abundance

Crop type had a significant effect ($p < 0.01$) on the prokaryotes absolute abundance, based on qPCR and when averaged across fertilizer types, but fertilizer type had no significant effect (Table S4). The prokaryotes abundance in the SM crop was significantly higher than in the VM ($p < 0.05$). However, there was no correlation between prokaryotes abundance and diversity.

3.4. Gene Function Prediction by Tax4Fun2

Two-way ANOVA analysis indicated that the gene abundances coding for *ammonium oxidation, carbamate kinase, glutamate dehydrogenase, nitrate reductase, nitrite oxidoreductase,* and *nitrite reductase (NO-forming)* were significantly influenced by the interaction effects of legume and fertilizer treatments (Figure 4, Table S1). Among fertilizer treatments, CM altered the abundance of many of the genes of interest. In particular, the MM crop (with CM) showed the highest abundance of *glutamate dehydrogenase, carbamate kinase, ammonium oxidation,* and *nitrate reductase (NO-forming)* genes. However, other fertilizer treatments (Ctr and CF) did not show a significant difference among legume treatments, although the abundance of glutamate dehydrogenase and *nitrite reductase (NO-forming)* genes was higher in the mixed cropping systems than in SM. *Nitrogenase* gene abundance, which was expected to be affected by the absence of legumes, did not show any significant difference between legume and fertilizer treatments.

3.5. Soil Chemical Properties

A significant increase in soil pH was observed in the CM treatment (Table 3) compared to other treatments, regardless of legume treatments. The CM also had a high concentration of salt-based ions, such as potassium, sodium, and calcium (Table 1). Additionally, the application of CM significantly increased NO_3^--N concentration ($p < 0.001$), but there was no significant difference in NH_4^+-N concentration between treatments.

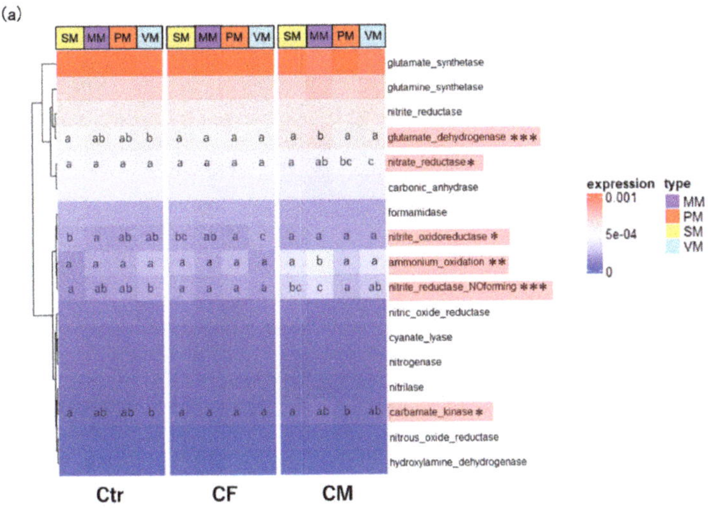

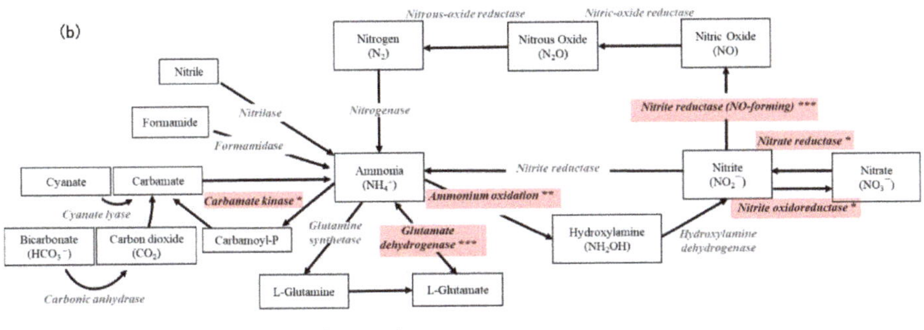

Figure 4. (**a**) Predicted prokaryotic genes coding N metabolism pathway from KEGG database. Asterisk show *P* value of two-way ANOVA interaction for legume and fertilizer treatments (* $p < 0.05$, ** $p < 0.01$, and *** $p < 0.001$). Different letters in the table indicate significant ($p < 0.05$) pairwise differences among legume treatments. Individual KEGG codes are described in Table S1. (**b**) N metabolism pathway (ko00910) and enzymes based on Kyoto Encyclopedia. The genes highlighted in red indicate significant pathways for the two-way ANOVA interaction.

3.6. Legume Nodulation and Plant Biomass

A higher number of nodules was observed in PM than in other legume treatments, regardless of fertilizer treatment (Table S3). With CF application in particular, more leguminous nodulation was observed in PM and MM. However, legume nodulation counts did not correlate with plant biomass. With CM application, MM had the highest biomass (13.9 g) among mixed cropping systems (10.1 and 9.5 g in VM and PM, respectively).

Table 3. Soil pH, nitrate, and ammonium content after plant cultivation.

Treatment	pH (H$_2$O)	NO$_3^-$-N (mg kg^{-1})	NH$_4^+$-N (mg kg^{-1})
SM			
Ctr	6.7 ± 0.03	8.2 ± 1.6	4.7 ± 0.30
CF	6.7 ± 0.04	7.1 ± 0.4	3.4 ± 0.75
CM	7.6 ± 0.09	15.9 ± 2.4	5.4 ± 1.22
MM			
Ctr	6.7 ± 0.02	10.9 ± 4.2	4.2 ± 2.64
CF	6.7 ± 0.02	6.3 ± 3.3	4.7 ± 1.37
CM	7.3 ± 0.10	19.7 ± 6.2	5.1 ± 2.48
PM			
Ctr	6.7 ± 0.02	12.8 ± 2.2	6.0 ± 2.80
CF	6.6 ± 0.02	7.9 ± 3.2	5.7 ± 0.98
CM	7.2 ± 0.12	35.0 ± 19.3	2.6 ± 0.37
VM			
Ctr	6.7 ± 0.02	10.6 ± 3.2	4.5 ± 1.35
CF	6.7 ± 0.03	6.7 ± 3.1	4.9 ± 0.68
CM	7.2 ± 0.08	30.8 ± 13.2	5.4 ± 2.22
Two-way ANOVA		*p*	
Legume	<0.001	0.45	0.92
Fertilizer	<0.001	<0.001	0.90
Legume × fertilizer	<0.01	0.91	0.39

The expressed plant species included single maize, SM; maize cropped with velvet bean (*Mucuna pruriens* (L.) DC.), MM; maize cropped with common bean (*Phaseolus vulgaris* L.), PM; maize cropped with cowpea (*Vigna unguiculate* (L.) Walp.), VM. Fertilizer applied with the control, Ctr; chemical fertilizer, CF; carbonized chicken manure, CM. Two-way ANOVA was performed to examine the effects of the interactions between plant species and fertilizer treatments. *p* values are shown at the bottom of the table.

4. Discussion

4.1. Changes in Microbial Diversity and Community after Biochar Application

Among the fertilizer treatments, CM clearly changed the structure of the prokaryotic community (Figures 1 and 2). The application of CM also altered soil pH (Table 3), which is well-known to strongly influence microbial community and abundance in soils [37,38]. Overall, *Thaumarchaeota*, *Verrucomicrobia*, *Chloroflexi*, and *Planctomycetes* were more abundant under CM application. In previous studies, the abundance of *Thaumarchaeota* and *Verrucomicrobia* was positively correlated with soil pH [39,40], and *Chloroflexi* and *Plancto-mycetes* became more abundant with the enrichment of labile carbon and aromatic compounds [41]. Thus, the chemical property and carbon content of CM might explain the larger impacts on soil prokaryotic communities when compared to chemical fertilizer.

The degree that differences in community composition increased with CM application varied among legume treatments. It is possible that the amount and quality of plant root exudates affect soil microbiomes. Some chemical compounds (e.g., aromatic organic acids) released from plant roots can enrich specific microbial species and activities that are beneficial for plant growth [42]. Furthermore, the physical properties of biochar itself may influence plant exudate availability on microbial community and diversity in the intercropping system because of absorption of the root exudates onto biochar's surface or its porosity. Liao et al. [43] indicated that the presence of biochar stimulated the assimilation of plant-delivered carbon by members of *Firmicutes* and *Bacteroidetes* in a legume-based intercropping system. In support of this previous study, *Bacteroidetes* increased in PM and VM under CM application in our study (Table S2). Thus, biochar might increase or decrease specific groups of prokaryotes when applied to particular legume species.

Under CM application, the prokaryotic community became more diverse in PM crops but less diverse in MM crops (Figure 1). This result contradicts a previous study that demonstrated that biochar application has a positive effect on soil microbial diversity in mixed crops [44]. With PM treatment, members of *Proteobacteria* and *Bacteroidetes* (i.e.,

Betaproteobacteria, Gammaproteobacteria, Deltaproteobacteria, and *Sphingobacteriia*) that were significantly positively correlated with Shannon and Simpson diversity indices became more abundant. Both phyla are highly abundant in rhizosphere soil [45,46]. In contrast, CM applied to MM crops increased the relative abundances of *Thaumarchaeota, Verrucomicrobia, Chloroflexi,* and *Planctomycetes,* which were negatively correlated with diversity indices. Thus, our results indicate that specific combinations of legumes and maize that are used with biochar can be vital for determining the diversity of prokaryotes in soils.

4.2. Soil Microbial Functions Related to the N Cycle

Functional prediction analysis with Tax4Fun2 indicated that CM treatment enhanced N metabolism functioning (e.g., *ammonium oxidation* and *nitrite reductase (NO-forming)*), especially in MM treatment (Figure 4). Biochar application was similarly reported to enhance ammonia-oxidizing archaea and bacteria in rotated cropland [47], as well as increase *nitrite reductase* gene abundance [48]. Soil pH has been considered a critical environmental factor impacting *ammonium oxidation* and *nitrite reduction,* with an optimal range of pH 7 to 8 [49]. Therefore, in the present study, the increase in soil pH through biochar application (pH 6 to 7; Table 3) was a possible cause of enhanced N-cycling. In addition, we used biochar made of chicken manure that contains a higher N content (4.06%) compared to biochar made from other materials such as sugar cane bagasse (1.8%) and rice straw (1.3%) [50]. Thus, we note that the biochar used in the current study altered not only soil pH and physical conditions, but also N availability in soils (Table 3).

Within CM treatments, there was enhanced *ammonium oxidation, nitrite reductase (NO-forming), carbamate kinase,* and *glutamate dehydrogenase,* particularly in MM compared to the other legume treatments. A higher abundance of those genes indicates a faster conversion of organic-N to ammonium or nitrate, which then becomes available for use by the plant. As N is a primary element for plant growth, we observed the highest plant biomass in MM treatment (Table S3). Our results are consistent with previous studies showing the improvement of crop productivity in the maize/velvet bean intercropping system compared to mono-cropping systems [23,24]. These studies indicated that weed reduction by cover crop can cause increased harvest; however, to our knowledge, little information is available regarding the interaction of velvet bean used in intercropping systems and prokaryotic N-cycling function. It should be mentioned that Tax4Fun2 is not an actual measurement but a functional prediction tool based on 16S rRNA gene sequences. Future work on shotgun metagenomics would be necessary to examine these predicted observations. However, the present study may provide basic knowledge for further metagenomic studies of soil prokaryotes and N-cycling gene expression in mixed cropping systems.

While comparing the community compositions at the phylum level, *Thaumarchaeota, Chloroflexi,* and *Planctomycetes* were more abundant in CM-applied soils than other fertilizer treatments (Table 2). Among the legume treatments, these taxa were most abundant in MM. *Thaumarchaeota,* archaea ubiquitously present in a wide variety of ecosystems, may contribute to the increase in ammonia-oxidizing functioning in a soil community [51]. In addition, members of *Planctomycetes* can perform anaerobic oxidation of ammonium to di-nitrogen via the anammox pathway, which might correspond to greater denitrification functioning [52]. Moreover, *Chloroflexi* is often associated with carbohydrate and amino acid degradation [53]. Thus, consistent with the hypotheses, the use of CM in a specific legume-combined mixed cropping system (i.e., MM) enhanced the N-cycling gene abundance with an increase in microbial taxa involved in N-cycling.

Similar to previous studies, CF treatment facilitated legume nodulation in MM and PM more than in other fertilizer treatments due to its phosphorus and potassium content [54] (Table S3). In leguminous nodules, N-fixation is performed by members of the *Alphaproteobacteria* and *Betaproteobacteria* classes (e.g., *Rhizobium spp.* and *Paraburkholderia spp.*) [55]. However, we found no significant differences in *nitrogenase* gene abundance or in the abundance of *Alphaproteobacteria* and *Betaproteobacteria* between fertilizer types.

4.3. Effects of the Legume Varieties and Biochar Application on Soil Microbes

In this study, the variability of legume species significantly changed the effect of biochar application on the diversity, community, and N cycling function of soil prokaryotes. A recent study indicated that after biochar application, the microbial community was strongly affected by plant species [21,56], in agreement with our observation. Research on eight different legume accessions and the rhizosphere microbiome showed only 0.7% OTUs that were shared across all treatments, indicating a strong legume species-specific effect on the rhizosphere microbial community [57]. In addition, different legume species are known to have different types of genes that host specific rhizobium because of their co-evolution with symbiotic microorganisms [15].

In mixed cropping systems, the competition and complementary relationship of nutrients between cereals and legumes are key to promote microbial activity, which is important for P and N acquisition. For example, nutrient deficiency caused by mixed cropping systems facilitated the nodulation of legume and mycorrhizal transfers of P and N [58,59]. Additionally, coexisting maize can facilitate the nodule formation of legumes with flavonoids (signaling compounds for rhizobia) contained in root residues [60]. Biochar has a positive influence on the maize root exudate production and promotes the N fixation and N transfer from legume to maize [61], suggesting the complex effect of biochar application in the association with legume species and maize on the rhizosphere microbial community. Thus, future investigations should consider the effect of biochar in the context of the combination of crop species.

5. Conclusions

Our results demonstrated that the choice of legume species in the intercropping system is an important factor controlling the effect of biochar application on soil microbial diversity, community, and functions. The combination of biochar and common bean enhanced microbial diversity with the increase in the abundance of *Proteobacteria* and *Bacteroidetes*. On the other hand, biochar combined with velvet bean altered the soil microbial genes for *ammonia oxidation* and *nitrite reductase (NO-forming)*, but not the prokaryotic diversities. The increase in the relative abundances of *Thaumarchaeota*, *Planctomycetes*, *Verrucomicrobia*, and *Chloroflexi* was identified as a potential cause for the alteration of the N-cycling functional genes in velvet bean treatments. These results suggest that consideration of the legume-species-specific effect is necessary to optimize the positive effect of biochar on microbial diversity and functions.

Supplementary Materials: The following supporting information can be downloaded at: https://www.mdpi.com/article/10.3390/agriculture12101548/s1, Table S1: Selected genes were tested by two-way ANOVA, Table S2: Relative abundances (%) of phylum and classes, Table S3: Dry matter of plant shoots and roots, Table S4: 16S rRNA gene abundance, Figure S1: Rarefaction curve of all samples, Figure S2: Nonmetric multidimensional scaling plot based on Bray–Curtis dissimilarity for each fertilizer treatments.

Author Contributions: Conceptualization, Y.U. and A.K.; methodology, A.K. and Y.U; software, A.K.; validation, A.K. and Y.M.M.; investigation, A.K.; data curation, Y.M.M.; writing—original draft preparation, A.K.; writing—review and editing, Y.M.M. and Y.U; funding acquisition, Y.U. All authors have read and agreed to the published version of the manuscript.

Funding: This work was supported by JSPS KAKENHI grant numbers 21J22147, 21H02324, 18KK0183, and 18H02310.

Institutional Review Board Statement: Not applicable.

Data Availability Statement: Sequence data were deposited in the Sequence Read Archive of the National Center for Biotechnology Information (NCBI) under accession number PRJNA743765.

Conflicts of Interest: The authors declare no conflict of interest.

References

1. Rohr, J.R.; Barrett, C.B.; Civitello, D.J.; Craft, M.E.; Delius, B.; DeLeo, G.A.; Hudson, P.J.; Jouanard, N.; Nguyen, K.H.; Ostfeld, R.S.; et al. Emerging Human Infectious Diseases and the Links to Global Food Production. *Nat. Sustain.* **2019**, *2*, 445–456. [CrossRef] [PubMed]
2. Le Provost, G.; Thiele, J.; Westphal, C.; Penone, C.; Allan, E.; Neyret, M.; van der Plas, F.; Ayasse, M.; Bardgett, R.D.; Birkhofer, K.; et al. Contrasting Responses of Above- and Belowground Diversity to Multiple Components of Land-Use Intensity. *Nat. Commun.* **2021**, *12*, 3918. [CrossRef] [PubMed]
3. Wagg, C.; Hautier, Y.; Pellkofer, S.; Banerjee, S.; Schmid, B.; van der Heijden, M.G. Diversity and Asynchrony in Soil Microbial Communities Stabilizes Ecosystem Functioning. *eLife* **2021**, *10*, e62813. [CrossRef] [PubMed]
4. Peixoto, R.S.; Voolstra, C.R.; Sweet, M.; Duarte, C.M.; Carvalho, S.; Villela, H.; Lunshof, J.E.; Gram, L.; Woodhams, D.C.; Walter, J.; et al. Harnessing the Microbiome to Prevent Global Biodiversity Loss. *Nat. Microbiol.* **2022**, 1–10. [CrossRef]
5. Nannipieri, P.; Ascher, J.; Ceccherini, M.T.; Landi, L.; Pietramellara, G.; Renella, G. Microbial Diversity and Soil Functions. *Eur. J. Soil Sci.* **2003**, *54*, 655–670. [CrossRef]
6. Li, X.; Mu, Y.; Cheng, Y.; Liu, X.; Nian, H. Effects of Intercropping Sugarcane and Soybean on Growth, Rhizosphere Soil Microbes, Nitrogen and Phosphorus Availability. *Acta Physiol. Plant.* **2013**, *7*, 1113–1119. [CrossRef]
7. Lian, T.; Mu, Y.; Jin, J.; Ma, Q.; Cheng, Y.; Cai, Z.; Nian, H. Impact of Intercropping on the Coupling between Soil Microbial Community Structure, Activity, and Nutrient-Use Efficiencies. *PeerJ* **2019**, *7*, e6412. [CrossRef]
8. Yu, L.; Tang, Y.; Wang, Z.; Gou, Y.; Wang, J. Nitrogen-Cycling Genes and Rhizosphere Microbial Community with Reduced Nitrogen Application in Maize/Soybean Strip Intercropping. *Nutr. Cycl. Agroecosyst.* **2019**, *113*, 35–49. [CrossRef]
9. Horner, A.; Browett, S.S.; Antwis, R.E. Mixed-Cropping Between Field Pea Varieties Alters Root Bacterial and Fungal Communities. *Sci. Rep.* **2019**, *9*, 16953. [CrossRef]
10. Granzow, S.; Kaiser, K.; Wemheuer, B.; Pfeiffer, B.; Daniel, R.; Vidal, S.; Wemheuer, F. The Effects of Cropping Regimes on Fungal and Bacterial Communities of Wheat and Faba Bean in a Greenhouse Pot Experiment Differ between Plant Species and Compartment. *Front. Microbiol.* **2017**, *8*, 902. [CrossRef]
11. Li, H.; Shen, J.; Zhang, F.; Marschner, P.; Cawthray, G.; Rengel, Z. Phosphorus Uptake and Rhizosphere Properties of Intercropped and Monocropped Maize, Faba Bean, and White Lupin in Acidic Soil. *Biol. Fertil. Soils* **2010**, *46*, 79–91. [CrossRef]
12. Zhang, N.N.; Sun, Y.M.; Li, L.; Wang, E.T.; Chen, W.X.; Yuan, H.L. Effects of Intercropping and Rhizobium Inoculation on Yield and Rhizosphere Bacterial Community of Faba Bean (*Vicia faba* L.). *Biol. Fertil. Soils* **2010**, *46*, 625–639. [CrossRef]
13. Baumann, D.T.; Bastiaans, L.; Kropff, M.J. Competition and Crop Performance in a Leek–Celery Intercropping System. *Crop Sci.* **2001**, *41*, 764–774. [CrossRef]
14. Zhang, F.; Li, L. Using Competitive and Facilitative Interactions in Intercropping Systems Enhances Crop Productivity and Nutrient-Use Efficiency. *Plant Soil* **2003**, *248*, 305–312. [CrossRef]
15. Chang, D.; Gao, S.; Zhou, G.; Deng, S.; Jia, J.; Wang, E.; Cao, W. The Chromosome-Level Genome Assembly of Astragalus Sinicus and Comparative Genomic Analyses Provide New Resources and Insights for Understanding Legume-Rhizobial Interactions. *Plant Commun.* **2022**, *3*, 100263. [CrossRef]
16. Chen, Q.-L.; Ding, J.; Zhu, D.; Hu, H.-W.; Delgado-Baquerizo, M.; Ma, Y.-B.; He, J.-Z.; Zhu, Y.-G. Rare Microbial Taxa as the Major Drivers of Ecosystem Multifunctionality in Long-Term Fertilized Soils. *Soil Biol. Biochem.* **2020**, *141*, 107686. [CrossRef]
17. Hartmann, M.; Frey, B.; Mayer, J.; Mäder, P.; Widmer, F. Distinct Soil Microbial Diversity under Long-Term Organic and Conventional Farming. *ISME J.* **2015**, *9*, 1177–1194. [CrossRef]
18. Chen, C.; Zhang, J.; Lu, M.; Qin, C.; Chen, Y.; Yang, L.; Huang, Q.; Wang, J.; Shen, Z.; Shen, Q. Microbial Communities of an Arable Soil Treated for 8 Years with Organic and Inorganic Fertilizers. *Biol. Fertil. Soils* **2016**, *52*, 455–467. [CrossRef]
19. Francioli, D.; Schulz, E.; Lentendu, G.; Wubet, T.; Buscot, F.; Reitz, T. Mineral vs. Organic Amendments: Microbial Community Structure, Activity and Abundance of Agriculturally Relevant Microbes Are Driven by Long-Term Fertilization Strategies. *Front. Microbiol.* **2016**, *7*, 1446. [CrossRef]
20. Yu, L.; Homyak, P.M.; Kang, X.; Brookes, P.C.; Ye, Y.; Lin, Y.; Muhammad, A.; Xu, J. Changes in Abundance and Composition of Nitrifying Communities in Barley (*Hordeum vulgare* L.) Rhizosphere and Bulk Soils over the Growth Period Following Combined Biochar and Urea Amendment. *Biol. Fertil. Soils* **2020**, *56*, 169–183. [CrossRef]
21. Kochanek, J.; Soo, R.M.; Martinez, C.; Dakuidreketi, A.; Mudge, A.M. Biochar for Intensification of Plant-Related Industries to Meet Productivity, Sustainability and Economic Goals: A Review. *Resour. Conserv. Recycl.* **2022**, *179*, 106109. [CrossRef]
22. Liu, G.; Chen, L.; Jiang, Z.; Zheng, H.; Dai, Y.; Luo, X.; Wang, Z. Aging Impacts of Low Molecular Weight Organic Acids (LMWOAs) on Furfural Production Residue-Derived Biochars: Porosity, Functional Properties, and Inorganic Minerals. *Sci. Total Environ.* **2017**, *607–608*, 1428–1436. [CrossRef]
23. Akobundu, I.O.; Udensi, U.E.; Chikoye, D. Velvetbean (*Mucuna* Spp.) Suppresses Speargrass (Imperata Cylindrica (L.) Raeuschel) and Increases Maize Yield. *Int. J. Pest Manag.* **2000**, *46*, 103–108. [CrossRef]
24. Cardoso, E.J.B.N.; Nogueira, M.A.; Ferraz, S.M.G. Biological N-2 Fixation and Mineral N in Common Bean-Maize Intercropping or Sole Cropping in Southeastern Brazi. *Exp. Agric.* **2007**, *43*, 319–330. [CrossRef]
25. Dahmardeh, M.; Ghanbari, A.; Baratali, S.; Mahmood, R. Effect of Intercropping Maize (*Zea mays* L.) With Cow Pea (*Vigna unguiculata* L.) on Green Forage Yield and Quality Evaluation. *Asian J. Plant Sci.* **2009**, *8*, 235–239. [CrossRef]

26. Chan, K.Y.; Zwieten, L.V.; Meszaros, I.; Downie, A.; Joseph, S.; Chan, K.Y.; Zwieten, L.V.; Meszaros, I.; Downie, A.; Joseph, S. Using Poultry Litter Biochars as Soil Amendments. *Aust. J. Soil Res.* **2008**, *46*, 437–444. [CrossRef]
27. Rajkovich, S.; Enders, A.; Hanley, K.; Hyland, C.; Zimmerman, A.R.; Lehmann, J. Corn Growth and Nitrogen Nutrition after Additions of Biochars with Varying Properties to a Temperate Soil. *Biol. Fertil. Soils* **2012**, *48*, 271–284. [CrossRef]
28. Géant, C.B.; Francine, S.B.; Adrien, B.N.; Wasolu, N.; Mulalisi, B.; Espoir, M.B.; Jean, M.M.; Antoine, K.L.; Gustave, M.N. Optimal Fertiliser Dose and Nutrients Allocation in Local and Biofortified Bean Varieties Grown on Ferralsols in Eastern Democratic Republic of the Congo. *Cogent Food Agric.* **2020**, *6*, 1805226. [CrossRef]
29. Susila, A.D.; Prasetyo, T.; Kartika, J.G. *Optimum Fertilizer Rate for Yard Long Bean* (Vigna unguilata L.) *Production in Ultisol Jasinga*, Sustainable Agriculture and Natural Resource Management (SANREM) Knowledgebase, Virginia Tech: Blacksburg, VA, USA, 2008.
30. Hu, F.; Zhao, C.; Feng, F.; Chai, Q.; Mu, Y.; Zhang, Y. Improving N Management through Intercropping Alleviates the Inhibitory Effect of Mineral N on Nodulation in Pea. *Plant Soil* **2017**, *412*, 235–251. [CrossRef]
31. Lithourgidis, A.S.; Vlachostergios, D.N.; Dordas, C.A.; Damalas, C.A. Dry Matter Yield, Nitrogen Content, and Competition in Pea–Cereal Intercropping Systems. *Eur. J. Agron.* **2011**, *34*, 287–294. [CrossRef]
32. Bolyen, E.; Rideout, J.R.; Dillon, M.R.; Bokulich, N.A.; Abnet, C.C.; Al-Ghalith, G.A.; Alexander, H.; Alm, E.J.; Arumugam, M.; Asnicar, F.; et al. Reproducible, Interactive, Scalable and Extensible Microbiome Data Science Using QIIME 2. *Nat. Biotechnol.* **2019**, *37*, 852–857. [CrossRef] [PubMed]
33. Wemheuer, F.; Taylor, J.A.; Daniel, R.; Johnston, E.; Meinicke, P.; Thomas, T.; Wemheuer, B. Tax4Fun2: Prediction of Habitat-Specific Functional Profiles and Functional Redundancy Based on 16S RRNA Gene Sequences. *Environ. Microbiome* **2020**, *15*, 11. [CrossRef] [PubMed]
34. Kanehisa, M.; Goto, S. KEGG: Kyoto Encyclopedia of Genes and Genomes. *Nucleic Acids Res.* **2000**, *28*, 27–30. [CrossRef] [PubMed]
35. He, X.; Yin, H.; Fang, C.; Xiong, J.; Han, L.; Yang, Z.; Huang, G. Metagenomic and Q-PCR Analysis Reveals the Effect of Powder Bamboo Biochar on Nitrous Oxide and Ammonia Emissions during Aerobic Composting. *Bioresour. Technol.* **2021**, *323*, 124567. [CrossRef] [PubMed]
36. Wang, J.; Long, Z.; Min, W.; Hou, Z. Metagenomic Analysis Reveals the Effects of Cotton Straw–Derived Biochar on Soil Nitrogen Transformation in Drip-Irrigated Cotton Field. *Environ. Sci. Pollut. Res.* **2020**, *27*, 43929–43941. [CrossRef] [PubMed]
37. Fierer, N.; Jackson, R.B. The Diversity and Biogeography of Soil Bacterial Communities. *Proc. Natl. Acad. Sci. USA* **2006**, *103*, 626–631. [CrossRef]
38. Santoyo, G.; Pacheco, C.H.; Salmerón, J.H.; León, R.H. The Role of Abiotic Factors Modulating the Plant-Microbe-Soil Interactions: Toward Sustainable Agriculture. A Review. *Span. J. Agric. Res.* **2017**, *15*, 13. [CrossRef]
39. Lu, X.; Seuradge, B.J.; Neufeld, J.D. Biogeography of Soil Thaumarchaeota in Relation to Soil Depth and Land Usage. *FEMS Microbiol. Ecol.* **2017**, *93*, fiw246. [CrossRef]
40. Shen, C.; Ge, Y.; Yang, T.; Chu, H. Verrucomicrobial Elevational Distribution Was Strongly Influenced by Soil PH and Carbon/Nitrogen Ratio. *J. Soils Sediments* **2017**, *17*, 2449–2456. [CrossRef]
41. Sheng, Y.; Zhu, L. Biochar Alters Microbial Community and Carbon Sequestration Potential across Different Soil PH. *Sci. Total Environ.* **2018**, *622–623*, 1391–1399. [CrossRef]
42. Zhalnina, K.; Louie, K.B.; Hao, Z.; Mansoori, N.; da Rocha, U.N.; Shi, S.; Cho, H.; Karaoz, U.; Loqué, D.; Bowen, B.P.; et al. Dynamic Root Exudate Chemistry and Microbial Substrate Preferences Drive Patterns in Rhizosphere Microbial Community Assembly. *Nat. Microbiol.* **2018**, *3*, 470–480. [CrossRef] [PubMed]
43. Liao, H.; Li, Y.; Yao, H. Biochar Amendment Stimulates Utilization of Plant-Derived Carbon by Soil Bacteria in an Intercropping System. *Front. Microbiol.* **2019**, *10*, 1361. [CrossRef] [PubMed]
44. Imparato, V.; Hansen, V.; Santos, S.S.; Nielsen, T.K.; Giagnoni, L.; Hauggaard-Nielsen, H.; Johansen, A.; Renella, G.; Winding, A. Gasification Biochar Has Limited Effects on Functional and Structural Diversity of Soil Microbial Communities in a Temperate Agroecosystem. *Soil Biol. Biochem.* **2016**, *99*, 128–136. [CrossRef]
45. Chaparro, J.M.; Badri, D.V.; Vivanco, J.M. Rhizosphere Microbiome Assemblage Is Affected by Plant Development. *ISME J.* **2014**, *8*, 790–803. [CrossRef]
46. Wolińska, A.; Kuźniar, A.; Zielenkiewicz, U.; Izak, D.; Szafranek-Nakonieczna, A.; Banach, A.; Błaszczyk, M. Bacteroidetes as a Sensitive Biological Indicator of Agricultural Soil Usage Revealed by a Culture-Independent Approach. *Appl. Soil Ecol.* **2017**, *119*, 128–137. [CrossRef]
47. Zhang, H.; Sun, H.; Zhou, S.; Bai, N.; Zheng, X.; Li, S.; Zhang, J.; Lv, W. Effect of Straw and Straw Biochar on the Community Structure and Diversity of Ammonia-Oxidizing Bacteria and Archaea in Rice-Wheat Rotation Ecosystems. *Sci. Rep.* **2019**, *9*, 9367. [CrossRef] [PubMed]
48. Su, X.; Wang, Y.; He, Q.; Hu, X.; Chen, Y. Biochar Remediates Denitrification Process and N2O Emission in Pesticide Chlorothalonil-Polluted Soil: Role of Electron Transport Chain. *Chem. Eng. J.* **2019**, *370*, 587–594. [CrossRef]
49. Nicol, G.W.; Leininger, S.; Schleper, C.; Prosser, J.I. The Influence of Soil PH on the Diversity, Abundance and Transcriptional Activity of Ammonia Oxidizing Archaea and Bacteria. *Environ. Microbiol.* **2008**, *10*, 2966–2978. [CrossRef]
50. Chan, K.Y.; Xu, Z. Biochar: Nutrient Properties and Their Enhancement. In *Biochar for Environmental Management*; Routledge: Abingdon, UK, 2009; ISBN 978-1-84977-055-2.
51. Brochier-Armanet, C.; Gribaldo, S.; Forterre, P. Spotlight on the Thaumarchaeota. *ISME J.* **2012**, *6*, 227–230. [CrossRef]

52. Fuerst, J.A.; Sagulenko, E. Beyond the Bacterium: Planctomycetes Challenge Our Concepts of Microbial Structure and Function. *Nat. Rev. Microbiol.* **2011**, *9*, 403–413. [CrossRef]
53. Bayer, K.; Jahn, M.T.; Slaby, B.M.; Moitinho-Silva, L.; Hentschel, U. Marine Sponges as Chloroflexi Hot Spots: Genomic Insights and High-Resolution Visualization of an Abundant and Diverse Symbiotic Clade. *mSystems* **2018**, *3*, e00150-18. [CrossRef] [PubMed]
54. Sangakkara, U.R.; Hartwig, U.A.; Nösberger, J. Soil Moisture and Potassium Affect the Performance of Symbiotic Nitrogen Fixation in Faba Bean and Common Bean. *Plant Soil* **1996**, *184*, 123–130. [CrossRef]
55. Moulin, L.; Munive, A.; Dreyfus, B.; Boivin-Masson, C. Nodulation of Legumes by Members of the β-Subclass of Proteobacteria. *Nature* **2001**, *411*, 948–950. [CrossRef] [PubMed]
56. Azeem, M.; Sun, D.; Crowley, D.; Hayat, R.; Hussain, Q.; Ali, A.; Tahir, M.I.; Jeyasundar, P.G.S.A.; Rinklebe, J.; Zhang, Z. Crop Types Have Stronger Effects on Soil Microbial Communities and Functionalities than Biochar or Fertilizer during Two Cycles of Legume-Cereal Rotations of Dry Land. *Sci. Total Environ.* **2020**, *715*, 136958. [CrossRef]
57. Pérez-Jaramillo, J.E.; de Hollander, M.; Ramírez, C.A.; Mendes, R.; Raaijmakers, J.M.; Carrión, V.J. Deciphering Rhizosphere Microbiome Assembly of Wild and Modern Common Bean (*Phaseolus vulgaris*) in Native and Agricultural Soils from Colombia. *Microbiome* **2019**, *7*, 114. [CrossRef] [PubMed]
58. Zhang, H.; Wang, X.; Gao, Y.; Sun, B. Short-Term N Transfer from Alfalfa to Maize Is Dependent More on Arbuscular Mycorrhizal Fungi than Root Exudates in N Deficient Soil. *Plant Soil* **2020**, *446*, 23–41. [CrossRef]
59. Hu, J.; Li, M.; Liu, H.; Zhao, Q.; Lin, X. Intercropping with Sweet Corn (*Zea mays* L. Var. Rugosa Bonaf.) Expands P Acquisition Channels of Chili Pepper (*Capsicum annuum* L.) via Arbuscular Mycorrhizal Hyphal Networks. *J. Soils Sediments* **2019**, *19*, 1632–1639. [CrossRef]
60. Li, B.; Li, Y.-Y.; Wu, H.-M.; Zhang, F.-F.; Li, C.-J.; Li, X.-X.; Lambers, H.; Li, L. Root Exudates Drive Interspecific Facilitation by Enhancing Nodulation and N2 Fixation. *Proc. Natl. Acad. Sci. USA* **2016**, *113*, 6496–6501. [CrossRef]
61. Liu, L.; Wang, Y.; Yan, X.; Li, J.; Jiao, N.; Hu, S. Biochar Amendments Increase the Yield Advantage of Legume-Based Intercropping Systems over Monoculture. *Agric. Ecosyst. Environ.* **2017**, *237*, 16–23. [CrossRef]

Article

Adaptation and High Yield Performance of Honglian Type Hybrid Rice in Pakistan with Desirable Agricultural Traits

Muhammad Ashfaq [1,*], Renshan Zhu [2], Muhammad Ali [1], Zhiyong Xu [3], Abdul Rasheed [1], Muhammad Jamil [1], Adnan Shakir [1] and Xianting Wu [2,*]

1   Faculty of Agricultural Sciences, University of the Punjab, Lahore 54590, Pakistan
2   Department of Genetics, College of Life Sciences, Wuhan University, Wuhan 430072, China
3   Wuhan Yansheng Agriculture Technology Ltd., Wuhan 430076, China
*   Correspondence: ashfaq.iags@pu.edu.pk (M.A.); xiantwu@whu.edu.cn (X.W.)

Abstract: Honglian type cytoplasmic male sterility (CMS) is one of the three known major CMS types of rice (*Oryza sativa* L.) commercially used in hybrid rice seed production. Hybrid rice generated by the Honglian type CMS is a special group of hybrid rice, having distinct agricultural characteristics. The main objective of the study was to screen out the Honglian hybrid rice adapted for growing in Pakistan based on desirable traits. Different Honglian-type hybrid rice varietieswere tested locally in different locations in Pakistan based on various desirabletraits. Three Honglian types of hybrids (HP1, HP2, HP3) performed well, had better agricultural traits and showed high yield potential over the check variety. Different qualitative and quantitative traits were studied to conclude the advantages of these varieties for Pakistani local adaptation evaluations. Forty-eight SSR markers were used to study the genetic diversities of the hybrids. Nine selected polymorphic SSR markers (RM-219, RM-236, RM-274, RM-253, RM-424, RM-567, RM-258, RM-481, RM-493) showed genetic variations among Honglian hybrid rice varieties through PCR analysis. In 2019 and 2020, the increment of the yield potential of HP1, HP2 and HP3 was better (+43.90%, +35.44%, +37.13% and +30.91%, +33.37%, +33.62%, respectively, in both years)than the check variety KSK-133. All the desirable traits were analyzed through Principal Component Analysis (PCA). The principal components with more than one eigenvalue showed more variability. The average variability of 74.78% was observed among genotypes and their desirable traits in both years. National Uniform Yield Trial (NUYT) and Distinctness, Uniformity, Stability (DUS) trials are being conducted under the supervision of National Coordinated Rice (NCR) and Federal Seed Certification and Registration Department (FSCRD), Government of Pakistan. In the 2020 trial, the average yield of 104 rice varieties/hybrids was 8608 kg/ha; HP1, HP2 and HP3 (8709 kg/ha, 8833 kg/ha, and 9338 kg/ha, respectively) were all higher than the average yield, and HP3 yield was higher than over check varieties (D-121, Guard-53). In the 2021 trial, the average yield of 137 varieties was 7616 kg/ha; the HP1 yield (7863 kg/ha) was higher than the average overcheck varieties/hybrids. Various qualitative and quantitative traits showed desirable genetic diversity among the rice hybrids. It was also observed that, under higher temperatures, the seeds setting rate of Honglian-type hybrid rice was stable, which is the guarantee for stable yield and rice production in Pakistan. Moreover, it was considerably better, suggesting that Honglian-type hybrid rice varieties can be grown in Pakistan because they are less risky under climate change, especially the global warming challenges.

Keywords: rice; hybrid; desirable; agricultural traits; adaptation; Honglian

---

Citation: Ashfaq, M.; Zhu, R.; Ali, M.; Xu, Z.; Rasheed, A.; Jamil, M.; Shakir, A.; Wu, X. Adaptation and High Yield Performance of Honglian Type Hybrid Rice in Pakistan with Desirable Agricultural Traits. *Agriculture* **2023**, *13*, 242. https://doi.org/10.3390/agriculture13020242

Academic Editor: Gaoneng Shao

Received: 1 November 2022
Revised: 9 January 2023
Accepted: 12 January 2023
Published: 19 January 2023

1. Introduction

Rice (*Oryza sativa* L) is an important food grain crop of the World population. Low yield and other undesirable traits are the common problems of rice crops that cause huge yield losses every year. Therefore, yield loss and undesirable agricultural traits cause increasing problems in agricultural practices, and selection for high-yielding rice varieties with good quality becomes an urgent challenge for plant breeding programs. Rice is the leading staple food for both China and Pakistan, and more than half of the world's population depends on the rice crop [1,2]. It is grown under various agro-ecological conditions in different subtropical and tropical countries, including Pakistan and India. To fulfill the future food demand of the ever-increasing world population there is an urgent need to take necessary steps to increase this crop's productivity [3]. Crop improvement programs also depend on the use of germplasm resources available in various parts of the world. The improvement and expansion of world supply will also depend on developing and improving rice varieties with higher yield potential and on several conventional and biotechnological methods for developing high-yielding varieties having resistance against biotic and abiotic stresses [4].

Honglian Type hybrid rice belongs to the three-line hybrid rice category, which is internationally recognized as one of the three cytoplasmic male sterility types of hybrid rice, with "Wild Abortive Type" by Academician Prof. Yuan Longping and "Baotai Type" by the Japanese group. The Baotai type is not used in indica hybrid rice, while the wild abortive type is only used in indica rice because of its sporophyte sterile type [5–7]. The honglian type belongs to the gametophyte sterile type, which can be used in indica and japonica rice [8]. In the agricultural field trial, when a single cytoplasmic type is popularized in a large area, the homogenization of the matched varieties may lead to weak resistance to some diseases and cause severe loss or crop failures. The Honglian type is another crucial genetic resource besides wild abortive type in China. Developing Honglian-type hybrid rice can enrich the genetic diversity of rice varieties and reduce the potential threats to agricultural practices caused by single cytoplasm type growth, which is a significant guarantee of the food security for both China and the world!

Hybrid rice technology has been developed and utilized in China for more than half a century, and it is one of the main technologies used to feed population and secure the country's food safety. Hybrid rice technology has a substantial yield and enhances the yield production by 25% more than inbred rice varieties [9].

The success of the hybrids depends upon a combination of desirable traits, high yield, adaptability, stability, distinctness, uniformity, novelty and various allelic interactions [10,11]. High yield and desirable traits are essential for crop breeding and enhance agricultural production under stressful environments [12–16]. In the current scenario food, security and sustainable agricultural productions are very crucial for peoples of all the nations. However, world food security is threatened by multiple factors, i.e., increasing population, high temperature, heat stress, climate change, loss of arable lands, urbanization, lack of acclimatization and increasing demands of food and feed. Rice meets the requirements of 21% of the total calorie intake of the world population and up to 76% of that of Southeast Asia [17,18]. Enhancement of rice production would have a great impact and would be very helpful in world food security. This will only become possible by the introduction of new high yielding varieties with good agricultural traits, high adaptability, high fertility rate under stress environment and approaches to produce them.Rice genetics and genomics have been advancing over the last decade and have an important role for the development of new varieties since the determination of rice genome sequence [19,20]. However, the increase of yield per hectare of rice in China and other Southeast Asian countries is too slow to meet future demands. Genetic diversity, the Honglian type of hybrid rice and different techniques are available to achieve the desired outcome.

The rice suffers from extensive stress to sustain full yield potential due to biotic and abiotic factors [21], the main constraints of the yield in South Asia and other countries cause severe loss. Therefore, it is advantageous to select those cultivars which can tolerate multiple stresses at the same time. Different strategies will be fruitful in managing the abiotic stresses in such environments. For the last few decades, rice breeding programs and other techniques of rice provided knowledge and resources to understand the various problems, including overcoming the problem of hybrid sterility; the discovery of cytoplasmic male sterility system and restorer genes for the production of new Honglian type of hybrid rice;

the increase of yield with the development of mapping populations; and the exploitation of heterosis, genetic resources and identification of new QTLs.

On the other hand, breaking undesirable linkages between abiotic stress and plant height, abiotic stress and earliness, abiotic stress and low yield and other quality traits could also be helpful for the development of resistant varieties [22,23].

The main characteristics are prominent genetic diversity, broad hybridization selectivity and excellent comprehensive agronomic traits. Honglian-type hybrid rice can be used as a mid-season rice in the Yangtze River basin of China, and an early-season rice and late-season rice in south China. Due to its heat tolerance characteristic, it is also suitable for spreading in tropical or subtropical regions along "the Belt and Road" countries such as in South Asian and African countries. In the past two decades, old varieties generated from the second-generation sterile lines of Honglian Type CMS have been widely spread and welcomed by these countries, and their growth area was enlarged every year. Now, new Honglian type varieties with better quality, higher yield and wider adaptabilities have been selected from the third-generation to fifth-generation sterile lines and developed to carry out adaptation tests in Pakistan. These breakthrough varieties of Honglian Type hybrid rice are generated based on the solid embodiment of the scientific concepts, aiming at combinations between "high yield, high quality, wide adaptability and ecological balance", which are the leading concepts for guiding the direction of hybrid rice breeding. The main objective of the study was to screen the Honglian type of hybrid rice on the basis of adaptability, genetic diversity and high yield performance in different locations of Pakistan.

2. Materials and Methods

2.1. Plant Materials

A set of 10 hybrids, including HP1, HP2, HP3, HP4, HP5, HP6, HLR-006, WR-1906, Guard53 (check variety), D-121(check variety) and one open-pollinated variety, KSK-133 (originated from Wuhan University China, Yuan Longping High-Tech Agriculture Co., LTD China WINALL Hi-Tech Seed Co., LTD China and Rice Research Institute, Kala Shah Kaku, Pakistan), were tested in the year 2019-20 in different locations (Gujranwala, Lahore, Pakpattan) of Pakistan under Randomized Complete Block Design (RCBD). Different qualitative and quantitative parameters and seed morphological characters were studied to see the diversity among rice lines. In the year 2021 some of these hybrids were tested in National Uniform Yield Trials (NUYT) at various locations in Pakistan, with sowing date (23 June 2020), transplanting date (18 July 2020) and harvesting date (28 October 2020). In the first year of the trials, seven different sites were selected for testing the hybrid varieties, and in the second year ten different sites were selected for testing the hybrids (See Supplementary Tables for details).

2.2. Traits Measurement

Various agro morphological traits (seed length, width, thickness, length–width ratio, curling%, bursting%, cooked grain length, brown rice, milled rice, head rice and yield/ha) were measured with the help of a meter rod, scale and weighing balance at physiological maturity of each rice hybrid/line. The seed length–width ratio was measured with the help of the following equation in millimeters. These traits showed significant differences among the rice genotypes based on their origin and genetic diversity.

$$\text{Seed Length Width Ratio} = \frac{\text{Seed Length (mm)}}{\text{Seed Width (mm)}}$$

2.3. Seed Morphological Traits

Various seed morphological traits (seed length, seed width, seed thickness, seed length–width ratio and 1000 grain weight) were measured with the help of a Digital Vernier Caliper (Jinhua Longtai Tools Co., LTD. Zhejiang, China, IP-67) and weighing balance (Locosc Ningbo Precision Technology Co., LTD. Ningbo, Zhejiang, China, LP-7610). Three lots of ten seeds data of each genotype were selected randomly. The seed morphology of each genotype is shown in Figure 1. The humidity of the rice paddy was measured with the help of a grain moisture meter (Model number: FG-506, Kett Japanese Company).

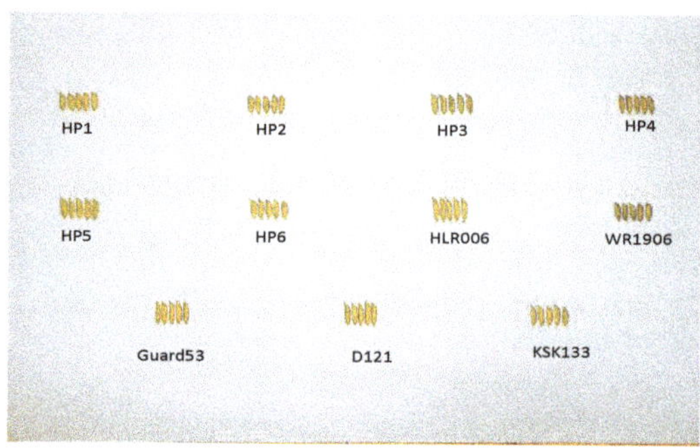

Figure 1. Seed morphology of ten rice hybrids and one open pollinated variety.

2.4. Principal Component Analysis, Variance and Correlation

Principal Component Analysis (PCA), variability and correlation of the rice hybrids were studied based on measured traits. On the basis of all the measured traits, three hybrids were selected for further genetic analysis in different sites in Pakistan to check the stability and adaptability in changing climatic conditions.PCA is a technique used for large datasets, presenting more variability among the genotypes and traits and also minimizing information loss. Desirable traits of all the hybrids were also analyzed by using Principal Component Analysis (PCA) to determine the genetic variability for these traits. Mean values of all the recorded traits of the hybrids/genotypes were used for PCA analysis.

2.5. NUYT and DUS Trials

Based on in-house trial results, three hybrids, HP1, HP2 and HP3, were selected for further analysis in NUYT (National Uniform Yield Trials) and DUS (Distinctness, Uniformity and Stability) trials in the year 2020-21, along with other hybrids and check varieties in different locations, i.e., the Soil Salinity Research Institute (SSRI PindiBhatian), Bahawalnagar, the Rice Research Institute Kala Shah Kaku (RRI KSK), the Rice Research Institute, Kala Shah Kaku (RRI KSK Sialkot), the Rice Research Institute (RRI Dokri), the Pakistan Agricultural Research Council (PARC KSK), the National Institute for Biotechnology and Genetic Engineering (NIBJE), the Nuclear Institute for Agriculture and Biology, (NIAB), Agriculture Research Institute (ARI Usta Muhammad), the Agriculture Research Institute, Dera Ismail Khan (ARI DI Khan), the Guard Rice Golarchi, Sun Crop, Meskay and Femtee Trading Company (FMTC) Shikarpur, the Emkay Farooqabad, Dokri Larkana, Soil Salinity Research Institute (SRRI Thatha) and Tara Crops of Pakistan based on yield and quality parameters under Randomized Complete Block Design (RCBD) with three replications. A total of 104 hybrids in 2020 and 137 hybrids in 2021 (including Honglian hybrid rice and check varieties) were tested under different ecological zones in Pakistan.

2.6. DNA Extraction and Quality Analysis

The leaf samples of the three selected hybrids (HP1, HP2 and HP3) were collected and stored at 4 °C. All the samples were collected at tillering stage. A CTAB method [24] was used to extract DNA. A nanodrop spectrophotometer (ND-1000) was used to check the quality and quantity of DNA at 260 and 280 nm. In all samples, DNA with good quality (concentration higher than 100 ng/µL) was used in PCR analysis.

2.7. DNA Fingerprinting and PCR Analysis

A set of 48 primers was used for PCR analysis [25] that covered almost the whole rice genome. Among them, nine primers were highly polymorphic, showing vast genetic differences in the tested rice hybrids. These polymorphic primers showed the highest genetic diversity among the rice hybrids (Table 1). A total of 100 ng DNA of each hybrid was used in the experiment. The PCR amplification reaction was set as heating 94 °C for 4 min, followed by 30 cycles of denaturation at 94 °C for 1 min, annealing at 55 °C for 1 min, extension at 72 °C for 1 min and final extension at 72 °C for 10 min. The 6% gel was used in this experiment, to see more clear bands of each hybrid. Furthermore, PCR samples were scanned on a gel documentation system to see the genetic differences among the rice hybrids. These SSR primers were used by genotyping and identifying rice hybrids/varieties using the SSR marker method [26] (NY/T 1433-2014).

Table 1. SSR markers showed variability among rice hybrids based on various genotypic traits.

S. No.	Primer Name	Chromosomal Location	Annealing Temp °C	Primer Sequence (5′–3′)	Fluorescence	Product Size	Nature of Polymorphism
1	RM-219	9	55	F:cgtcggatgatgtaaagcct R:catatcggcattcgcctg	FAM	194–215	Polymorphic
2	RM-236	7	55	F:cttacagagaaacggcatcg R:gctggtttgtttcaggttcg	VIC	151–166	Polymorphic
3	RM-274	5	55	F:cctcgcttatgagagcttcg R:cttctccatcactcccatgg	V1C	149–162	Polymorphic
4	RM-253	6	55	F:tccttcaagagtgcaaaacc R:gcattgtcatgtcgaagcc	PET	133–142	Polymorphic
5	RM-424	2	55	F:tttgtggctcaccagttgag R:tggcgcattcatgtcatc	NED	240–280	Polymorphic
6	RM-567	4	55	F:atcagggaaatcctgaaggg R:ggaaggagcaatcaccactg	PET	248–260	Polymorphic
7	RM-258	10	55	F:tgctgtatgtagctcgcacc R:tggcctttaaagctgtcgc	FAM	128–146	Polymorphic
8	RM-481	7	55	F:tagctagccgattgaatggc R: ctccacctcctatgttgttg	FAM	146–165	Polymorphic
9	RM-493	1	55	F: tagctccaacaggatcgacc R:gtacgtaaacgcggaaggtg	VIC	210–264	Polymorphic

2.8. Statistical Analysis

The Principal Component Analysis (PCA) and correlation were analyzed by using SAS (Statistical Analysis System) version 9.2 [27] to see the genetic variability in rice hybrids. The average data of all the hybrids were calculated based on the mean values. Statistical significance is the determination of relationships between two or more variables for the prevalence of significant variance for all studied traits that implicates the usefulness of the rice hybrids for genetics analysis at level of significance of 1% and 5%.

3. Results

3.1. In-House Yield Trials

In-house yield performance data of the rice hybrids HP1, HP2 and HP3 were developed by the University of Punjabat different locations (Gujranwala, Lahore, Pakpattan) in Punjab, Pakistan. All the parameters were taken at the time of maturity. In all the locations, the yield was measured in tons/ha. All the varieties performed very well over check variety in all the locations (Table 2).

Table 2. Yield comparison of hybrid rice varieties with check variety.

Sr. No	Variety Name	Origin	Average Yield tons/ha 2019	Average Yield tons/ha 2020	% Increase/Decrease With Check Variety 2019	% Increase/Decrease With Check Variety 2020
1	HP1	China	12.75	10.63	+43.90%	+30.91%
2	HP2	China	12	10.83	+35.44%	+33.37%
3	HP3	China	12.15	10.85	+37.13%	+33.62%
4	HP4	China	9.98	10.10	+12.64%	+24.38%
5	HP5	China	10.6	9.85	+19.63%	+21.30%
6	HP6	China	10.22	10.5	+15.34	+29.31%
7	HLR006	China	8.14	8.10	−8.12%	−0.24%
8	WR1906	China	9.81	10.2	+10.72	+25.61%
9	Guard53	China	9.91	10.25	+11.85%	+26.23%
10	D121	China	10.40	10.35	+17.38	+27.46%
11	KSK133	Pakistan	8.86	8.12	-	-

3.2. Genetic Diversity Study of Honglian Type Hybrid Rice

Forty-eight SSR markers were used to measure the genetic diversity of Honglian type rice. The varieties showed maximum genetic diversity concerning their specific markers, product size, polymorphism and chromosomal locations. Finally, nine polymorphic SSR markers were selected to determine the genetic diversity of Honglian type rice HP1, HP2, HP3 that showed the most remarkable genetic diversity. The RM-236 and RM-424 were found to be more appropriate for HP1, HP2 and HP3 that showed more diversity among them, and very useful for traits variation (Supplementary Table S3).

3.3. Principal Component Analysis (PCA) with Respect to Yield and Other Traits

Principal component analysis was used to determine the phenotypic diversity under adaptability trials 2020–2021. All the hybrids showed variability according to their quality and yield contributing traits. The Honglian hybrid rice varieties (HP1, HP2, HP3) performed excellently among the full hybrids and over-check varieties under different ecological zones, having significant differences among them. The quality traits, i.e., seed length, width, thickness, length–width ratio, stickiness, curling %, bursting %, cooked grain length, brown rice, milled rice, head rice and yield per hectare of HP1 and HP3, were higher over the check varieties (D-121, Guard-53). The results are shown in Table 3. The average seed length was 7.02 mm, the cooked grain length 11.01 mm and other quality traits were higher over check varieties in both years. The yield per hectare of HPI (7863 kg/ha) was higher over check varieties in 2021 (Supplementary Table S2).

Table 3. Yield and quality traits comparison of hybrids with check varieties in adaptability trials 2020-21.

Adaptability Trials in the Year 2020													
Sr. No	Variety	Length (mm)	Width (mm)	L/W Ratio (mm)	Thickness (mm)	Stickiness	Curling %	Bursting %	C.G.L. (mm)	Brown Rice (gm)	Milled Rice (gm)	Head Rice Recovery %	Yield Kg/hac
1	HP1	7.01	2.04	3.44	1.82	sticky	2	6	10.2	81	77.3	53.8	8709
2	HP2	6.73	2.06	3.27	1.74	sticky	5	16	9.8	80	75	53.3	8833
3	HP3	6.8	2.11	3.22	1.76	sticky	2	4	10.3	84.4	79.4	61.2	9338
4	D-121	6.72	1.98	3.4	1.75	sticky	4	7	10.8	80	73.3	56.4	9171
5	Guard-53	6.9	2.08	3.32	1.77	sticky	3	8	10.3	81.6	74.5	62.3	8395
Adaptability trials in the year 2021													
Sr. No	Variety	Length (mm)	Width (mm)	L/W ratio (mm)	Thickness (mm)	Stickiness	Curling %	Bursting %	C.G.L. (mm)	Brown rice (gm)	Milled rice (gm)	Head rice recovery %	Yield Kg/hac
1	HP1	7.3	2.13	3.43	1.81	sticky	4	12	12.9	82.03	76.6	51.67	7863
2	HP2	6.94	2.09	3.32	1.76	sticky	5	4	11.1	81	74.56	60	7288
3	HP3	6.6	1.75	3.77	1.78	sticky	25	11	10.2	80.13	74.66	64.67	7387
4	D-121	6.96	2.01	3.46	1.78	sticky	4	12	11.5	81.63	74.86	51	7518
5	Guard-53 (1)	6.66	2.02	3.30	1.77	sticky	8	4	9.7	82.86	76.46	65.30	7341

The yield per hectare (9338 kg/ha) was measured as higher over control. Based on the yield performance, the variety was positioned at number four in adaptability trials among 104 hybrids (Supplementary Table S1). Similarly, HP3 performance was higher over check varieties in the year 2020.

In the year 2020, adaptability trials of 102 hybrids and two check hybrids were conducted in different locations, i.e., the Rice Research Institute Kala Shah Kaku(RRI, KSK), the Rice Research Institute (RRI, Dokri), the Pakistan Agricultural Research Council (PARC, KSK), Guard Rice; Golarchi, Four Brothers Multan, Emaky Sheikhupura, Chaudhry Khair Din (CKD), and Dera GhaziKhan, in the country regarding yield and other quality parameters. Based on in-house trials, three hybrids (HP1, HP2 and HP3) were selected for further evaluation in NUYT and DUS trials. Our hybrids showed excellent performance concerning yield and various quality parameters over check varieties (D-121, Guard-53). The quality traits of seed length (7.01 mm; 6.8 mm), width (2.04 mm; 2.11 mm), thickness (1.82 mm; 1.76 mm), length–width ratio (3.44 mm; 3.22 mm), brown rice (81 gm; 84.4 gm) and milled rice (77.3 gm; 79.4 gm) of HP1 and HP3 were higher than the check varieties. The results showed that the yield kg per hectare of HP1 (8709), HP2 (8833) and HP3 (9338) was more than the check varieties, i.e., D-121 (9171) and Guard-53 (8395). The almost average performance of our hybrid varieties was higher than the check varieties, and all the information was mentioned in Table 3 and Supplementary Table S1.

In 2021, 135 hybrids and two check hybrids were evaluated in ten different locations in Pakistan under adaptability trials. Our hybrids (HP1, HP2 and HP3) showed significant differences to check varieties (D-121 and Guard-53) in yield and various quality parameters (Table 3, Supplementary Table S2). Almost all the traits showed significant variation among each other and with check varieties. The average performance of HP hybrid characteristics was higher than the check varieties in almost all the traits studied. HP1 seed length (7.3 mm), width (2.13 mm), thickness (1.81 mm), length–width ratio (3.43 mm), cooked grain length (12.9 mm), brown rice (82.03 gm), milled rice (76.6 gm) and grain yield 7863 kg/ha was higher than both check hybrids.

The principal components also showed variability among the entire set of genotypes and their contributing traits. Those principal components have more than one eigenvalue that shows more variability and has more importance in selection criteria. In 2020, four components with more than one eigenvalue contributed to variation in a collective 67.43%. Similarly, in 2021, six principal components showed a maximum variation of 82.16%. Such results showed significant differences among genotypes and traits. Some components had significant positive effects on various quality and seed parameters. The components had more considerable positive effects for the selection of promising genotypes. The eigenvectors with positive values with their respective traits showed more variation in genotypes and studied traits (Table 4).

Table 4. Eigenvalue, variation and cumulative% variability of various yield and quality parameters of hybrid rice.

Traits	PC	2020			2021		
		Eigenvalue	Variation%	Cumulative %	Eigenvalue	Variation %	Cumulative %
Length	PC1	2.48	22.63	22.62	3.18	22.78	22.78
Width	PC2	2.13	19.45	42.07	2.60	18.58	41.36
Thickness	PC3	1.49	13.56	55.63	1.92	13.72	55.08
L/W ratio	PC4	1.29	11.79	67.43	1.57	11.23	66.32
Curling %	PC5	0.96	8.77	76.21	1.18	8.45	74.77
Bursting %	PC6	0.71	6.46	82.67	1.03	7.38	82.16
C.G.L (mm)	PC7	0.57	5.24	87.92	0.90	6.44	88.60
Brown rice %	PC8	0.47	4.29	92.21	0.66	4.77	93.38
Milled rice %	PC9	0.39	3.63	95.85	0.51	3.69	97.08
Head rice %	PC10	0.26	2.42	98.27	0.40	2.88	99.97
Yield/ha	PC11	0.18	1.72	100	0.003	0.026	100

Scree plot, biplot and traits variation among the significant principal components numbers (PC1, PC2) had maximum variation between them and in comparison with other numbers. The scree plot describes the association between eigenvalues and cumulative% variability. This showed the variation among genotypes and their traits.

Some of the traits showed positive significant association with each other that was very important for enhancing the various qualitative and quantitative traits of hybrids under the adaptability trials 2020–2021. Seed length had positive significant associations with length–width ratio (r = 0.5159*), thickness (r = 0.3907*) and cooked grain length (r = 0.5027*). A significant positive association was observed between thickness with cooked grain length and yield/hectare (r = 0.4059*; r = 0.3091*). Brown rice was associated with milled rice and head rice in adaptability trials 2020 (Table 5). In 2021, a significant positive association of seed width was observed with length–width ratio, thickness and bursting% (r = 3054*; r =3051*; r = 0.1679*). The bursting%, cooked grain length, brown rice% and yield/hectare had a significant positive association with curling% (Table 6). Correlations among the significant traits were beneficial for selecting better hybrids and determining the interrelationship between the traits. According to the National Uniform Yield Trial (NUYT), Distinctness, Uniformity and Stability (DUS) confirmed that HP1 and HP3 hybrid varieties performed very well in both years over check varieties (D-121, Guard-53) in various quality and yield parameters. In the first year of the trials, seven different sites were selected for testing the hybrid varieties, and in the second year, ten different sites were selected for testing the hybrids. In all locations, HP hybrids showed excellent results in all country zones.

Table 5. Association of various morphological, quality and yield traits of hybrid rice in adaptability trials 2020.

Variables	Length (mm)	Width (mm)	L/W Ratio	Thickness (mm)	Curling (%)	Bursting (%)	C.G. L (mm)	Brown Rice (%)	Milled Rice (%)	Head Rice (%)	Yield/ha
Length (mm)	1.00										
Width (mm)	0.0961	1.00									
L/W Ratio	0.5159 *	0.2458 *	1.00								
Thickness (mm)	0.3907 *	−0.2190 *	−0.2532 *	1.00							
Curling (%)	−0.1223	−0.0416	0.1757	−0.3379 *	1.00						
Bursting (%)	−0.1830	−0.1154	−0.1700	−0.0676	0.3539 *	1.00					
C.G. L (mm)	0.5027 *	0.0173	0.2155 *	0.4059 *	−0.2315 *	−0.1600	1.00				
Brown Rice (%)	−0.0942	−0.1156	−0.0335	0.0936	0.0885	0.1505	−0.1284	1.00			
Milled Rice (%)	−0.0687	−0.2450 *	−0.1475	0.1706	0.0446	−0.0940	−0.0961	0.6637 *	1.00		
Head Rice (%)	−0.2902 *	−0.1262	−0.1737	−0.0459	−0.0626	−0.1414	−0.1488	0.4243 *	0.4807 *	1.00	
Yield/ha	0.0122	−0.0263	−0.2354 *	0.3091 *	−0.0846	−0.0175	0.1135	0.1474	0.1423	−0.0296	1.00

* showed the Level of significance. $p < 0.05$= * (Statistically significant) and *p* represent the probability value.

Table 6. Association of various morphological, quality and yield traits of hybrid rice in adaptability trials 2021.

Variables	Length (mm)	Width (mm)	L/W Ratio	Thickness (mm)	Curling (%)	Bursting (%)	C.G. L (mm)	Brown Rice (%)	Milled Rice (%)	Head Rice (%)	Yield/ha
Length (mm)	1.00										
Width (mm)	−0.6068 *	1.00									
L/W Ratio	−0.3201 *	0.3054 *	1.00								
Thickness (mm)	−0.3201 *	0.3051 *	1.0000 *	1.00							
Curling (%)	0.0586	0.1062	−0.1481	−0.1482	1.00						
Bursting (%)	−0.0227	0.1679 *	−0.1625	−0.1626	0.4136 *	1.00					
C.G. L (mm)	0.0576	−0.1179	0.0670	0.0670	0.1951 *	−0.8077 *	1.00				
Brown Rice (%)	0.1666	0.0248	−0.1328	−0.1328	0.5975 *	0.1447	0.2168 *	1.00			
Milled Rice (%)	−0.1234	0.1154	0.1646	0.1646	−0.1375	−0.2631 *	0.1943 *	−0.1055	1.00		
Head Rice (%)	0.0471	−0.0213	−0.0600	−0.0600	−0.0176	−0.1663	0.1665	−0.0196	0.4282 *	1.00	
Yield/ha	−0.0831	0.0222	0.0193	0.0193	0.1981*	0.2262*	−0.1145	0.0637	−0.1348	−0.0656	1.00

* showed the Level of significance. $p < 0.05$= * (Statistically significant) and *p* represent the probability value.

3.4. DNA Analysis

Following the SSR marker method for identifying rice varieties (NY/T1433-2014), an experiment was conducted to analyze HP1, HP2 and HP3 hybrid rice varieties; a total of 48 primer pairs were used for this purpose. Among them, nine pairs of markers could clearly distinguish the three varieties and showed more remarkable genetic diversity (Figure 2). No.1, No.2, No.3, No.4, No.5, No.6, and No.7 primers distinguished HP1 from the other two varieties (HP2, HP3). On the other hand, No.2 and No.5 primers distinguished HP2 from the other two varieties (HP1, HP3) and No.5, No.8, and No.9 primers distinguished HP3 from the other two varieties (HP1 and HP2).

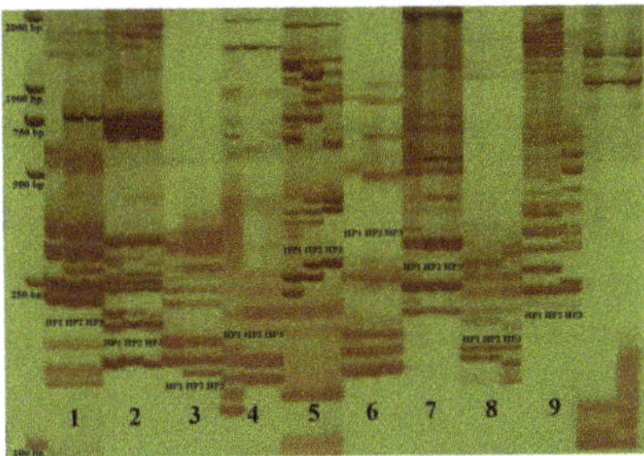

Figure 2. Genetic diversity of elite Honglian type of hybrid rice varieties. Markers (RM219, RM236, RM274, RM253, RM424) distinguished HP1 with HP2 and Hp3. Markers RM236 and RM424 distinguished HP2 with HP1 and HP3. Markers RM424, RM481 and RM493 distinguished HP3 with HP1 and HP2.

4. Discussion

4.1. Genetic Studies and Characteristics of HonglianType Hybrid Rice

Crop improvement is followed by genetic diversity, with various characteristics screened. In this study, different hybrids were studied in 2019 under in-house trials for yield and yield-related traits and their adaptation to changing climatic conditions. With respect to yield, all the Honglian hybrid rice varieties showed more than 30% more yield potential over check variety. As Honglian hybrid rice varieties have desirable yield and yield related traits, i.e., high yield, high quality, heat tolerant, drought tolerant, high tillering ability, high fertility and good grain quality parameters, etc., PCA and correlation analysis displayed significant genetic differences among the genotypes along with all the desired traits indicating the existence of variability [28,29]. The association of different rice traits and patterns of influence on the grain yield of rice was investigated. Such types of evaluation are very important to determine the direct effects of various traits on yield to determine the selection criteria for high grain yield. We found that some of traits have greater value, including seed length, seed width, seed thickness, curling%, bursting %, cooked grain length, head rice recovery %, etc., which have accounted for high grain yield. The positive associations of yield with other desirable traits were found to be significant, providing the information for selecting desirable rice hybrids/genotypes which are more favorable to acclimatize to changing environments. Positive significant genetic differences and correlation studies among the genotypes, along with desired traits, provide information to the researcher for the better selection of genotypes [30,31].

Genetic variation of traits is essential for selection and other breeding applications [32] Similarly, in previous studies, the Honglian type of hybrid rice displayed good performance in various Southeast Asian countries on the basis of various desirable traits [33]. The rice hybrids showed significant differences on the basis of various morphological traits. It was found that the average performance of HP1 and HP3 was excellent over check varieties in both years with respect to yield and yield related traits. An NUYT test provides excellent information regarding the adaptability of HP hybrids over a wide range of environments in all the locations of Pakistan, and is also helpful for the food security of China and the world's population due to its high yield and other desired traits [34]. This was a significant step towards selecting good hybrids in different ecological zones, providing information regarding the suitability of hybrid seed production and technology transfer of Honglian hybrid rice to farmers, students and scientists. With the advancement of new molecular breeding techniques and genomics technology, two-line hybrid CMS and three-line hybrid CMS systems can be adopted to improve crops. Hence, the CMS systems of hybrid rice have great scope and are helpful for the enhancement of productivity [35,36]. Molecular markers are significant for determining genetic diversity and developing new hybrids [37,38]. DNA fingerprint results showed that Honglian type varieties are highly variable from the traditional indica rice, which will support as an alternative germplasm for breeding selection, which is useful for modifying the local varieties in Pakistan. Some of the markers have been used in previous studies to determine the genetic variation among the Honglian type of hybrid rice and their genetic traits [37].

4.2. Correlation Studies

Correlation analysis allows us toobtain information on the relationship between variables, i.e., dependent and independent. It helps plant breeders to better understand the relationship of the yield-related traits, which will lead to the selection of genotypes with desired characteristics [39]. Knowledge about correlation coefficients is essential because the grain yield and other quantitative traits are influenced by many factors [40,41]. Correlation is one of the best methods to find out the relations among various traits and to lead the way for the frequency of traits and the compulsory screening to be measured in improving traits, for instance, grain yield [31,42].

4.3. Principal Component Analysis (PCA) Studies

The hybrid varieties were also analyzed using PCA to compare characteristics of HP varieties over check varieties. Principal components with more than one eigenvalue showed more variability among the traits studied for each genotype. Those principal components with more than one eigenvalue showed a collective variation of 67.42% in the year 2020. The PC1 had 22.62%, PC2 showed 19.45%, PC3 exhibited 13.56%, and PC4 had 11.79% variability between the rice varieties and their various traits in 2020. The variance and eigenvalue associated with principal components decreased gradually and stopped at 0.18%. In the year 2021, the first six components showed maximum variability of 82.13% (Table 4).

Based on the principal eigenvalue components, variation is considered to be very important in screening and selecting the rice hybrids/varieties [43]. Principal components and eigenvalues that showed genetic differences among genotypes are shown in Figure 3. Higher eigen values showed more variability and helped to select parents and improve the varieties through breeding [44]. In biplot principal component analysis, almost all traits positively affected their respective genotypes except curling% and bursting % (Figure 4). This showed the diversity of the genotypes, along with the desired characteristics. This information will be beneficial for the further screening of genotypes for developing a new plant population and starting a new breeding program. In adaptability trials, traits showed variability with the positive effect and their principal components (Figure 5), which were considered more important in selecting diverse genotypes. Such genotypic and phenotypic characteristics could be utilized in a breeding program for the screening and developing of new plant populations [45,46]. The study was equally beneficial for the best interests of breeders and researchers to utilize in their research in the best interest of the community and economy of the country.

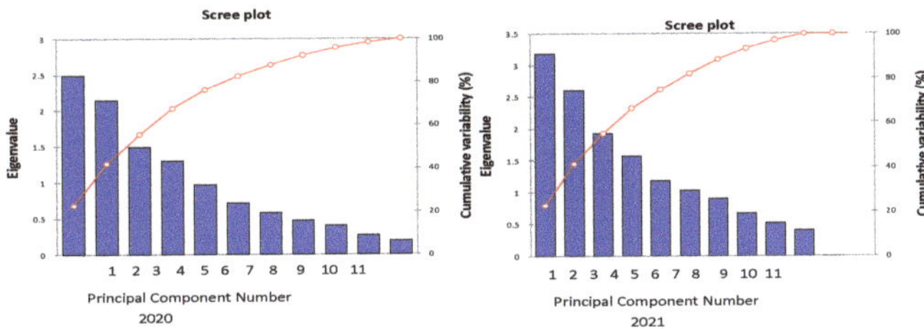

Figure 3. Scree plot of principal component analysis rice hybrids between their eigen values, number of principal component and accumulative variability% under adaptability trials 2020–2021.

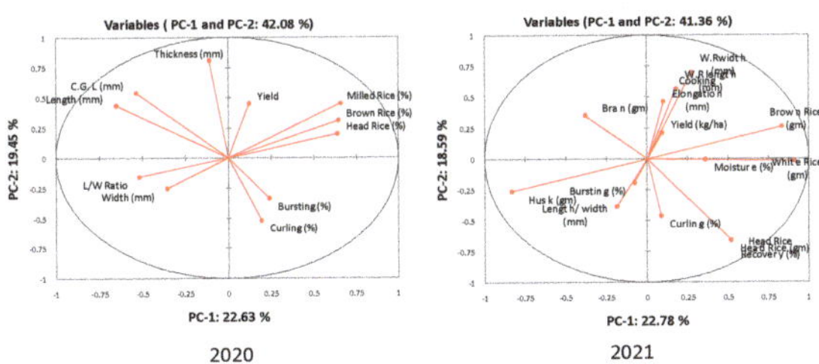

Figure 4. Traits variation on the basis of their major principal components under adaptability trials 2020–2021.

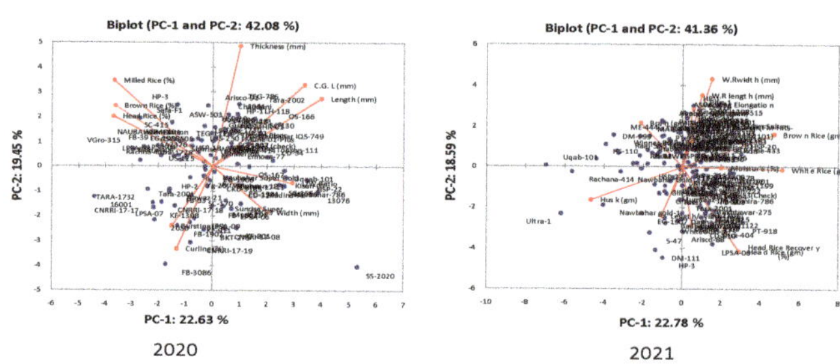

Figure 5. Biplot PCA analysis of various hybrid rice varieties and their traits association with principal component numbers under adaptability trials 2020–2021.

Honglian hybrid rice (HP1, HP3) yield is higher than the check variety in continuous year inspections and different locations, and its yield is higher because it is heat tolerant and widely adaptive, so the yield is comparably stable in different environments. Moreover, its seeds setting rate, tiller numbers, and spikes per plants are higher than the other varieties, which is the solid base for the higher yield, and this is critical for the agriculturally safe production in Pakistan, especially facing the challenges of climate change and sharp environmental variations in the future.Exploiting the heterosis of the hybrids had great importance for improving high grain yield in rice and other traits [47]. The parental material of Honglian hybrid rice had a strong restoration ability in the cytoplasmic male sterility system (CMS) for developing new three-line hybrids [48–51].

4.4. Honglian Type Hybrid Rice Research Importance and Future Prospects

"Food and safety come as the first". Rice is the staple food for more than 50% of the world population and its demand is increasing day by day. Therefore, to gradually improve wide-ranging rice production capacity and meet the inflexible demands for food consumption globally is a major strategic issue concerning food security, social stability, people's health, and economic developments. Wuhan University is one of the leading teams in the field of hybrid rice research in China. The team made many research achievements and developed a series of Honglian type hybrid rice varieties with excellent characteristics such as high quality, high yield, wide adaptability, multi-diseases/insect resistance and environmental friendliness. The main characteristics are prominent genetic diversity, wide hybridization selectivity and excellent comprehensive agronomic traits [52].

Several novel techniques could be beneficial for improving hybrid rice breeding techniques for developing new hybrid varieties. To advance the application of hybrid rice, genome editing technology based on clustered regularly interspaced palindromic repeats (CRISPR), the Cas9 system, has been extensively used to improve crops [53,54]. Another technique, de novo domestication of allotetraploid rice, could be very fruitful to stabilize heterosis and other desirable traits in hybrid rice which show more adaptation in changing environments, such asin other polyploidy species, i.e.,wheat, triticale, cotton, tobacco and strawberry [55–57]. Genetic resources in hybrid rice breeding will be very desirable in the future to develop breeding techniques for improving crops and enhancing the selection process of superior hybrid lines.

In this regard, the University of the Punjab and Wuhan University, China have plans for the further testing of Honglian-type hybrid rice varieties and hybrid seed production in Pakistan under different climatic conditionsto achieve maximum results in the rice field through technology transformation techniques. Different plant populations (F_2, RILs, NILs, etc.) will be developed by using diverse germplasm (Figure 6) to produce elite breeding plant varieties to meet the food requirements of the world's population.

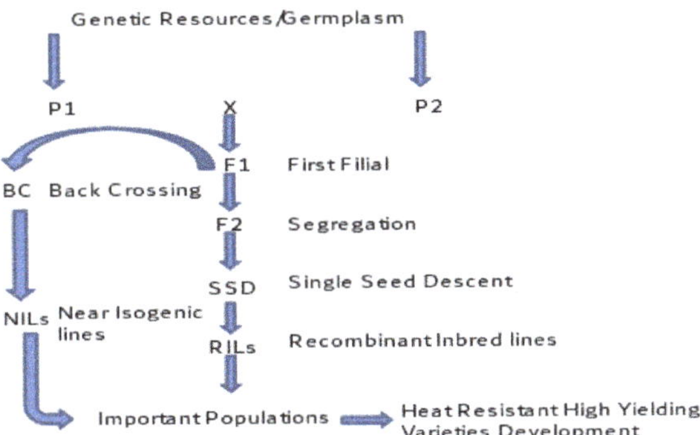

Figure 6. Genetic resources are a vital component for the improvement and development of new hybrid rice high-yielding varieties.

5. Conclusions

The present findings conclusively demonstrate that the yield of 9338 kg/ha of HP3 in 2020 and 7863 kg/ha of HP1 in 2021 were higher than the average yield of all the hybrids tested in both years and over check varieties. We found that some of the traits, i.e., seed length, width, thickness, cooked grain length, brown rice, milled rice and yield per hectare of HP3 and HP1, respectively, were greater than the check varieties (D-121, Guard-53). These hybrids and their parent material could be further used for the development of new distinct uniform homozygous plant populationson the basis of the desired characteristics. The performance of Honglian type hybrid rice was more stable and considerably better than other hybrids/varieties in high temperature locations, which indicates its better adaptation and acclimatization in various ecological zones of Pakistan. Finally, we summarize that HP1,HP2 and HP3, as the new generations of Honglian type hybrids with advanced characteristics, are beneficial for introduction into Pakistan for future development and industrialization. The NUYT and DUS analysis convincingly showed that they are suitable to grow in Pakistan. Based on the historical economical contributions of the old varieties of Honglian type hybrid rice, it is promising that development and industrialization of these new varieties would contribute well to Pakistan and benefit the people both in China and Pakistan. In the current scenario, this type of study and the genetic material could be very useful for the production of high yielding varieties that would be more fruitful to the farmers community and strengthen the country's economy.

Supplementary Materials: The following are available online at https://www.mdpi.com/article/10.3390/agriculture13020242/s1, Table S1: (Part-I): Means of Paddy Yield (kg/ha) of Rice Hybrids Evaluated in NUYT during Kharif, 2020, 2020, (Part-II). Quality data of NUYT 2020 Hybrids Rice; Table S2: (Part-I): Mean of Paddy Yield (kg/ha) of Rice Hybrids Evaluated in NUYT during Kharif, 2021, (Part-II) Quality Characteristics of Rice Hybrids Evaluated in NUYT 2021; Table S3: SSR markers used for rice hybrids for their Genetic diversity study.

Author Contributions: M.A. (Muhammad Ashfaq) conceived the idea of the study and wrote up the manuscript. R.Z. is the principal researcher and breeder of the Honglian type hybrid rice and provided the material. Z.X. carried out the SSR marker analysis. M.A. (Muhammad Ali) and A.R. helped to organize the experimental material for further data recording. M.J. and A.S. helped with data analysis and supported the write-up and review of the manuscript. X.W. designed the experiment and helped a lot in the completion/write-up of the research work. All authors have read and agreed to the published version of the manuscript.

Funding: This research was funded by the Ministry of Science and Technology (MOST), China (National Key Research and Development Project: 2021YFE0101000) and the Pakistan Science Foundation (PSF/CRP/18th Protocol (11) under the international collaboration project supported by both governments. The APC was funded by the collaborative project.

Institutional Review Board Statement: Not applicable.

Informed Consent Statement: Not applicable.

Data Availability Statement: The data that support the findings of this are available in the main manuscript and the supplementary files.

Acknowledgments: The authors would like to thank the Wuhan University, China and the University of Punjab for supporting and facilitating the joint project on Honglian-type hybrid rice. The Ministry of Science and Technology of both countries and the Pakistan Science Foundation (PSF) supported financially to complete this research work in a timely manner.

Conflicts of Interest: The authors declare no conflict of interest.

References

1. Seck, P.A.; Diagne, A.; Mohanty, S.; Wopereis, M.C. Crops that feed the world 7: Rice. *Food Secur.* **2012**, *4*, 7–24. [CrossRef]
2. FAO. *The Future of Food and Agriculture—Alternative Pathways to 2050*; License: CCBY-NC-SA3.0IGO; FAO: Rome, Italy, 2018; 224p
3. Ram, S.G.; Thiruvengadam, V.; Vinod, K.K. Genetic diversity among cultivars, landraces and wild relatives of rice as revealed by microsatellite markers. *J. Appl. Genet.* **2007**, *48*, 337–345. [CrossRef]
4. Khush, G. What it will take to Feed 5.0 Billion Rice consumers in 2030. *Plant Mol. Biol.* **2005**, *59*, 1–6. [CrossRef] [PubMed]
5. Ren, G.J.; Yan, L.A.; Xie, H.A. Retrospective and perspective on indica three-line hybrid rice breeding research in China. *Chin. Sci. Bull.* **2016**, *61*, 3748–3760. [CrossRef]
6. Yang, Z.A. Retrospects and prospects on the development of japonica hybrid rice in the north of China. *Acta Agron. Sin.* **1998**, *24*, 840–846.
7. Li, S.; Yang, D.; Zhu, Y. Characterization and use of male sterility in hybrid rice breeding. *J. Integr. Plant Biol.* **2007**, *49*, 791–804. [CrossRef]
8. Zhang, S.; Haung, X.; Han, B. Understanding the genetic basis of rice heterosis: Advances and prospects. *Crop J.* **2021**, *9*, 688–692. [CrossRef]
9. Verma, R.L.; Katara, J.L.; Sah, R.P.; Sarkar, S.; Azaharudin, T.P.; Patra, B.C.; Sahu, R.K.; Mohapatra, S.D.; Mukherjee, A.; Manish, K.; et al. CRMS 54A and CRMS 55A: Cytoplasmic male sterile lines with enhanced seed producibility. *NRRI Newsl. Jan. March* **2018**, *39*, 1–14.
10. Takai, T.; Arai-Sanoh, Y.; Iwasawa, N.; Hayashi, T.; Yoshinaga, S.; Kondo, M. Comparative Mapping Suggests Repeated Selection of the Same Quantitative Trait Locus for High Leaf Photosynthesis Rate in Rice High-Yield Breeding Programs. *Crop Sci.* **2012**, *52*, 2649–2658. [CrossRef]
11. Xie, F.; Zhang, J. Shanyou 63: An elite mega rice hybrid in China. *Rice* **2018**, *11*, 17. [CrossRef]
12. Boyce, C.K.; Jung, E.L. Plant evolution and climate over geological timescales. *Annu. Rev. Earth Planet. Sci.* **2017**, *45*, 61–87. [CrossRef]
13. Mittler, R.; Blumwald, E. Genetic engineering for modern agriculture: Challenges and perspectives. *Annu. Rev. Plant Biol.* **2010**, *61*, 443–462. [CrossRef] [PubMed]
14. Madan, P.; Jagadish, S.; Craufurd, P.; Fitzgerald, M.; Lafarge, T.; Wheeler, T. Effect of elevated CO_2 and high temperature on seed-set and grain quality of rice. *J. Exp. Bot.* **2012**, *63*, 3843–3852. [CrossRef]
15. Mohammed, A.R.; Tarpley, L. Effects of high night temperature and spikelet position on yield-related parameters of rice (*Oryza sativa* L.) plants. *Eur. J. Agron.* **2010**, *33*, 117–123. [CrossRef]
16. Morita, S.; Wada, H.; Matsue, Y. Countermeasures for heat damage in rice grain quality under climate change. *Plant Prod. Sci.* **2016**, *19*, 1–11. [CrossRef]
17. Fitzgerald, M.A.; McCouch, S.R.; Hall, R.D. Not just a grain of rice: The quest for quality. *Trends Plant Sci.* **2009**, *14*, 133–139. [CrossRef] [PubMed]
18. Miura, K.; Ashikari, M.; Matsuoka, M. The role of QTLs in the breeding of high-yielding rice. *Trends Plant Sci.* **2011**, *16*, 319–326. [CrossRef]
19. Goff, S.A.; Ricke, D.; Lan, T.H.; Presting, G.; Wang, R.; Dunn, M.; Glazebrook, J.; Sessions, A.; Oeller, P.; Varma, H.; et al. A draft sequence of the rice genome (*Oryza sativa* L. ssp. *japonica*). *Science* **2002**, *296*, 92–100. [CrossRef]
20. International Rice Genome Sequencing Project. The map-based sequence of the rice genome. *Nature* **2005**, *436*, 793–800. [CrossRef]
21. Waddington, S.R.; Li, X.; Dixon, J.; Hyman, G.; De Vicente, M.C. Getting the focus right: Production constraints for six major food crops in Asian and African farming systems. *Food. Secur.* **2010**, *2*, 27–48. [CrossRef]
22. Kamoshita, A.; Babu, R.C.; Boopathi, N.; Fukai, S. Phenotypic and genotypic analysis of drought-resistance traits for development of rice cultivars adapted to rainfed environments. *Field Crops Res.* **2008**, *109*, 1–23. [CrossRef]
23. Vikram, P.; Swamy, B.P.M.; Dixit, S.; Trinidad, J.; Sta, M.T.; Cruz, P.C.; Maturan, M.; Amante, M.; Kumar, A. Linkages and interactions analysis of major effect drought grain yield QTLs in rice. *PLoS ONE* **2016**, *11*, e0151532. [CrossRef] [PubMed]
24. Murray, M.G.; Thompson, W.F. Rapid isolation of high molecular weight plant DNA. *Nucl. Acids Res.* **1980**, *8*, 4321–4326. [CrossRef]

25. Panaud, O.; Chen, X.; McCouch, S. Development of microsatellite markers and characterisation of simple sequence length polymorphism (SSLP) in rice (*Oryza sativa* L.). *Mol. Gen. Gene* **1996**, *252*, 597–607.

26. MOA. *Protocol for Identifcation of Rice Varieties-SSR Marker Method*; NY/T 1433-2014; China Agriculture Press: Beijing, China, 2014.

27. SAS Institute Inc. *SAS®9.2 Enhanced Logging Facilities*; SAS Institute Inc.: Cary, NC, USA, 2008.

28. Dutta, P.; Dutta, P.N.; Borua, P.K. Morphological Traits as Selection Indices in Rice: A Statistical View. *Univ. J. Agric. Res.* **2013**, *1*, 85–96.

29. Konate, A.K.; Zongo, A.; Kam, H.; Sanni, A.; Audebert, A. Genetic variability and correlation analysis of rice (*Oryza sativa* L.) inbred lines based on agro-morphological traits. *Afr. J. Agric. Res.* **2016**, *11*, 3340–3346.

30. Ogunbayo, S.A.; Sié, M.; Ojo, D.K.; Sanni, K.A.; Akinwale, M.G.; Toulou, B.; Shittu, A.; Idehen, E.O.; Popoola, A.R.; Daniel, I.O.; et al. Genetic variation and heritability of yield and related traits in promising rice genotypes (*Oryza sativa* L.). *J. Plant Breed. Crop Sci.* **2014**, *6*, 153–159.

31. Paswan, S.K.; Sharma, V.; Singh, V.K.; Ahmad, A.; Deepak. Genetic Variability Studies for Yield and Related Attributes in Rice Genotypes (*Oryza sativa* L.). *Res. J. Agric. Sci.* **2019**, *5*, 750–752.

32. Tiwari, D.N.; Tripathi, S.R.; Tripathi, M.P.; Khatri, N.; Bastola, B.R. Genetic Variability and Correlation Coefficients of Major Traits in Early Maturing Rice under Rainfed Lowland Environments of Nepal. *Adv. Agric.* **2019**, *2019*, 5975901. [CrossRef]

33. Zhu, R.S.; Yu, J.H.; Ding, J.P. Research and utilization of honglian type hybrid rice. *Hybrid Rice* **2010**, *25*, 12–16. (In Chinese)

34. Li, J.; Xin, Y.; Yuan, L. Hybrid Rice Technology Development, Ensuring China's Food Security. In *IFPRI Discussion Paper 00918*; International Food Policy Research Institute: Washington, DC, USA, 2009; China National Hybrid Rice Research and Development Center: Changsha, China, 2009; Available online: www.ifpri.org/pubs/otherpubs.htm#dp (accessed on 30 October 2022).

35. Rao, Y.; Li, Y.; Qian, Q. Recent progress on molecular breeding of rice in China. *Plant Cell Rep.* **2014**, *33*, 551–564. [CrossRef] [PubMed]

36. Rout, D.; Jena, D.; Singh, V.; Kumar, M.; Arsode, P.; Singh, P.; Lalkatara, J.; Samantaray, S.; Verma, R. Hybrid Rice Research, Current Status and Prospects. In *Recent Advances in Rice Reserach*; IntechOpen: London, UK, 2020; Available online: https://www.intechopen.com/chapters/73489 (accessed on 30 October 2022). [CrossRef]

37. Chen, L.K.; Yan, X.C.; Dai, J.H.; Chen, S.P.; Liu, Y.Z.; Wang, H.; Chen, Z.Q.; Guo, T. Significant association of the novel Rf4-targeted SNP marker with the restorer for WA-CMS in different rice backgrounds and its utilisation in molecular screening. *J. Integr. Agric.* **2017**, *16*, 2128–2135. [CrossRef]

38. Tang, H.W.; Luo, D.P.; Zhou, D.G.; Zhang, Q.Y.; Tian, D.S.; Zheng, X.M.; Chen, L.T.; Liu, Y.G. The rice restorer Rf4 for wild-abortive cytoplasmic male sterility encodes a mitochondrial-localised PPR protein that functions in reduction of WA352 transcripts. *Mol. Plant* **2014**, *7*, 1497–1500. [CrossRef]

39. Khatun, K.M.; Hanafi, M.M.; Yusop, M.R.; Wong, M.Y.; Salleh, F.M.; Ferdous, J. Genetic variation, heritability, and diversity analysis of upland rice (*Oryza sativa* L.) genotypes based on quantitative traits. *BioMed Res. Inter.* **2015**, *2015*, 290861.

40. Hossain, S.; Haque, M.; Rahman, J. Genetic variability, correlation and path coefficient analysis of morphological traits in some extinct local Aman rice (*Oryza sativa* L). *J. Rice. Res.* **2015**, *4*, 1–6. [CrossRef]

41. Girma, B.T.; Kitil, M.A.; Banje, D.G.; Biru, H.M.; Serbessa, T.B. Genetic variability study of yield and yield related traits in rice (*Oryza sativa* L.) genotypes. *Adv. Crop Sci. Technol.* **2018**, *6*, 381–387.

42. Kamana, B.; Ankur, P.; Subarna, S.; Prasad, K.B.; Koshraj, U. Genetic variability, correlation and path analysis of rice genotypes in rainfed condition at Lamjung, Nepal. *Russ. J. Agric. Socio Econ. Sci.* **2019**, *8*, 274–280.

43. Sharma, R.C.; Chaudhary, N.K.; Ojha, B.; Yadav, L.; Pandey, M.P.; Shrestha, S.M. Variation in rice landraces adapted to the lowlands and hills in Nepal. *Plant Genet. Res.* **2007**, *5*, 120–127. [CrossRef]

44. Shahidullah, S.M.; Hanafi, M.M.; Ashrafuzzaman, M.; Ismail, M.R.; Salam, M.A. Phenological characters and genetic divergence in aromatic rices. *Afr. J. Biotechnol.* **2009**, *8*, 3199–3207.

45. Sanni, K.; Fawole, I.; Guei, R.; Ojo, D.; Somado, E.; Tia, D.; Ogunbayo, S.; Sanchez, I. Geographical patterns of phenotypic diversity in Oryza sativa landraces of Côte d'Ivoire. *Euphytica* **2008**, *160*, 389–400. [CrossRef]

46. Seetharam, K.; Thirumeni, S.; Paramasvum, K. Estimation of genetic diversity in rice (*Oryza sativa* L.) genotypes using SSR markers and morphological characters. *Afr. J. Biotechnol.* **2009**, *8*, 2050–2059.

47. Dai, Z.; Lu, Q.; Luan, X.; Ouyang, L.; Guo, J.; Liang, J.; Zhu, H.; Wang, W.; Wang, S.; Zeng, R.; et al. Development of a platform for breeding by design of CMS restorer lines based on an SSSL library in rice (*Oryza sativa* L.). *Breed. Sci.* **2016**, *66*, 768–775. [CrossRef] [PubMed]

48. Chen, R.; Feng, Z.; Zhang, X.; Song, Z.; Cai, D. A New Way of Rice Breeding: Polyploid Rice Breeding. *Plants* **2021**, *10*, 422. [CrossRef]

49. Awad-Allah, M.M.A.; Elekhtyar, N.M.; El-Abd, M.A.-E.-M.; Abdelkader, M.F.M.; Mahmoud, M.H.; Mohamed, A.H.; El-Diasty, M.Z.; Said, M.M.; Shamseldin, S.A.M.; Abdein, M.A. Development of New Restorer Lines Carrying Some Restoring Fertility Genes with Flowering, Yield and Grains Quality Characteristics in Rice (*Oryza sativa* L.). *Genes* **2022**, *13*, 458. [CrossRef] [PubMed]

50. Qian, Q.; Guo, L.; Smith, S.M.; Li, J. Breeding high-yield superior quality hybrid super rice by rational design. *Natl. Sci. Rev.* **2016**, *3*, 283–294. [CrossRef]

51. Li, P.; Zhou, H.; Yang, H.; Xia, D.; Liu, R.; Sun, P.; Wang, Q.; Gao, G.; Zhang, Q.; Wang, G.; et al. Genome-Wide Association Studies Reveal the Genetic Basis of Fertility Restoration of CMS-WA and CMS-HL in xian/indica and aus Accessions of Rice (*Oryza sativa* L.). *Rice* **2020**, *13*, 1–12. [CrossRef]

52. Gao, C. Genome engineering for crop improvement and future agriculture. *Cell* **2021**, *184*, 1621–1635. [CrossRef]
53. Chen, K.; Wang, Y.; Zhang, R.; Zhang, H.; Gao, C. CRISPR/Cas genome editing and precision plant breeding in agriculture. *Annu. Rev. Plant Biol.* **2019**, *70*, 667–697. [CrossRef]
54. Kumlehn, J.; Pietralla, J.; Hensel, G.; Pacher, M.; Puchta, H. The CRISPR/Cas revolution continues: From efficient gene editing for crop breeding to plant synthetic biology. *J. Integr. Plant Biol.* **2018**, *60*, 1127–1153. [CrossRef]
55. Yu, H.; Lin, T.; Meng, X.; Du, H.; Zhang, J.; Liu, G.; Chen, M.; Jing, Y.; Kou, L.; Li, X.; et al. A route to de novo domestication of wild allotetraploid rice. *Cell* **2021**, *184*, 1156–1170. [CrossRef]
56. Cheng, F.; Wu, J.; Cai, X.; Liang, J.; Freeling, M.; Wang, X. Gene retention, fractionation and subgenome differences in polyploid plants. *Nat. Plants* **2018**, *4*, 258–268. [CrossRef] [PubMed]
57. Sattler, M.; Carvalho, C.; Clarindo, W. The polyploidy and its key role in plant breeding. *Planta* **2016**, *243*, 281–296. [CrossRef] [PubMed]

agriculture

MDPI

Article

Influence of Transgenic (*Bt*) Cotton on the Productivity of Various Cotton-Based Cropping Systems in Pakistan

Muhammad Waseem Riaz Marral [1], Fiaz Ahmad [2], Sami Ul-Allah [3,4,*], Atique-ur-Rehman [1], Shahid Farooq [5] and Mubshar Hussain [1,6,*]

1   Department of Agronomy, Bahauddin Zakariya University, Multan 60800, Pakistan
2   Physiology/Chemistry Section, Central Cotton Research Institute, Multan 60800, Pakistan
3   College of Agriculture, BZU Bahadur Sub-Campus, Layyah 31200, Pakistan
4   College of Agriculture, University of Layyah, Layyah 31200, Pakistan
5   Department of Plant Protection, Faculty of Agriculture, Harran University, Sanlıurfa 63050, Turkey
6   School of Veterinary and Life Sciences, Murdoch University, 90 South Street, Murdoch, WA 6150, Australia
*   Correspondence: samipbg@bzu.edu.pk (S.U.-A.); mubashir.hussain@bzu.edu.pk (M.H.)

Abstract: Cotton (*Gossypium hirsutum* L.) is an important fiber crop in Pakistan with significant economic importance. Transgenic, insect-resistant cotton (carrying a gene from *Bacillus thuringiensis* (*Bt*)) was inducted in the cotton-based cropping systems of Pakistan during 2002, and is now sown in >90% of cotton fields in the country. However, concerns are rising that *Bt* cotton would decrease the productivity of winter crops (sown after cotton), leading to decreased system productivity. This two-year field study determined the impacts of transgenic (*Bt*) and non-transgenic (non-*Bt*) cotton genotypes on the productivities of winter crops (i.e., wheat, Egyptian clover, and canola), and the overall productivities of the cropping systems including these crops. Four cotton genotypes (two *Bt* and two non-*Bt*) and three winter crops (i.e., wheat, Egyptian clover, and canola) were included in the study. Nutrient availability was assessed after the harvest of cotton and winter crops. Similarly, the yield-related traits of cotton and winter crops were recorded at their harvest. The productivities of the winter crops were converted to net economic returns, and the overall economic returns of the cropping systems with winter crops were computed. The results revealed that *Bt* and non-*Bt* cotton genotypes significantly ($p < 0.05$) altered nutrient availability (N, P, K, B, Zn, and Fe). However, the yield-related attributes of winter crops were not affected by cotton genotypes, whereas the overall profitability of the cropping systems varied among the cotton genotypes. Economic analyses indicated that the *Bt* cotton–wheat cropping system was the most profitable, with a benefit–cost ratio of 1.55 in the semi-arid region of Pakistan. It is concluded that *Bt* cotton could be successfully inducted into the existing cropping systems of Pakistan without any decrease to the overall productivity of the cropping system.

Keywords: *Gossypium hirsutum*; Egyptian clover; canola; wheat; economic analyses

Citation: Marral, M.W.R.; Ahmad, F.; Ul-Allah, S.; Atique-ur-Rehman; Farooq, S.; Hussain, M. Influence of Transgenic (*Bt*) Cotton on the Productivity of Various Cotton-Based Cropping Systems in Pakistan. *Agriculture* **2023**, *13*, 276. https://doi.org/10.3390/agriculture13020276

Academic Editor: Martin Weih

Received: 26 December 2022
Revised: 18 January 2023
Accepted: 20 January 2023
Published: 23 January 2023

1. Introduction

Cotton (*Gossypium hirsutum* L.) is the most important fiber crop grown in Pakistan. It serves as a foundation for Pakistan's economy [1]. Pest infestations exert negative effects on cotton productivity, as >160 insect pests infest cotton at various growth stages [2,3]. Subsequently, cotton farmers incur significant amounts of money on pesticides to combat the pest infestation [2,4,5]. Farmers in Pakistan spend ~$300 million every year on pest control, and most of the pesticides (~80%) are sprayed in the cotton crop [6,7]. The extensive use of pesticides causes significant environmental and human health hazards [8,9]. The cultivation of transgenic, insect resistance crops can decrease the use of pesticides [10].

The use of genetically modified (GM) crops is drawing significant attention in Pakistan, due to their insect resistance abilities [11]. Transgenic cotton (commonly known as *Bt* cotton) is one of the greatest examples of GM crops, which are bollworms-resistant [12]. *Bt* cotton can reduce the damage caused by specific insects, and enhance crop productivity because of its resistance to bollworms [13]. The reduced use of insecticides and lower pest infestation are two major benefits of GM crops [14–16]. Numerous studies have shown that GM crops with *Bt* Cry proteins are resistant to various lepidopterans [17,18]. *Bt* cotton could help farmers in reducing the use of pesticides [2,5,19]. Therefore, *Bt* cotton is considered to be environmentally friendly [20,21].

There are increasing concerns on the potential negative consequences of GM crops on the soil health and productivity of other crops [22–24]. Numerous studies have revealed that the cultivation of GM crops increases the levels of *Bt* toxins in the soil [25–27]. *Bt* toxins are naturally present in the soil, and the recurrent planting of GM crops raises their concentration in the soil and alters the composition and behavior of soil microorganisms [28,29]. *Bt* toxins produced by GM crops enter the soil and could exert negative impacts on the productivity of other crops [30]. The cultivation of the *Bt* cotton line 'GK19' increased the accumulation of *Bt* proteins in the soil under salinity stress [31]. Increased levels of *Bt* toxins might exert negative impacts on agroecosystems [29] and alter the chemical composition of the root zone [27,32]. Similarly, toxins–microbe interactions significantly influence soil properties and nutrient availability. The changing rhizosphere conditions due to the cultivation of GM crops increased soil phosphorus (P) availability [33]. Furthermore, the cultivation of GM crops increases the available N and the oxidative metabolism in soil because of the enhanced activities of urease and dehydrogenase enzymes [34].

The induction of *Bt* cotton in the cotton-based cropping systems of Pakistan is being criticized due to the associated negative impacts. However, its introduction had favorable consequences on the environment and on farmers' health [35]. Cases of pesticide poisoning in farmers have decreased in China, India, and Pakistan after the introduction of *Bt* cotton [36,37]. However, the impacts of *Bt* cotton cultivation on the productivity of subsequent winter crops and the overall productivity of cropping systems have been less explored in Pakistan. Cotton is followed by wheat crop in the cotton-based cropping systems of Pakistan [38]. However, the recurrent cultivation of the same crops caused significant weed and insect infestation problems. It is suggested to include alternative crops (other than wheat) in the cotton-based cropping system of the country [38]. However, the impacts of *Bt* cotton on the productivity of alternative crops are still unknown.

This study assessed the impacts of *Bt* cotton cultivation on nutrient availability, the yield-related traits of different crops (wheat, Egyptian clover, and canola) sown after cotton, and the overall productivity of the cropping systems, including these crops. It was hypothesized that the cultivation of *Bt* cotton would reduce the yield of winter crops; however, the overall productivity of the cropping systems would not be affected. It was further hypothesized that cotton-based cropping system with *Bt* genotypes would have higher economic returns compared to those having non-*Bt* genotypes.

2. Materials and Methods

2.1. Experimental Site

This field study was conducted at CCRI (Central Cotton Research Institute), Multan (30.2° N, 71.43° E, and 122 meters above sea level), Pakistan, during 2016–2017 and 2017–2018. The soil of the experimental site was analyzed to assess the nutrient availability and the physico-chemical characters before and after the experiments. The soil texture was silt–clay–loam. The soil physico-chemical properties are given in Table 1.

2.2. Experimental Treatments

This experiment consisted of two factors, i.e., cotton genotypes (*Bt* and non-*Bt*) and winter crops. The *Bt* genotypes included in the study were 'CIM-616' (Bt_1) and 'GH Mubarik' (Bt_2), while the non-*Bt* genotypes were 'CIM-620' (non-Bt_1) and 'N-414' (non-Bt_2). Similarly, wheat (*Triticum aestivum* L.), Egyptian clover (*Trifolium alexandrinum* L.), and canola (*Brassica napus*) were winter crops that were sown after cotton harvest. The cultivation of winter crops resulted in three possible cropping systems, i.e., cotton–wheat, cotton–Egyptian clover, and cotton–canola. These systems may further be classified into to *Bt* and non-*Bt* cotton-based cropping systems. The *Bt* and non-*Bt* cotton genotypes were sown in May, whereas winter crops were sown in November following the harvest of the cotton crop (Table 2). The experiment was laid out according to two factorial design, where cotton genotypes were kept in main plots (6 m × 10 m), whereas winter crops were randomized in the sub-plots (2 m × 10 m). All treatments had three replications, and the experiment was repeated for two years. The sub-plots were regarded as being experimental units, and each experimental unit had three replications, as described above.

Table 1. Physicochemical characteristics of the soil in the experimental area before the initiation of the experiments.

Soil Properties	Unit	2016–2017	2017–2018
Organic matter content	%	0.59	0.56
Total nitrogen (N)	kg ha^{-1}	22.12	22.23
Available phosphorus (P)	kg ha^{-1}	18.02	18.08
Available potassium (K)	kg ha^{-1}	245.15	249.15
pH		8.17	8.19
EC	dS m^{-1}	4.96	5.00
Silt	%	54.15	54.00
Sand	%	25.75	26.10
Clay	%	20.10	19.90

Table 2. The production practices used for the cultivation of cotton and winter crops used in study.

Crops Name	Genotype Name	Planting Time *	Seed Rate (kg ha^{-1})	Fertilizer NPK (kg ha^{-1})	R × R (cm)	P × P (cm)	Harvesting Time
Cotton	GH Mubarik and CIM-616 (*Bt*) CIM-620 and CIM-554 (non-*Bt*)	08 and 10 May	25	250-175-125 (*Bt*) 200-145-100 (non-*Bt*)	75	20	Last picking in October
Wheat	Galaxy-2013	13 and 16 November	125	130-100-62	25		21 and 23 April
Canola	Hyola-420	12 and 13 November	5	90-60-50	30	4-5	6 and 10 April
Egyptian clover	Anmol berseem	9 and 11 November	25	22-115-0			Last cutting in April

Different dates in planting and harvesting time column indicate the dates in first and second years of the experiment, R × R = row to row spacing, P × P = plant to plant spacing. *, the first and second dates denote the planting time of the respective crop during 1st and 2nd year of the study, respectively.

2.3. Crop Husbandry

Cotton and winter crops were planted following the recommendations in production technology provided by the local agriculture extension department (https://www.agripunjab.gov.pk/, accessed on 12 January 2015). The recommendations followed in the study are given in Table 2. Irrigation was applied according to the moisture needs of the crops. All crop protection measures were taken to protect crops from insect and disease infestation. All crops were harvested when they reached physiological maturity.

2.4. Data Collection

2.4.1. Soil Properties

Particle size distribution was determined using the hydrometer method [39]. Soil EC was recorded using a digital EC meter following the standard procedures detailed by Dellavalle [40]. A digital pH meter was used to measure the soil pH from a saturated soil paste [40].

2.4.2. Nutrient Availability

Soil nitrogen (N) availability was measured spectrophotometrically with a segmented-flow system. The phosphorus (P) was determined using the vanadomolybdate method, potassium (K) through flame photometry, and zinc (Zn) and iron (Fe) through atomic absorption spectrophotometry [41]. Soil organic matter was measured using a loss-on-ignition protocol, as introduced by Hoogsteen et al. [42].

2.4.3. Weed Infestation

Weed infestation was evaluated 45 days after the sowing of cotton and winter crops. The weeds present in a 1 m^2 quadrat were counted from each experimental unit at three different places. The weeds were identified, grouped into narrow and broad-leaved, and their densities were computed. The density of each experimental unit was averaged from different locations within a replication.

2.5. Morphological and Yield-Related Traits

2.5.1. Cotton

The number of sympodial and monopodial branches were counted from 10 randomly selected plants in each experimental unit and averaged. The weights of 10 opened bolls from a single plant were measured using a sensitive balance from 10 randomly selected plants in each experimental unit, and averaged. Three manual pickings were performed from each experimental unit to record the seed cotton yield. The seed cotton yields of three pickings were added and converted to seed cotton yield per hectare, using a unitary method. The cotton stalks were harvested and left in the field for two weeks. Afterwards, the stalks were weighed to record the total biomass (biological yield), and they were expressed in kg ha^{-1}. The harvest index was estimated by dividing the seed cotton yield to biological yield, and this was expressed as a percentage.

2.5.2. Wheat

The number of productive (spike-bearing) tillers present in a 1 m^2 area were counted. The lengths of 10 randomly selected spikes were recorded from four central rows in each treatment and averaged. The number of grains were counted from 10 randomly selected spikes and averaged. The weight of 1000 grains from three random samples in each treatment was measured using a sensitive balance. The grain yield per plot was measured on a sensitive balance when the seed reached the required moisture level, i.e., 12%, and converted to t ha^{-1}.

2.5.3. Canola

The number of siliques per plant were counted from three randomly selected plants in each experimental unit. Ten random siliques were opened, and the number of grains in them were counted and averaged. The weight (g) of 1000 seeds from randomly sampled seeds per plot was measured on a sensitive balance. The mature crop was harvested, sundried, and threshed manually to record the seed yield, which was converted into kg ha^{-1}. The biological yield was recorded by weighing the total above ground biomass harvested from four central rows of each experimental plot, and this was converted into kg ha^{-1}.

2.5.4. Egyptian Clover

All plants within the experimental unit were harvested during each cut, and weighed to record the fresh forage yield. A pre-weighed amount of fresh forage was oven-dried, and fresh forage yield was converted into dry forage yield using a unitary method. The fresh and dry forage yields were converted into t/ha. Crude protein was measured via a near-infrared spectroscopy system [43].

2.6. Economic Analysis

The profit abilities of different cotton genotypes via winter crops interactions were computed following CIMMYT [44]. The input costs and outputs obtained in monetary terms were calculated. The costs regarding land rent, irrigation, labor, seeds, fertilizers, pesticides, sowing, harvesting, etc., were computed. The existing market prices of the produce were used to compute the gross income. These expenses were deducted from the gross income to obtain the net income. The benefit–cost ratio (BCR) was computed by dividing the net economic returns with the expenses incurred.

2.7. Statistical Analysis

The collected data for the nutrient availability, weed density, and yield-related parameters of different crops were checked for normality using the Shapiro-Wilk normality test [45]. The parameters having non-normal distributions were normalized using the Arcsine transformation technique to meet the normality assumption of the Analysis of Variance (ANOVA). The differences among the years were tested, which were significant; therefore, the data of both years were analyzed, presented, and interpreted separately. A two-way ANOVA was used to test the significance among the treatments, and the means were compared using a least significant difference post hoc test at 95% probability, where ANOVA denoted significant differences [46]. The interactive effects of cotton genotypes and winter crops were significant for most of the studied traits. Therefore, the interactive effects were presented and interpreted. A one-way ANOVA was used to analyze the data on the yield-related traits of cotton. All statistical computations were performed on SPSS statistical software, version 21.0 [47].

3. Results

3.1. Nutrient Availability

Soil nutrient availability and organic matter content were significantly altered by cotton genotypes via winter crops interaction during both years (Table 3). Wheat sown after the *Bt* cotton genotype 'GH-Mubarik' had the highest available N during each year, while canola sown after the non-*Bt* genotypes 'N-414' and 'CIM-620' resulted in the lowest available N during both years (Table 3). Egyptian clover sown after the non-*Bt* cotton genotype 'CIM-620' during the first year, and the *Bt* genotype 'CIM-616' during the second year, recorded the highest P, whereas the lowest values were recorded for wheat sown after the non-*Bt* genotype 'N-414' during both years (Table 3). The Egyptian clover sown after the non-*Bt* genotype 'N-414' resulted in the highest available K, which was statistically similar to the wheat cultivation after the *Bt* genotype 'GH-Mubarik'. The lowest available K was noted for canola sown after the *Bt* cotton genotypes during the first year, and the non-*Bt* genotype 'N-414' during the second year (Table 3). Egyptian clover following the non-*Bt* genotypes resulted in the highest available Zn during both years, while the lowest Zn was recorded for canola sown after the *Bt* genotype 'GH-Mubarik' during both years (Table 3). The interactive effects of wheat and the non-*Bt* genotype 'CIM-620' resulted in the highest available Fe during each year, while the lowest Fe was recorded for Egyptian clover sown after the non-*Bt* genotype 'CIM-620' during first year, and canola sown after the non-*Bt* genotype 'N-414' during the second year (Table 3). The highest organic matter content was recorded for wheat sown after the *Bt* genotypes during both years, while canola sown after the non-*Bt* genotypes resulted in the lowest value of soil organic matter during both years (Table 3).

Table 3. The influences of cotton genotypes from winter crops interaction on nutrient availability and soil organic matter contents in the soil after the harvest of winter crops.

Treatments	2016–2017			2017–2018		
	Wheat	**Egyptian Clover**	**Canola**	**Wheat**	**Egyptian Clover**	**Canola**
	Available nitrogen (kg ha^{-1})					
CIM-616 (Bt_1)	0.17 ± 0.001 a–c	0.15 ± 0.003 c–e	0.14 ± 0.001 de	0.18 ± 0.003 ab	0.16 ± 0.005 b–d	0.16 ± 0.001 b–d
GH-Mubarik (Bt_2)	0.19 ± 0.003 a	0.14 ± 0.002 de	0.16 ± 0.001 b–d	0.19 ± 0.002 a	0.16 ± 0.004 b–d	0.17 ± 0.003 a–c
CIM-620 (NBt_1)	0.18 ± 0.002 ab	0.15 ± 0.001 c–e	0.14 ± 0.004 de	0.18 ± 0.001 ab	0.14 ± 0.003 cd	0.14 ± 0.002 d
N-414 (NBt_2)	0.16 ± 0.004 b–d	0.15 ± 0.002 c–e	0.13 ± 0.002 e	0.18 ± 0.002 ab	0.16 ± 0.002 b–d	0.16 ± 0.002 b–d
LSD ($p \leq 0.05$)	0.020			0.020		
	Available phosphorous (kg ha^{-1})					
CIM-616 (Bt_1)	19.36 ± 0.02 a–e	19.50 ± 0.03 a–c	19.38 ± 0.02 a–e	19.40 ± 0.01 a–c	19.59 ± 0.07 a	19.28 ± 0.04 b–d
GH-Mubarik (Bt_2)	19.20 ± 0.04 de	19.52 ± 0.02 ab	19.24 ± 0.05 de	19.10 ± 0.02 de	19.42 ± 0.04 ab	19.14 ± 0.06 de
CIM-620 (NBt_1)	19.28 ± 0.06 c–e	19.58 ± 0.07 a	19.30 ± 0.04 b–e	19.18 ± 0.06 de	19.48 ± 0.02 ab	19.20 ± 0.05 c–e
N-414 (NBt_2)	19.18 ± 0.04 e	19.40 ± 0.03 a–d	19.40 ± 0.03 a–d	19.08 ± 0.05 e	19.30 ± 0.03 a–d	19.30 ± 0.04 a–d
LSD ($p \leq 0.05$)	0.10			0.12		
	Available potassium (kg ha^{-1})					
CIM-616 (Bt_1)	394 ± 6.1 b–d	400 ± 4.3 ab	394 ± 3.3 e	402 ± 2.2 a–c	404 ± 2.2 a–c	406 ± 3.2 a–c
GH-Mubarik (Bt_2)	388 ± 5.3 de	396 ± 6.1 a–c	394 ± 3.2 e	408 ± 6.1 a	402 ± 3.1 a–c	404 ± 2.6 a–c
CIM-620 (NBt_1)	390 ± 3.4 c–e	398 ± 3.3 ab	386 ± 8.3 de	400 ± 3.4 bc	402 ± 3.4 a–c	402 ± 2.7 a–c
N-414 (NBt_2)	392 ± 2.4 b–e	402 ± 1.2 a	390 ± 4.5 c–e	400 ± 3.3 bc	406 ± 4.3 ab	400 ± 2.3 c
LSD ($p \leq 0.05$)	7.58			6.30		
	Available zinc (kg ha^{-1})					
CIM-616 (Bt_1)	1.46 ± 0.01 d	1.60 ± 0.03 b	1.44 ± 0.03 de	1.56 ± 0.04 cd	1.60 ± 0.02 bc	1.48 ± 0.02 ef
GH-Mubarik (Bt_2)	1.46 ± 0.02 d	1.58 ± 0.02 bc	1.40 ± 0.02 e	1.50 ± 0.04 de	1.62 ± 0.02 b	1.42 ± 0.03 f
CIM-620 (NBt_1)	1.58 ± 0.02 bc	1.66 ± 0.01 a	1.44 ± 0.04 c	1.60 ± 0.03 bc	1.68 ± 0.04 a	1.56 ± 0.04 cd
N-414 (NBt_2)	1.58 ± 0.01 bc	1.68 ± 0.02 a	1.56 ± 0.02 bc	1.58 ± 0.02 bc	1.68 ± 0.05 a	1.58 ± 0.02 bc
LSD ($p \leq 0.05$)	0.04			0.06		
	Available iron (kg ha^{-1})					
CIM-616 (Bt_1)	7.62 ± 0.12 d–f	7.42 ± 0.10 fg	7.72 ± 0.09 c–e	7.68 ± 0.11 de	7.78 ± 0.13 c–e	7.78 ± 0.10 c–e
GH-Mubarik (Bt_2)	7.82 ± 0.14 b–e	7.68 ± 0.11 de	7.84 ± 0.11 b–d	7.90 ± 0.12 bc	8.04 ± 0.17 ab	7.80 ± 0.14 c–e
CIM-620 (NBt_1)	8.14 ± 0.19 a	7.32 ± 0.09 g	7.94 ± 0.11 a–c	8.20 ± 0.16 a	7.96 ± 0.11 bc	7.80 ± 0.13 c–e
N-414 (NBt_2)	8.04 ± 0.11 ab	7.56 ± 0.14 ef	7.60 ± 0.09 ef	7.96 ± 0.10 bc	7.84 ± 0.12 b–d	7.62 ± 0.12 e
LSD ($p \leq 0.05$)	0.24			0.20		
	Soil organic matter (%)					
CIM-616 (Bt_1)	0.59 ± 0.02 a	0.53 ± 0.04 c–e	0.51 ± 0.02 de	0.62 ± 0.01 ab	0.62 ± 0.01 ab	0.59 ± 0.01 cd
GH-Mubarik (Bt_2)	0.58 ± 0.03 ab	0.53 ± 0.04 c–e	0.51 ± 0.02 de	0.60 ± 0.02 bc	0.63 ± 0.01 a	0.60 ± 0.01 bc
CIM-620 (NBt_1)	0.57 ± 0.04 a–c	0.52 ± 0.03 de	0.49 ± 0.01 ef	0.59 ± 0.02 cd	0.59 ± 0.01 cd	0.57 ± 0.01 d
N-414 (NBt_2)	0.58 ± 0.02 ab	0.54 ± 0.04 b–d	0.45 ± 0.02 f	0.58 ± 0.01 cd	0.58 ± 0.02 cd	0.60 ± 0.01 bc
LSD ($p \leq 0.05$)	0.04			0.03		

Means followed by different letters significantly differ ($p \leq 0.05$) from each other. The values of different traits are means of three replications ± standard errors of means. NS = non-significant; *Bt* = transgenic genotypes; *NBt* = conventional or non-transgenic genotypes.

3.2. Weed Density

The densities of broadleaved, narrow-leaved, and total weeds were significantly affected by the interactive effect of cotton genotypes and winter crops. Overall, *Bt* genotypes recorded lesser weed infestation compared with the non-*Bt* genotypes included in the current study (Figure 1).

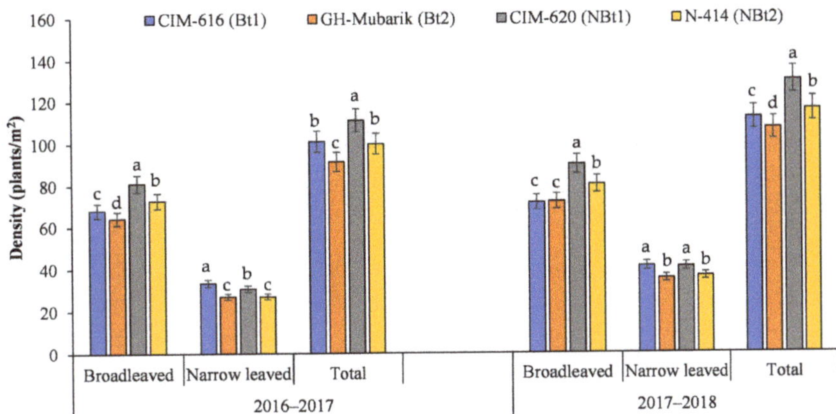

Figure 1. The densities (±standard errors of means, n = 3) of broadleaved, narrow-leaved, and total weed species recorded in different cotton genotypes. Here, Bt and NBt stand for *Bt* and non-*Bt* genotypes.

Similarly, the wheat crop had the highest density of narrow-leaved, broadleaved, and total weeds during both years (Figure 2). Wheat sown after the non-*Bt* genotype 'CIM-620' had the highest density of broad-leaved weeds, whereas the lowest density was observed in Egyptian clover and canola crops sown after both *Bt* genotypes (Table 4).

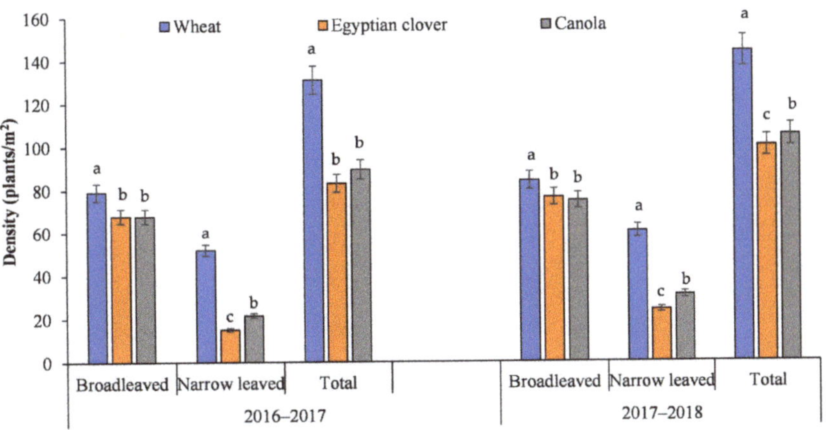

Figure 2. The impacts of different winter crops on weed density (±standard errors of means, n = 3) in different winter crops included in the study.

Similarly, the lowest density of narrow-leaved weeds was recorded for Egyptian clover sown after the *Bt* genotype 'GH-Mubarik' and non-*Bt* genotype 'CIM-620' during both years, whereas the highest density was noted for wheat sown after the *Bt* genotype 'CIM-616' and the non-*Bt* genotype 'CIM-620' during both years (Table 4). Likewise, the highest density of total weeds was recorded in wheat crop sown after the non-*Bt* genotype 'CIM-620', whereas the lowest density was recorded in Egyptian clover sown after the *Bt* genotypes during both years (Table 4).

Table 4. Interactive effects of different cotton genotypes and winter crops on the densities of broadleaved, narrow-leaved, and total weeds recorded in winter crops.

Treatments	2016–2017			2017–2018		
	Wheat	**Egyptian Clover**	**Canola**	**Wheat**	**Egyptian Clover**	**Canola**
	Broadleaved weeds density (m^{-2})					
CIM-616 (Bt_1)	81.3 ± 3.1 b	61.0 ± 3.1 f	62.3 ± 2.8 f	83.0 ± 3.4 cd	66.7 ± 3.9 g	65.0 ± 3.1 g
GH-Mubarik (Bt_2)	68.7 ± 4.3 e	60.3 ± 2.4 f	64.0 ± 2.9 f	74.3 ± 2.6 ef	69.3 ± 5.2 fg	72.7 ± 3.4 ef
CIM-620 (NBt_1)	90.3 ± 3.4 a	77.7 ± 2.8 bc	75.3 ± 2.4 cd	97.0 ± 6.3 a	88.7 ± 2.8 b	83.7 ± 5.1 bc
N-414 (NBt_2)	76.0 ± 2.7 cd	72.7 ± 3.3 de	69.7 ± 1.4 e	82.0 ± 4.7 cd	81.0 ± 3.0 cd	78.0 ± 3.1 de
LSD ($p \leq 0.05$)	4.42			5.57		
	Narrow-leaved weeds density (m^{-2})					
CIM-616 (Bt_1)	57.0 ± 2.2 a	16.0 ± 3.6 e	27.3 ± 2.1 c	64.0 ± 4.6 a	24.0 ± 3.4 e	35.3 ± 4.4 c
GH-Mubarik (Bt_2)	48.0 ± 3.3 b	12.7 ± 3.4 ef	20.3 ± 3.4d	55.7 ± 3.3 b	22.0 ± 3.3 e	28.0 ± 3.2 d
CIM-620 (NBt_1)	56.0 ± 4.5 a	11.7 ± 2.6 f	24.3 ± 2.9 c	66.3 ± 4.1 a	21.7 ± 4.5 e	34.3 ± 4.0 c
N-414 (NBt_2)	46.7 ± 2.6 b	20.0 ± 2.9 d	14.3 ± 1.7 ef	55.7 ± 4.8 b	28.0 ± 3.7 d	25.0 ± 4.2 de
LSD ($p \leq 0.05$)	3.47			3.53		
	Total weeds density (m^{-2})					
CIM-616 (Bt_1)	138 ± 6 b	77.0 ± 8 h	89.7 ± 3 f	147 ± 10 b	90.7 ± 5 h	100 ± 5 g
GH-Mubarik (Bt_2)	116 ± 5 d	73.0 ± 11 i	84.3 ± 2 g	130 ± 11 d	91.3 ± 6 h	100 ± 6 g
CIM-620 (NBt_1)	146 ± 10 a	89.3 ± 4 f	99.7 ± 8 e	163 ± 19 a	110 ± 8 f	118 ± 11 e
N-414 (NBt_2)	122 ± 4 c	92.7 ± 5 f	84.0 ± 4 g	137 ± 6 c	109 ± 7 f	103 ± 5 g
LSD ($p \leq 0.05$)	3.84			5.43		

Means followed by different letters significantly differ ($p \leq 0.05$) from each other. The values of different traits are means of three replications $\pm$ standard errors of means. NS = non-significant; *Bt* = transgenic genotypes; N*Bt* = conventional or non-transgenic genotypes.

3.3. Yield-Related Attributes of Cotton

Various yield-related traits of cotton were significantly affected by genotypes, except for the number of monopodial branches and the harvest index (Table 5).

The *Bt* genotype 'CIM-616' and non-*Bt* genotype 'CIM-620' recorded the highest number of sympodial branches during the first year, whereas both *Bt* genotypes recorded the highest number of sympodial branches during the second year. The non-*Bt* genotypes had the lowest number of sympodial branches during the second year (Table 5). The *Bt* genotype 'CIM-616' recorded the highest seed cotton yield during the first year, while it was not affected by genotypes during the second year (Table 5).

3.4. Yield-Related Attributes of Wheat

Yield-related traits of wheat crop were significantly impacted by different cotton genotypes with some exceptions, i.e., grain yield during the first year, and harvest index during both years. Overall, the wheat crop planted after the non-*Bt* genotypes recorded higher values for the number of productive tillers, number of grains per spike, 1000-grain weight, and grain and biological yields (Table 6). Wheat sown after the *Bt* genotypes recorded lower values for the number of productive tillers, number of grains per spike, 1000-grain weight, and grain, and biological yields, compared to non-*Bt* genotypes during both years (Table 6).

Table 5. The influences of different *Bt* and non-*Bt* genotypes on yield-related attributes of cotton crop.

Treatments	2016	2017	2016	2017	2016	2017
	Monopodial Branches (Plant^{-1})		Sympodial Branches (Plant^{-1})		Boll Weight (g)	
CIM-616 (*Bt₁*)	1.78 ± 0.2	1.78 ± 0.3	23.6 ± 0.9 b	26.0 ± 1.1 a	3.2 ± 0.1 a	3.2 ± 0.04 a
GH-Mubarik (*Bt₂*)	1.67 ± 0.3	1.67 ± 0.4	22.3 ± 0.8 b	25.6 ± 1.2 a	3.1 ± 0.1 ab	3.1 ± 0.02 b
CIM-620 (*NBt₁*)	1.89 ± 0.3	1.89 ± 0.2	25.0 ± 1.4 a	23.0 ± 0.8 b	3.0 ± 0.05 bc	3.0 ± 0.08 b
N-414 (*NBt₂*)	1.67 ± 0.4	1.67 ± 0.3	22.0 ± 1.1 bc	24.0 ± 1.2 b	2.9 ± 0.04 c	2.9 ± 005 c
LSD ($p \leq 0.05$)	NS	NS	1.49	1.29	0.15	0.07
	Seed cotton yield (kg ha^{-1})		Harvest index (%)			
CIM-616 (*Bt₁*)	2892 ± 141 a	2832 ± 221	32.7 ± 2.12	32.4 ± 1.9		
GH-Mubarik (*Bt₂*)	2685 ± 123 b	2635 ± 213	29.5 ± 2.21	29.4 ± 3.1		
CIM-620 (*NBt₁*)	2645 ± 129 b	2563 ± 303	33.9 ± 2.39	31.2 ± 2.2		
N-414 (*NBt₂*)	2613 ± 147 b	2570 ± 309	32.3 ± 2.53	31.1 ± 2.6		
LSD ($p \leq 0.05$)	150.14	NS	NS	NS		

Means followed by different letters significantly differ ($p \leq 0.05$) from each other. The values of different traits are means of three replications ± standard errors of means. NS = non-significant; *Bt* = transgenic genotypes; *NBt* = conventional or non-transgenic genotypes.

Table 6. Yield-related parameters of wheat crop sown after the harvest of transgenic and non-transgenic cotton genotypes.

Treatments	2016–2017	2017–2018	2016–2017	2017–2018	2016–2017	2017–2018
	Productive Tillers (m^{-2})		Grains (Spike^{-1})		1000-Grain Weight (g)	
CIM-616 (*Bt₁*)	189 ± 15 ab	191 ± 17 ab	55.7 ± 1.9 b	56.0 ± 2.0 b	36.2 ± 1.8 c	36.8 ± 1.4 c
GH-Mubarik (*Bt₂*)	181 ± 11 b	176 ± 14 b	53.3 ± 1.7 b	53.9 ± 1.8 c	37.9 ± 1.7 bc	37.9 ± 1.6 bc
CIM-620 (*NBt₁*)	197 ± 10 a	201 ± 14 a	59.5 ± 1.2 a	58.8 ± 1.4 a	40.2 ± 1.5 a	40.7 ± 1.3 a
N-414 (*NBt₂*)	202 ± 12 a	202 ± 16 a	59.1 ± 1.4 a	58.1 ± 1.6 a	39.8 ± 1.6 ab	39.6 ± 2.3 ab
LSD ($p \leq 0.05$)	14.4	14.4	2.4	1.7	2.0	1.8
	Grain yield (t ha^{-1})		Biological yield (t ha^{-1})		Harvest index (%)	
CIM-616 (*Bt₁*)	5.82 ± 0.8	5.95 ± 0.2 b	17.6 ± 0.5 bc	15.7 ± 0.4 bc	33.1 ± 1.2	37.8 ± 2.1
GH-Mubarik (*Bt₂*)	5.98 ± 0.7	5.92 ± 0.1 b	17.1 ± 0.6 c	15.3 ± 0.5 c	34.9 ± 1.6	38.8 ± 2.2
CIM-620 (*NBt₁*)	6.30 ± 0.7	6.26 ± 0.2 a	18.2 ± 0.6 ab	16.4 ± 0.6 ab	34.6 ± 1.8	38.2 ± 2.1
N-414 (*NBt₂*)	6.21 ± 0.8	6.31 ± 0.2 a	18.7 ± 0.5 a	16.9 ± 0.5 a	33.2 ± 2.0	37.4 ± 2.4
LSD ($p \leq 0.05$)	NS	0.25	0.86	0.86	NS	NS

Means followed by different letters significantly differ ($p \leq 0.05$) from each other. The values of different traits are means of three replications ± standard errors of means. NS = non-significant; *Bt* = transgenic genotypes; *NBt* = conventional or non-transgenic genotypes.

3.5. Yield-Related Attributes of Canola

Yield-related traits of canola were not affected by different cotton genotypes except for seed yield during the first year, where the crop sown after non-*Bt* genotypes recorded higher values compared to the *Bt* genotypes (Table 7).

Table 7. Yield-related traits of canola sown after different transgenic and non-transgenic cotton genotypes.

Treatments	2016–2017	2017–2018	2016–2017	2017–2018	2016–2017	2017–2018
	Siliques (Plant^{-1})		Seeds (Silique^{-1})		1000-Seed Weight (g)	
CIM-616 (Bt_1)	106 ± 22	105 ± 7 ab	26.7 ± 2.3	26.9 ± 3.7	2.77 ± 0.3	2.73 ± 0.4
GH-Mubarik (Bt_2)	103 ± 12	102 ± 9 b	24.0 ± 3.4	25.0 ± 4.0	2.90 ± 0.4	2.85 ± 0.3
CIM-620 (NBt_1)	109 ± 6	109 ± 7 ab	25.2 ± 3.6	25.8 ± 3.1	2.87 ± 0.6	2.90 ± 0.2
N-414 (NBt_2)	112 ± 16	113 ± 8 a	26.3 ± 4.1	27.0 ± 3.3	2.83 ± 0.5	2.93 ± 0.5
LSD ($p \leq 0.05$)	NS	8.4	NS	NS	NS	NS
	Biological yield (kg ha^{-1})		Seed yield (kg ha^{-1})		Harvest index (%)	
CIM-616 (Bt_1)	4800 ± 343	5271 ± 234	1650 ± 212 b	1797 ± 158	34.4 ± 2.2	34.1 ± 3.4
GH-Mubarik (Bt_2)	5132 ± 412	5070 ± 267	1700 ± 223 b	1833 ± 123	33.2 ± 3.1	36.2 ± 2.1
CIM-620 (NBt_1)	5233 ± 345	5345 ± 312	1950 ± 201 a	1850 ± 112	37.4 ± 2.6	34.7 ± 3.3
N-414 (NBt_2)	4876 ± 321	5478 ± 434	1900 ± 198 a	1900 ± 121	39.1 ± 2.8	34.7 ± 3.1
LSD ($p \leq 0.05$)	NS	NS	197.7	NS	NS	NS

Means followed by different letters significantly differ ($p \leq 0.05$) from each other. The values of different traits are means of three replications $\pm$ standard errors of means. NS = non-significant; *Bt* = transgenic genotypes; N*Bt* = conventional or non-transgenic genotypes.

3.6. Yield-Related Attributes of Egyptian Clover

The yield-related traits of Egyptian clover were significantly affected by different cotton genotypes during both years (Table 8). The crop sown after non-*Bt* genotypes recorded higher values for the total forage yield, dry matter content, and crude protein during both years of the study, compared to the *Bt* genotypes (Table 8).

Table 8. The influence of various cotton varieties on yield-related traits of Egyptian clover.

Treatments	Fresh Forage Yield (t ha^{-1})		Dry forage Yield (t ha^{-1})		Crude Protein (%)	
	2016–2017	2017–2018	2016–2017	2017–2018	2016–2017	2017–2018
CIM-616 (Bt_1)	28.3 ± 1.21 b	30.7 ± 1.98 b	2.91 ± 0.11 b	3.62 ± 0.09 b	21.0 ± 1.2 b	20.6 ± 1.7 b
GH-Mubarik (Bt_2)	28.2 ± 1.26 b	32.0 ± 2.02 b	2.97 ± 0.17 b	3.72 ± 0.16 ab	20.2 ± 1.6 b	20.3 ± 2.4 b
CIM-620 (NBt_1)	34.1 ± 2.34 a	34.8 ± 1.12 a	3.50 ± 0.21 a	3.87 ± 0.11 a	24.0 ± 2.1 a	22.3 ± 1.6 ab
N-414 (NBt_2)	32.3 ± 2.31 a	33.2 ± 1.63 ab	3.35 ± 0.18 a	3.76 ± 0.12 ab	23.7 ± 1.9 a	23.6 ± 1.2 a
LSD ($p \leq 0.05$)	1.96	2.63	0.18	0.17	2.44	2.52

Means followed by different letters significantly differ ($p \leq 0.05$) from each other. The values of different traits are means of three replications $\pm$ standard errors of means. NS = non-significant; *Bt* = transgenic genotypes; N*Bt* = conventional or non-transgenic genotypes.

3.7. Economic Returns/System Productivity

Economic analysis showed that the *Bt* genotypes–wheat cropping system resulted in the highest net benefits, whereas the non-*Bt* genotypes–canola cropping system recorded the lowest net benefits (Table 9). Similarly, the *Bt* genotypes–wheat cropping system resulted in the highest benefit–cost ratio (BCR), while the non-*Bt* genotypes–canola cropping system resulted in the lowest BCR (Table 9).

Table 9. The impacts of different transgenic and non-transgenic cotton genotypes on system productivity of various cotton-based cropping systems.

Treatments	2016–2017				2017–2018			
	TE	GI	NI	BCR	TE	GI	NI	BCR
Bt_1 × Wheat	1563.59	2607.75	1044.16	1.67	1563.59	2531.58	967.99	1.62
Bt_2 × Wheat	1563.59	2516.55	952.96	1.61	1563.59	2423.59	860.00	1.55
NBt_1 × Wheat	1629.85	2572.23	942.38	1.58	1629.85	2466.93	837.09	1.51
NBt_2 × Wheat	1629.85	2564.07	934.22	1.57	1629.85	2491.97	862.12	1.53
Bt_1 × Canola	1500.76	1979.11	478.35	1.32	1500.76	2009.13	508.37	1.34
Bt_2 × Canola	1500.76	1910.59	409.83	1.27	1500.76	1923.63	422.87	1.28
NBt_1 × Canola	1567.02	1966.87	399.85	1.26	1567.02	1904.60	337.59	1.22
NBt_2 × Canola	1567.02	1925.86	358.85	1.23	1567.02	1926.66	359.65	1.23
Bt_1 × Egyptian clover	1621.02	1989.23	368.21	1.23	1621.02	2016.24	395.22	1.24
Bt_2 × Egyptian clover	1621.02	1893.59	272.57	1.17	1621.02	1957.23	336.21	1.21
NBt_1 × Egyptian clover	1687.28	2010.18	322.90	1.19	1687.28	1987.73	300.45	1.18
NBt_2 × Egyptian clover	1687.28	1953.95	266.68	1.16	1687.28	1954.70	267.42	1.16

BCR = benefit–cost ratio; Bt_1 = CIM-616; Bt_2 = GH-Mubarik; NBt_1 = CIM-620; NBt_2 = N-414; Bt = transgenic cotton; NBt = non-transgenic cotton; TE= total expenditure; GI= gross income; NI= net income, the values of TE, GI, and NI are in US$.

4. Discussion

The results of the current study indicated that nutrient availability, yield-related attributes of winter crops, and overall system productivity were significantly affected by the cotton genotypes. *Bt* cotton genotypes improved the nutrient availability and system productivity, which are directly linked to the higher fertilizer input and the better yield of the *Bt* genotypes. Increased fertilizer use after the introduction of *Bt* genotypes have increased crop yields because of better pest control [48]. Furthermore, farmers started applying more fertilizers after the induction of *Bt* genotypes in the existing cropping systems of Pakistan [49]. Cotton yields and related benefits may also be affected by numerous factors, such as changes to irrigation systems, crop production technologies, agronomic practices, farmer training, or weather fluctuations, etc. [48]. Earlier studies have reported that nutrient availability may vary across *Bt* and non-*Bt* genotypes because of the differences in their nutrient requirements and absorption [33]. Therefore, the results of the current study regarding nutrient availability are in agreement with the earlier studies.

The highest P, N, Zn, Fe, and organic matter contents were recorded from the soil cultivated with winter crops after the *Bt* genotypes. Wheat crop sown after *Bt* genotypes resulted in higher values of available N, Fe, and organic matter contents, whereas Egyptian clover following *Bt* genotypes resulted in higher P, K, and Zn contents. Nutrient uptake is dependent on plants and their genetic makeup. Several factors affect the nutrient uptake capacity of plants [50]. These factors include root surface area, and the type and quantity of root exudates released in the rhizosphere and microbial communities [50]. Moreover, the plant characteristics and interactive effect between roots and soil microorganisms also play significant roles in nutrient uptake [51]. The quantity of nutrients available to plants depends mainly upon their availability in the root zone [52]. Genetically modified (GM) crops can disrupt soil nutrient cycles due to changes in the root zone [53]. The quantity and quality of root exudates affects microbial activity, which alters the solubility of mineral or fixed P, and P availability [2,21,27]. The availability of soil nutrients is significantly affected by the cultivation of *Bt* cotton. Growing *Bt* cotton also decreased the available N and K, while increasing Zn and P [29,54]. However, our study indicated that the available N was increased in the soil cultivated with *Bt* genotypes. The *Bt* genotypes received higher amounts of nutrients, which can be linked to the increased nutrient availability. Similarly, the availabilities of K, Zn, and P also increased in the treatments with *Bt* genotypes. Higher root biomass-mediated exudation is responsible for the enhanced availability of Zn and Fe in *Bt* cotton cultivated soil, compared to non-*Bt* cotton [55,56].

Weed infestation exerts significant negative impacts on crop yields [57,58]. Weed infestation decreases cotton yield by 0.26 to 66%, depending on the weed species and their densities [59]. The cultural practices used in the existing cropping systems of Pakistan encourage the growth of several weed species [60]. The recurrent cultivation of *Bt* genotypes may result in the proliferation of specific weed species. Different genotypes significantly vary in their weed competitive ability [61]. Several earlier studies have reported that cotton genotypes significantly differ in their competitive ability with weeds [62,63]. The weed competitive abilities of these cultivars were linked with their potential to establish a crop canopy. The cultivars which developed a dense canopy in a shorter period were more competitive. However, these studies did not include any *Bt* genotypes. Low weed density was recorded in the soil cultivated with *Bt* genotypes in the current study. Weed competition of *Bt* genotypes in the current study could be linked with its quicker canopy development, compared to non-*Bt* cotton genotypes.

Wheat and Egyptian clover sown after non-*Bt* genotypes had better yield-related traits. However, the yield-related traits of canola were not affected by cotton genotypes. The lower yields of winter crops in the fields cultivated with *Bt* genotypes can be linked with the increased levels of *Bt* toxins in the soil and the higher nutrient consumption by cotton plants. Moreover, the improvement in the yield-related traits of winter crops in the fields cultivated with non-*Bt* genotypes can be linked to the low *Bt* toxin levels in these soils. It has been observed that toxins produced in the aerial parts and roots of *Bt* cotton may cause soil pollution upon their release [34,64,65]. *Bt* toxins released from the plants become absorbed or bound to the soil particles, and then they become safe from degradation from other microorganisms that are present in the soil [66]. The recurrent cultivation of *Bt* cotton in the same field increases the level of *Bt* toxins in the soil, which can change the activity and composition of the soil microbes and the soil biochemical nature [29,67–69].

The *Bt* cotton–wheat cropping system revealed the highest net income and benefit–cost ratio (BCR). The highest productivity of this system was due to a higher production of *Bt* cotton and wheat. The *Bt* genotypes produced the highest yield due to a lower rate of insect infestation, compared with non-*Bt* genotypes. However, wheat yield was higher in the fields cultivated with non-*Bt* genotypes, and a lesser number of sprays in *Bt* cotton for pest management decreased the input costs. This eventually reduced expenses, which resulted in a higher net income and BCR than with non-*Bt* cotton.

The results confirmed the hypothesis of the study, where *Bt* genotypes exerted negative impacts on the yield-related traits of winter crops (with some exceptions) and improved the overall system productivity. Therefore, *Bt* genotypes can be included in the cotton-based cropping systems without any decrease in the productivity and economic returns.

5. Conclusions

The results of the current study indicated that nutrient availability, weed infestation, the yield-related traits of winter crops, and the system productivity of various cotton-based cropping systems were significantly affected by *Bt* and non-*Bt* cotton genotypes. Overall, *Bt* genotypes had higher yields than non-*Bt* genotypes. The soil cultivated with *Bt* genotypes resulted in higher N, P, and Zn availabilities. The yield-related traits of winter crops were negatively affected by the *Bt* genotypes. Economic analysis indicated that *Bt* cotton, followed by wheat, resulted in the highest economic returns and benefit–cost ratios. Therefore, *Bt* cotton can be successfully inducted in the cotton–wheat cropping systems of semi-arid regions in Pakistan in order to obtain higher economic benefits.

Author Contributions: Conceptualization, S.U.-A., S.F. and M.H.; data curation, M.W.R.M. and F.A.; formal analysis, M.W.R.M.; investigation, M.W.R.M. and A.-u.-R.; methodology, S.U.-A., A.-u.-R., S.F. and M.H.; project administration, M.H.; resources, F.A. and M.H.; software, S.U.-A. and A.-u.-R.; supervision, M.H.; validation, S.F.; visualization, S.U.-A.; writing—original draft, M.W.R.M.; writing—review and editing, F.A., S.U.-A., S.F. and M.H. All authors have read and agreed to the published version of the manuscript.

Funding: This research received no external funding.

Institutional Review Board Statement: Not applicable.

Data Availability Statement: All data are within the manuscript.

Acknowledgments: This manuscript has been prepared from the Ph.D. thesis of the first author.

Conflicts of Interest: The authors declare no conflict of interest.

References

1. *GOP Economic Survey of Pakistan*; Economic Advisory Wing: Islamabad, Pakistan, 2021.
2. Kouser, S.; Spielman, D.J.; Qaim, M. Transgenic Cotton and Farmers' Health in Pakistan. *PLoS ONE* **2019**, *14*, e0222617. [CrossRef] [PubMed]
3. Naeem-Ullah, U.; Ramzan, M.; Bokhari, S.H.M.; Saleem, A.; Qayyum, M.A.; Iqbal, N.; Habib ur Rahman, M.; Fahad, S.; Saeed, S. Insect Pests of Cotton Crop and Management Under Climate Change Scenarios. In *Environment, Climate, Plant and Vegetation Growth*; Springer International Publishing: Cham, Switzerland, 2020; pp. 367–396.
4. Zehr, U.B. *Cotton: Biotechnological Advances*; Springer Science & Business Media: Cham, Switzerland, 2010; Volume 65, ISBN 3642047963.
5. Kouser, S.; Qaim, M. Valuing Financial, Health, and Environmental Benefits of Bt Cotton in Pakistan. *Agric. Econ.* **2013**, *44*, 323–335. [CrossRef]
6. Arshad, M.; Suhail, A.; Asghar, M.; Tayyib, M.; Hafeez, F. Factors Influencing the Adoption of Bt Cotton in the Punjab, Pakistan. *J. Agric. Soc. Sci.* **2007**, *11*, 19.
7. Khan, M.; Mahmood, H.Z.; Damalas, C.A. Pesticide Use and Risk Perceptions among Farmers in the Cotton Belt of Punjab, Pakistan. *Crop Prot.* **2015**, *67*, 184–190. [CrossRef]
8. Tooker, J.F.; Pearsons, K.A. Newer Characters, Same Story: Neonicotinoid Insecticides Disrupt Food Webs through Direct and Indirect Effects. *Curr. Opin. Insect Sci.* **2021**, *46*, 50–56. [CrossRef]
9. Ziółkowska, E.; Topping, C.J.; Bednarska, A.J.; Laskowski, R. Supporting Non-Target Arthropods in Agroecosystems: Modelling Effects of Insecticides and Landscape Structure on Carabids in Agricultural Landscapes. *Sci. Total Environ.* **2021**, *774*, 145746. [CrossRef]
10. Peshin, R.; Hansra, B.S.; Singh, K.; Nanda, R.; Sharma, R.; Yangsdon, S.; Kumar, R. Long-Term Impact of Bt Cotton: An Empirical Evidence from North India. *J. Clean. Prod.* **2021**, *312*, 127575. [CrossRef]
11. Rana, M.A. When Seed Becomes Capital: Commercialization of Bt Cotton in Pakistan. *J. Agrar. Chang.* **2021**, *21*, 702–719. [CrossRef]
12. Cheema, H.M.N.; Khan, A.A.; Noor, K. Bt Cotton in Pakistan. In *Genetically Modified Crops in Asia Pacific*; CSIRO Publishing: Clayton, Australia, 2021; p. 91.
13. Lv, N.; Liu, Y.; Guo, T.; Liang, P.; Li, R.; Liang, P.; Gao, X. The Influence of Bt Cotton Cultivation on the Structure and Functions of the Soil Bacterial Community by Soil Metagenomics. *Ecotoxicol. Environ. Saf.* **2022**, *236*, 113452. [CrossRef]
14. Smyth, S.J.; Kerr, W.A.; Phillips, P.W.B. Global Economic, Environmental and Health Benefits from GM Crop Adoption. *Glob. Food Sec.* **2015**, *7*, 24–29. [CrossRef]
15. Hutchison, W.D.; Burkness, E.C.; Mitchell, P.D.; Moon, R.D.; Leslie, T.W.; Fleischer, S.J.; Abrahamson, M.; Hamilton, K.L.; Steffey, K.L.; Gray, M.E.; et al. Areawide Suppression of European Corn Borer with Bt Maize Reaps Savings to Non-Bt Maize Growers. *Science* **2010**, *330*, 222–225. [CrossRef]
16. Kumar, K.; Gambhir, G.; Dass, A.; Tripathi, A.K.; Singh, A.; Jha, A.K.; Yadava, P.; Choudhary, M.; Rakshit, S. Genetically Modified Crops: Current Status and Future Prospects. *Planta* **2020**, *251*, 91. [CrossRef]
17. Di Lelio, I.; Barra, E.; Coppola, M.; Corrado, G.; Rao, R.; Caccia, S. Transgenic Plants Expressing Immunosuppressive DsRNA Improve Entomopathogen Efficacy against *Spodoptera littoralis* Larvae. *J. Pest Sci.* **2022**, *95*, 1413–1428. [CrossRef]
18. Katta, S.; Talakayala, A.; Reddy, M.K.; Addepally, U.; Garladinne, M. Development of Transgenic Cotton (Narasimha) Using Triple Gene Cry2Ab-Cry1F-Cry1Ac Construct Conferring Resistance to Lepidopteran Pest. *J. Biosci.* **2020**, *45*, 31. [CrossRef]
19. Klümper, W.; Qaim, M. A Meta-Analysis of the Impacts of Genetically Modified Crops. *PLoS ONE* **2014**, *9*, e111629. [CrossRef]
20. Kumar, S.; Chandra, A.; Pandey, K.C. *Bacillus thuringiensis* (Bt) Transgenic Crop: An Environment Friendly Insect-Pest Management Strategy. *J. Environ. Biol.* **2008**, *29*, 641–653.
21. Tokel, D.; Genc, B.N.; Ozyigit, I.I. Economic Impacts of Bt (*Bacillus thuringiensis*) Cotton. *J. Nat. Fibers* **2022**, *19*, 4622–4639. [CrossRef]

22. Halford, N.G.; Shewry, P.R. Genetically Modified Crops: Methodology, Benefits, Regulation and Public Concerns. *Br. Med. Bull.* **2000**, *56*, 62–73. [CrossRef]
23. Kuzma, J.; Grieger, K. Community-Led Governance for Gene-Edited Crops. *Science* **2020**, *370*, 916–918. [CrossRef]
24. Sendhil, R.; Nyika, J.; Yadav, S.; Mackolil, J.; Prashat, G.R.; Workie, E.; Ragupathy, R.; Ramasundaram, P. Genetically Modified Foods: Bibliometric Analysis on Consumer Perception and Preference. *GM Crops Food* **2022**, *13*, 65–85. [CrossRef]
25. Liu, J.; Liang, Y.; Hu, T.; Zeng, H.; Gao, R.; Wang, L.; Xiao, Y. Environmental Fate of Bt Proteins in Soil: Transport, Adsorption/Desorption and Degradation. *Ecotoxicol. Environ. Saf.* **2021**, *226*, 112805. [CrossRef] [PubMed]
26. Zhou, X.; Li, H.; Liu, D.; Hao, J.; Liu, H.; Lu, X. Effects of Toxin from *Bacillus thuringiensis* (Bt) on Sorption of Pb (II) in Red and Black Soils: Equilibrium and Kinetics Aspects. *J. Hazard. Mater.* **2018**, *360*, 172–181. [CrossRef] [PubMed]
27. Mandal, A.; Sarkar, B.; Owens, G.; Thakur, J.K.; Manna, M.C.; Niazi, N.K.; Jayaraman, S.; Patra, A.K. Impact of Genetically Modified Crops on Rhizosphere Microorganisms and Processes: A Review Focusing on Bt Cotton. *Appl. Soil Ecol.* **2020**, *148*, 103492. [CrossRef]
28. Stotzky, G. Persistence and Biological Activity in Soil of the Insecticidal Proteins from *Bacillus thuringiensis*, Especially from Transgenic Plants. *Plant Soil* **2005**, *266*, 77–89. [CrossRef]
29. Sarkar, B.; Patra, A.K.; Purakayastha, T.J.; Megharaj, M. Assessment of Biological and Biochemical Indicators in Soil under Transgenic Bt and Non-Bt Cotton Crop in a Sub-Tropical Environment. *Environ. Monit. Assess.* **2009**, *156*, 595–604. [CrossRef]
30. Fleming, D.; Musser, F.; Reisig, D.; Greene, J.; Taylor, S.; Parajulee, M.; Lorenz, G.; Catchot, A.; Gore, J.; Kerns, D.; et al. Effects of Transgenic *Bacillus thuringiensis* Cotton on Insecticide Use, Heliothine Counts, Plant Damage, and Cotton Yield: A Meta-Analysis, 1996–2015. *PLoS ONE* **2018**, *13*, e0200131. [CrossRef]
31. Luo, J.; Zhang, S.; Zhu, X.; Lu, L.; Wang, C.; LI, C.; Cui, J.; Zhou, Z. Effects of Soil Salinity on Rhizosphere Soil Microbes in Transgenic Bt Cotton Fields. *J. Integr. Agric.* **2017**, *16*, 1624–1633. [CrossRef]
32. Dunfield, K.E.; Germida, J.J. Impact of Genetically Modified Crops on Soil- and Plant-Associated Microbial Communities. *J. Environ. Qual.* **2004**, *33*, 806. [CrossRef]
33. Sarkar, B.; Patra, A.K.; Purakayastha, T.J. Transgenic Bt-Cotton Affects Enzyme Activity and Nutrient Availability in a Sub-Tropical Inceptisol. *J. Agron. Crop Sci.* **2008**, *194*, 289–296. [CrossRef]
34. Sun, C.X.; Chen, L.J.; Wu, Z.J.; Zhou, L.K.; Shimizu, H. Soil Persistence of *Bacillus thuringiensis* (Bt) Toxin from Transgenic Bt Cotton Tissues and Its Effect on Soil Enzyme Activities. *Biol. Fertil. Soils* **2007**, *43*, 617–620. [CrossRef]
35. Noman, A.; Bashir, R.; Aqeel, M.; Anwer, S.; Iftikhar, W.; Zainab, M.; Zafar, S.; Khan, S.; Islam, W.; Adnan, M. Success of Transgenic Cotton (*Gossypium hirsutum* L.): Fiction or Reality? *Cogent Food Agric.* **2016**, *2*, 1207844. [CrossRef]
36. Flachs, A. Transgenic cotton: High hopes and farming reality. *Nat. Plants* **2017**, *3*, 16212. [CrossRef]
37. Smyth, S.J. The Human Health Benefits from GM Crops. *Plant Biotechnol. J.* **2020**, *18*, 887–888. [CrossRef]
38. Matloob, A.; Aslam, F.; Rehman, H.U.; Khaliq, A.; Ahmad, S.; Yasmeen, A.; Hussain, N. Cotton-Based Cropping Systems and Their Impacts on Production. In *Cotton Production and Uses*; Springer: Singapore, 2020; pp. 283–310. ISBN 9789811514722.
39. Kroetsch, D.; Wang, C. Particle Size Distribution. *Soil Sampl. Methods Anal.* **2008**, *2*, 713–725.
40. Dellavalle, N.B. Determination of Soil-Paste PH and Conductivity of Saturation Extract. In *Reference Methods for Soil Analysis*; Soil and Plant Analysis Council, Inc.: Athens, GA, USA, 1992; pp. 40–43.
41. Cunniff, P.; AOAC International. *Official Methods of Analysis of AOAC International*, 16th ed.; AOAC International: Gaithersburg, MD, USA, 1997.
42. Hoogsteen, M.J.J.; Lantinga, E.A.; Bakker, E.J.; Groot, J.C.J.; Tittonell, P.A. Estimating Soil Organic Carbon through Loss on Ignition: Effects of Ignition Conditions and Structural Water Loss. *Eur. J. Soil Sci.* **2015**, *66*, 320–328. [CrossRef]
43. Jafari, A.; Connolly, V.; Frolich, A.; Walsh, E.J. A Note on Estimation of Quality Parameters in Perennial Ryegrass by near Infrared Reflectance Spectroscopy. *Ir. J. Agric. Food Res.* **2003**, *42*, 293–299.
44. CIMMYT. *From Agronomic Data to Farmer Recommendations: An Economics Workbook*; CIMMYT: Heroica Veracruz, Mexico, 1988; ISBN 9686127194.
45. Shapiro, S.S.; Wilk, M.B. An Analysis of Variance Test for Normality (Complete Samples). *Biometrika* **1965**, *52*, 591–611. [CrossRef]
46. Steel, R.; Torrei, J.; Dickey, D. *Principles and Procedures of Statistics a Biometrical Approach*; McGraw-Hill College: New York, NY, USA, 1997.
47. IBM, Inc. *SPSS Statistics for Windows (Version 20)*; IBM SPSS Inc.: Armonk, NY, USA, 2012; pp. 1–8.
48. Kranthi, K.R.; Stone, G.D. Long-Term Impacts of Bt Cotton in India. *Nat. Plants* **2020**, *6*, 188–196. [CrossRef]
49. Ahmad, I.; Zhou, G.; Zhu, G.; Ahmad, Z.; Song, X.; Hao, G.; Jamal, Y.; Ibrahim, M.E.H. Response of Leaf Characteristics of BT Cotton Plants to Ratio of Nitrogen, Phosphorus and Potassium. *Pak. J. Bot.* **2021**, *53*, 873–881. [CrossRef]
50. Marschner, P. *Marschner's Mineral Nutrition of Higher Plants*, 3rd ed.; Elsevier: Amsterdam, The Netherlands, 2011; ISBN 9780123849052.
51. Jones, M.L.M.; Wallace, H.L.; Norris, D.; Brittain, S.A.; Haria, S.; Jones, R.E.; Rhind, P.M.; Reynolds, B.R.; Emmett, B.A. Changes in Vegetation and Soil Characteristics in Coastal Sand Dunes along a Gradient of Atmospheric Nitrogen Deposition. *Plant Biol.* **2004**, *6*, 598–605. [CrossRef]
52. Marschner, P.; Fu, Q.; Rengel, Z. Manganese Availability and Microbial Populations in the Rhizosphere of Wheat Genotypes Differing in Tolerance to Mn Deficiency. *J. Plant Nutr. Soil Sci.* **2003**, *166*, 712–718. [CrossRef]

53. Hu, H.; Xie, M.; Yu, Y.; Zhang, Q. Transgenic Bt Cotton Tissues Have No Apparent Impact on Soil Microorganisms. *Plant Soil Environ.* **2013**, *59*, 366–371. [CrossRef]
54. O'Callaghan, M.; Glare, T.R.; Burgess, E.P.J.; Malone, L.A. Effects of Plants Genetically Modified for Insect Resistance on Nontarget Organisms. *Annu. Rev. Entomol.* **2005**, *50*, 271–292. [CrossRef]
55. Beura, K.; Rakshit, A. Effect of Bt Cotton on Nutrient Dynamics under Varied Soil Type. *Ital. J. Agron.* **2011**, *6*, e35. [CrossRef]
56. Kumari, S.; Manjhi, B.K.; Beura, K.S.; Rakshit, A. Decomposition Bt Cotton Residues Affecting Soil Microbial Activity under Varied Soils. *Int. J. Agric. Environ. Biotechnol.* **2015**, *8*, 359. [CrossRef]
57. Shahzad, M.; Farooq, M.; Hussain, M. Weed Spectrum in Different Wheat-Based Cropping Systems under Conservation and Conventional Tillage Practices in Punjab, Pakistan. *Soil Tillage Res.* **2016**, *163*, 71–79. [CrossRef]
58. Shahzad, M.; Hussain, M.; Jabran, K.; Farooq, M.; Farooq, S.; Gašparovič, K.; Barboricova, M.; Aljuaid, B.S.; El-Shehawi, A.M.; Zuan, A.T.K. The Impact of Different Crop Rotations by Weed Management Strategies' Interactions on Weed Infestation and Productivity of Wheat (*Triticum aestivum* L.). *Agronomy* **2021**, *11*, 2088. [CrossRef]
59. Tariq, M.; Abdullah, K.; Ahmad, S.; Abbas, G.; Rahman, M.H.; Khan, M.A. Weed Management in Cotton. In *Cotton Production and Uses*; Springer: Singapore, 2020; pp. 145–161.
60. Farkas, A. Soil Management and Tillage Possibilities in Weed Control. *Herbologia* **2006**, *7*, 9–23.
61. Mahajan, G.; Chauhan, B.S. The Role of Cultivars in Managing Weeds in Dry-Seeded Rice Production Systems. *Crop Prot.* **2013**, *49*, 52–57. [CrossRef]
62. Rezakhanlou, A.; Mirshekari, B.; Zand, E.; Farahvash, F.; Baghestani, M.A. Evaluation of Competitiveness of Cotton Varieties to Cocklebur (*Xanthium srumarium* L.). *J. Food Agric. Environ.* **2013**, *11*, 308–311.
63. Chandler, J.M.; Meredith, W.R. Yields of Three Cotton (*Gossypium hirsutum*) Cultivars as Influenced by Spurred Anoda (*Anoda cristata*) Competition. *Weed Sci.* **1983**, *31*, 303–307. [CrossRef]
64. Tabashnik, B.E.; Carrière, Y.; Dennehy, T.J.; Morin, S.; Sisterson, M.S.; Roush, R.T.; Shelton, A.M.; Zhao, J.-Z. Insect Resistance to Transgenic Bt Crops: Lessons from the Laboratory and Field. *J. Econ. Entomol.* **2003**, *96*, 1031–1038. [CrossRef]
65. Heckel, D.G. How Do Toxins from *Bacillus thuringiensis* Kill Insects? An Evolutionary Perspective. *Arch. Insect Biochem. Physiol.* 2020.
66. Tapp, H.; Stotzky, G. Insecticidal Activity of the Toxins from *Bacillus thuringiensis* Subspecies Kurstaki and Tenebrionis Adsorbed and Bound on Pure and Soil Clays. *Appl. Environ. Microbiol.* **1995**, *61*, 1786–1790. [CrossRef]
67. Rui, Y.-K.; Yi, G.-X.; Zhao, J.; Wang, B.-M.; Li, Z.-H.; Zhai, Z.-X.; He, Z.-P.; Li, Q.X. Changes of Bt Toxin in the Rhizosphere of Transgenic Bt Cotton and Its Influence on Soil Functional Bacteria. *World J. Microbiol. Biotechnol.* **2005**, *21*, 1279–1284. [CrossRef]
68. Wei, X.-D.; Zou, H.-L.; Chu, L.-M.; Liao, B.; Ye, C.-M.; Lan, C.-Y. Field Released Transgenic Papaya Affects Microbial Communities and Enzyme Activities in Soil. *Plant Soil* **2006**, *285*, 347–358. [CrossRef]
69. Fang, M.; Motavalli, P.P.; Kremer, R.J.; Nelson, K.A. Assessing Changes in Soil Microbial Communities and Carbon Mineralization in Bt and Non-Bt Corn Residue-Amended Soils. *Appl. Soil Ecol.* **2007**, *37*, 150–160. [CrossRef]

MDPI

St. Alban-Anlage 66

4052 Basel

Switzerland

Tel. +41 61 683 77 34

Fax +41 61 302 89 18

www.mdpi.com

Agriculture Editorial Office

E-mail: agriculture@mdpi.com

www.mdpi.com/journal/agriculture